# *The Mechanics of*
# *Thin Film Coatings*

**FIRST EUROPEAN COATING SYMPOSIUM ON**

# The Mechanics of Thin Film Coatings

*Leeds University, UK*
*19 – 22 September 1995*

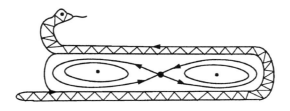

*Editors*

## P. H. Gaskell
## M. D. Savage
## J. L. Summers

University of Leeds, UK

**World Scientific**
*Singapore • New Jersey • London • Hong Kong*

*Published by*

World Scientific Publishing Co. Pte. Ltd.

P O Box 128, Farrer Road, Singapore 912805

*USA office:* Suite 1B, 1060 Main Street, River Edge, NJ 07661

*UK office:* 57 Shelton Street, Covent Garden, London WC2H 9HE

**Library of Congress Cataloging-in-Publication Data**
Biennial European Coating Symposium (1st : 1995 : Leeds, England)
    The mechanics of thin film coatings : proceedings of the First
    Biennial European Coating Symposium held in Leeds, 19th–22nd
    September 1995 / editors, P. H. Gaskell, M. D. Savage, J. L. Summers.
        p.    cm.
    Includes bibliographical references.
    ISBN 9810225431
    1. Coating processes -- Congresses.    I. Gaskell, P. H.    II. Savage, M. D.
    III. Summers, J. L. (Jonathan L.)    IV. Title.
    TP156.C57854    1995
    667'.9--dc20                                              96-13076
                                                                CIP

**British Library Cataloguing-in-Publication Data**
A catalogue record for this book is available from the British Library.

Printed in Singapore.

# CONTENTS

**Preface**

## SECTION 1 : Coating Process Fundamentals — 1

K. Ruschak and S. Weinstein (**Keynote Presentation**) — 3
*Mathematical modelling of fluid dies*

P. Bourgin and N. Tahiri — 23
*Generalised Jeffery-Hamel flow: application to high velocity coating*

A. Clarke — 32
*Recirculating flows in curtain coating*

K.N. Kumar, P. Murtagh and S. Subbiah — 42
*Numerical modelling of multilayer slot coating flows*

A. Münch — 52
*Towards a numerical simulation of thin liquid drops*

C. Richardson, P.H. Gaskell and M.D. Savage — 62
*New results in reverse roll coating*

## SECTION 2 : Instability and Coating Defects — 73

M.S. Carvalho and L.E. Scriven — 75
*Deformable roll coating: analysis of ribbing instability and its decay*

F. Durst, U. Lange and H. Raszillier — 85
*Minimization and control of random effects on film thickness uniformity by optimized design of coating die internals.*

A. T-L. Horng                                                                              95
*Chebyshev collocation method on solving stability of a liquid layer flow-*
*ing down an inclined plane*


I. Kliakhandler and G. Sivashinski                                                        107
*Kinetic instabilities in multilayer downflowing creeping films.*


Y.C. Severtson and C. Aidun                                                               117
*An inviscid mode of instability in two-layer flows*


## SECTION 3 : Experimental Investigations                                                **127**


P.M. Schweizer **(Keynote Presentation)**                                                 129
*Experimental methods for coating flows*


J-M. Buchlin, M. Decré, E. Gailly and P. Planquart                                        158
*Meniscus control by a string in a roll coating experiment*


J-M. Buchlin, M. Manna, M. Arnalsteen, M.L. Riethmuller and M.   168
Dubois
*Theoretical and experimental investigation of gas-jet wiping*


O. Cohu and A. Magnin                                                                     179
*Experimental investigations on roll coating with deformable rolls*


F. Cunha and M. Carbonaro                                                                 189
*Surface wave instability on aircraft de/anti-icing fluid films*


G. Innes, J.L. Summers and H.M. Thompson                                                  199
*Feed condition effects in forward and reverse roll coating*

# SECTION 4 : Roll and Gravure Coating 211

H. Benkreira, R. Patel, M. Naheem and J.M. Leclercq 213
*Film thickness and instabilities with forward gravure coating*

M.S. Carvalho and L.E. Scriven 221
*Deformable roll coating : modelling of steady flow in gaps and nips*

M. Kodama and K. Adachi 231
*Thin film forming in coating-nip flows*

S. Krauss, F. Durst and H. Raszillier 238
*A combined slot coating and roll coating technique*

# SECTION 5 : Spreading, Levelling, Surface Tension and Gravity Driven Flows 247

S. Kalliadasis 249
*Dynamics of liquid spreading on solid surfaces*

T.G. Myers 259
*Surface tension driven thin film flows*

W.S. Overdiep 269
*Application properties of decorative paints*

Y.D. Shikhmurzaev 279
*Spreading of liquids on dry and pre-wet solid surfaces*

S.K. Wilson and E.L. Terrill 288
*The dynamics of planar and axisymmetric holes in thin fluid layers*

# SECTION 6 : Rheological Effects in Coating Processes    299

D.C-H. Cheng (**Keynote Presentation**)    301
*A review of the role of rheology in coating processes*

C.K. Aidun and Y. Lu    348
*Dynamics of suspensions in coating flows*

C. Beriet, P.N. Bartlett, D.G. Chetwynd, J.W. Gardner and X. Liu    357
*Tribological properties of conducting polymer films for application in nanotechnology*

O.G. Harlen    366
*Simulation of viscoelastic flows*

# SECTION 7 : Novel Coating and Thin Film Problems    375

S.M. Bushnell-Watson M.R. Alexander, A.P. Ameen, W.M. Rainforth,    377
R.D. Short, F.R. Jones, W. Michaeli, M. Stollenwerk and J. Zabold
*The chemistry of thin film deposits formed from hexamethyldisoxane and hexamethyldisilazane plasmas*

M. Decré, J-M. Buchlin, J. Schmidt and M. Rabaud    387
*The onset of ribbing reconsidered: an experiment*

J. Lammers, S.B.G.M. O'Brien and M.N.M. Beerens    397
*Effects of evaporation during spin-coating*

**List of Delegates**    403

# PREFACE

Interest in coating process fundamentals has risen sharply in recent years; indeed coating research is now established as an inter-disciplinary pursuit incorporating analytical, computational and experimental methods. The purpose of this new series of biennial symposia is to provide a forum for the presentation and discussion of recent work together with an opportunity for the informal exchange of ideas and information between academics and industrialists.

This volume contains the papers presented at the First European Coating Symposium on The Mechanics of Thin Film Coatings held in Leeds, September 19-22, 1995. There are three Keynote presentations, given by experts in the field of experimental methods, rheology and the modelling of coating dyes. Included also are thirty contributions drawn from the following areas of coating research and application: instability and defects; spreading and levelling; flow visualisation; experimental methods; surface tension and gravity driven flows; rheological effects in coating processes. These proceedings should therefore, prove a valuable resource for both researchers and coating practitioners.

The papers are produced direct from lithographs of the authors' manuscripts and the editors do not accept responsibility for any erroneous comments or opinions.

Finally, the Symposium Committee wishes to acknowledge the generous financial support of industry – Autotype International Ltd., Horsell Graphic Industries Ltd., I.C.I Films, Ilford Ltd., Kodak Ltd., 3M U.K. – and sponsorship of the meeting by Barclays Bank plc, Silicon Graphics Inc. and The University of Leeds.

P. H. GASKELL
M. D. SAVAGE
J. L. SUMMERS
University of Leeds,
England, U.K.

# SECTION 1

# Coating Process

# Fundamentals

# MATHEMATICAL MODELING OF FLUID DIES

Kenneth J. Ruschak
Senior Technical Associate
and
Steven J. Weinstein
Technical Associate

Manufacturing Research and Engineering Organization
Eastman Kodak Company
Kodak Park 2/35
Rochester, New York 14652-3701

## SUMMARY

Dies for forming uniform liquid layers are basic to the coatings and plastics industries. The aim of this paper is to review and extend the hydrodynamics of flow in dies.

Flow distribution for specified geometries can usually be predicted using commercially available software packages to solve three-dimensional models. In many cases, simplified models can be effective as well.

Die design is the inverse of predicting die performance. Flow distribution is specified, and die geometry is determined. Die design is a constrained optimization problem requiring many predictions of flow distribution as geometric parameters are varied. The prediction of flow distribution may be just part of a more general model including the mechanical and thermal responses of the die. Simplified models for flow distribution are particularly advantageous in these circumstances.

Several aspects of the simplified equations for die design are reviewed and extended. These include the use of universal velocity profiles, the physical effects of inertia, the correct equations for the primary and secondary cavities, and the solution of the complete set of equations by perturbation methods. Additionally, the mechanism of operation of a two-cavity die is explained.

## 1. INTRODUCTION

Film-forming dies have internal flow chambers for forming liquid supplied in a tube into a uniform sheet. The flow chambers usually comprise cavities, slots, and holes. A uniform distribution is achieved primarily by partitioning flow resistance through and across the die. To force the liquid to the ends of the die, the resistance through the die must considerably exceed that across the die.

Attention here is focused on dies with internal cavities and slots. The liquid is usually fed to the center of a primary cavity running across the die. A primary slot whose height is one or two orders of magnitude less than the depth of the cavity offers a relatively high flow resistance. The cross-sectional area of the primary cavity is optionally tapered from center to end to prevent stagnant regions, even though reducted cross-sectional area increases flow resistance along the cavity. To even the flow distribution further, there may be a secondary cavity and slot.

A complementary strategy for achieving uniform flow distribution is locally adjusting the flow resistance through the die to compensate for pressure loss across the die. For a die with a primary cavity and slot, this can be done by tapering the length of the primary metering slot from center to end. Pressure loss in the primary cavity diminishes flow out the ends of the die. Reducing flow resistance by shortening the slot from center to end compensates for the pressure loss and maintains flow out the ends.

Flow resistances and compensatory geometric changes generally depend upon the flow rate and rheology of the liquid. Dies tightly designed for one flow rate and liquid may not perform well for other flow rates or liquids. Conservatively designed dies are generally more tolerant.

Models for die design are essential because the hardware is expensive and intricate, and there are too many geometric and rheological parameters for a purely experimental approach to be practical. Moreover, measuring flow distribution from a die is difficult.

Usually, three-dimensional computations of flow in dies can be done for specified geometries using commercially available software, at least for liquids that are no more than slightly elastic. The equations are nonlinear because of the inertial terms, and, in the case of non-Newtonian liquids, the viscous terms. Grid construction is complicated by the inherently disparate dimensions of die geometry, but once a die is discretized, different flow conditions are readily computed.. Grid refinement is essential, because high accuracy for flow distribution is required, and there can be several thousand unknowns. Constructing a grid and making several runs can take days.

Die design is the inverse of the problem of predicting flow distribution. Flow distribution is specified, and the optimum geometry is determined under constraints. There are always design contraints. For instance, certain geometric dimensions may be constrained because of space limitations. It is also common for the backpressure produced by the die to be limited. In some cases, fabrication tolerances and mechanical deformations of the die caused by internal pressure and nonisothermal temperature have to be considered. Solution of the optimization problem usually requires many iterations, and the optimizing routine needs to know how changes in the geometric variables affect flow distribution. Solution of the complete set of three-dimensional equations, no longer limited to fluid flow within preset boundaries, may not be effective or even possible. The need is well served, however, by simplified mathematical models that are rapidly computed.

The simplified equations are nonlinear because of the inertial terms and possibly the constitutive equation of the liquid. However, they are ordinary differential equations rather than partial differential equations and can be rapidly solved on common computers. In cases where the distribution of the liquid from the die must be highly uniform, the simplified equations can be linearized. The numerous advantages of linear equations include formal solution techniques and superposition.

In a single cavity die, or in the primary cavity of a two-cavity die, the main flow direction is along the cavity, and the results for rectilinear flow in conduits are commonly used. In the secondary cavity, however, the main flow direction is across the cavity. When inertia is important or the constitutive equation for the liquid is nonlinear, the main flow interacts with the weaker flow along the cavity. Consequently, results for flow in conduits are not directly applicable to secondary cavities, and appropriate flow equations must be derived.

Among the many analyses of flow in single-cavity dies are that for a T-die[1] and that for a linearly tapered coat hanger die[2]. Two-cavity dies have been analyzed[3-6], and the analog of a resistor network for multiple cavities has been proposed[5]. A single-cavity die has been analyzed with some linearization of the equations[7]. The three-dimensional nature of the flow in the secondary cavity has been recognized[3] but has not been treated in the context of a simplified model. Significant entry effects in single cavity dies have been found using three-dimensional numerical analysis[8].

In what follow, a two-cavity die is described and its performance for a viscous, Newtonian liquid estimated to reveal the basic operating principle. The simplified equations for die design under more general circumstances are then reviewed and extended. These equations are susequently solved for a two-cavity die by taking advantage of naturally occurring small parameters. The solution reveals the operating mechanism and makes clear the effect of coupling between the cavities. The original contributions are detailed in three papers submitted for publication[9-11].

Description of the flow geometry

Figure 1 is a schematic of a die comprising two cavities and slots. The die is symmetric about its center, and so only half is drawn. Liquid fed to the center of the die at a total flow rate of $2\,Q_e$ divides equally between the two sides by symmetry. The primary cavity has a cross-sectional area $A_i$ that optionally tapers from $A_e$ at its center to a nominal value at its end. A slot of uniform height $H_i$ connects the primary and secondary cavities. The length of the primary slot is optionally tapered from center to end. $L_e$ is the initial length of the slot at center. The secondary cavity has a constant cross-sectional area $A_o$, and the secondary slot has a constant height $H_o$ and length $L_o$. The cross-sectional shapes of the cavities usually have aspect ratios near unity to minimize flow resistance. The outer cavity cross section may expand gradually from its inlet to prevent flow separation and stagnant regions.

The length of the cavities, center-to-end, is $W$. The average flow rate per unit width exiting the die is $\langle q \rangle = Q_e / W$. For perfect flow distribution, the volumetric flow rate per unit width equals the average value everywhere.

## 2. RECTILINEAR FLOW IN CONDUITS

The simplified equations for the primary cavity result from overall balances of momentum and mass. They are best obtained by integrating the full set of equations over cross-sections of the cavities[10]. In this process, the gradients of pressure and

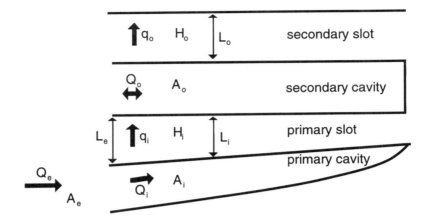

Figure 1. Schematic of a two-cavity die.

velocity in the cross section are lost, and as a result pressure and velocity profiles must be assumed. Trial functions for pressure and velocity profiles are typically obtained from profiles for developed flow in conduits of constant cross section. For Newtonian and power-law liquids and for die cross sections that vary in size but not shape, the velocity profile can be scaled to be independent of flow rate and cavity size. The flow equation for the primary cavity simplifies greatly as a result. A scaled velocity profile cannot be obtained for most other constitutive equations. For example, a scaled profile cannot be obtained for the Carreau rheological model.

For viscous, rectilinear flow of a power-law liquid in a conduit of constant cross section $A$, the momentum equation reduces to

$$\frac{\partial}{\partial Y}\left(\mu \frac{\partial V}{\partial Y}\right) + \frac{\partial}{\partial X}\left(\mu \frac{\partial V}{\partial X}\right) - \frac{\partial P}{\partial Z} = 0 \tag{1}$$

where $X$ and $Y$ are coordinates in the cross-section, $Z$ is the coordinate in the flow direction, and $V$ is the velocity. The viscosity depends upon the velocity gradients according to

$$\mu = M\left[\left(\frac{\partial V}{\partial Y}\right)^2 + \left(\frac{\partial V}{\partial X}\right)^2\right]^{\frac{n-1}{2}} \tag{2}$$

where parameter $M$ is the consistency coefficient and $n$ the power-law index. For a Newtonian liquid, $n = 1$, and for a shear-thinning liquid, $0 < n < 1$.

The scaled velocity profile is obtained from the solution to the following problem.

$$\frac{\partial}{\partial y}\left(\mu'\frac{\partial f}{\partial y}\right) + \frac{\partial}{\partial x}\left(\mu'\frac{\partial f}{\partial x}\right) + 1 = 0 \tag{3}$$

$$\mu' = \left[\left(\frac{\partial f}{\partial y}\right)^2 + \left(\frac{\partial f}{\partial x}\right)^2\right]^{\frac{n-1}{2}} \tag{4}$$

$$f = 0 \qquad \text{(boundary)} \tag{5}$$

Here $f$ is scaled velocity, and $x$ and $y$ are dimensionless coordinates defined by

$$x = \frac{X}{\sqrt{A}}, \qquad y = \frac{Y}{\sqrt{A}} \tag{6}$$

The domain in the $x$-$y$ plane has the shape of the conduit and a size giving an area of unity. From the solution to Eqs. (3)-(5), a viscous shape factor $\lambda$ and a momentum shape factor $\vartheta$ are defined by

$$\lambda \equiv \iint_A f \, dx \, dy \tag{7}$$

$$\vartheta \equiv \frac{1}{\lambda^2} \iint_A f^2 \, dx \, dy \tag{8}$$

The velocity profile and the viscosity are then determined by

$$V = \frac{1}{\lambda}\frac{Q}{A} f\left(\frac{X}{\sqrt{A}}, \frac{Y}{\sqrt{A}}\right) \tag{9}$$

$$\mu = \mu' \lambda^{1-n} \tag{10}$$

Comparing Eqs. (1) and (3) shows that the pressure gradient for rectilinear flow is given by

$$-\frac{\partial P}{\partial Z} = \frac{M Q^n}{\lambda^n A^{(3n+1)/2}} \tag{11}$$

A larger shape factor corresponds to a smaller pressure gradient. Eq. (8) becomes

$$\iint_A \rho V^2 dX dY = \vartheta \frac{\rho Q^2}{A} \tag{12}$$

The momentum shape factor is unity for a flat velocity profile and greater than unity for curved profiles.

These results for rectilinear flow in conduits are applicable to flow in the primary cavity of a die if the cross section of the cavity changes slowly enough that developed flow is maintained. Specifically, velocity gradients along the primary cavity are required to be negligibly small compared to those across the cavity:

$A_e / W^2 \ll 1$.

## 3. ESTIMATES FOR VISCOUS FLOW IN A TWO-CAVITY DIE

Detailed analysis of flow in a two-cavity die is prefaced by the following estimates for purely viscous flow of Newtonian liquid. In the primary cavity, the primary flow direction is lengthwise along the cavity, and the pressure gradient is that for rectilinear flow in conduits, Eq. (11). On the other hand, the primary flow direction in the secondary cavity is across the cavity. Nevertheless, for the special case of purely viscous flow of Newtonian liquid only, the same equation applies for pressure gradient along the cavity, because the governing Stokes equations are linear. When inertial terms or constitutive equations introduce nonlinearities, however, the primary flow across the cavity couples to the secondary flow along the cavity, and a flow equation accounting for the interaction have to be derived.

The equation for the pressure gradient in the primary cavity follows from Eq. (11)

$$\frac{dP_i}{dZ} = -\frac{\mu Q_i}{\lambda_i A_i^2} \tag{13}$$

Here $\lambda_i$ is the shape factor for the primary cavity. The variation of cross-sectional area along the cavity is taken to be

$$A_i = A_e \left(1 - \frac{Z}{W}\right)^\gamma \tag{14}$$

For the current estimate, $\gamma$ is 0.5 for simplicity. The equation of continuity is

$$\frac{dQ_i}{dZ} = -q_i \tag{15}$$

To estimate the pressure variation in the primary cavity, the expression for perfect flow distribution, following from Eq. (15) for $q_i = \langle q \rangle$, is entered in Eq. (13).

$$Q_i \approx Q_e \left(1 - \frac{Z}{W}\right) \tag{16}$$

The resulting pressure gradient is constant along the cavity.

The pressure drop across the slot between the cavities is that for fully developed viscous flow.

$$P_i - P_o = \frac{12\mu L_i q_i}{H_i^3} \tag{17}$$

Flow through the secondary slot is proportional to cavity pressure, and so the pressure in the secondary cavity can be considered constant if a highly uniform flow distribution is required. In that case, the derivative of Eq. (17) is

$$\frac{dP_i}{dZ} \approx \frac{12\mu}{H_i^3} \frac{d(L_i q_i)}{dZ} \tag{18}$$

Combining Eqs. (13) and (18) and integrating gives

$$L_i q_i \big|_0^W = -\frac{H_i^3 W^2}{12\lambda_i A_e^2} \langle q \rangle \tag{19}$$

The flow rate per unit width in the slot and the length of the slot are approximated by linear functions

$$q_i \approx \langle q \rangle + \Delta q_i \left(\frac{1}{2} - \frac{Z}{W}\right) \tag{20}$$

$$L_i \approx L_e - \Delta L \frac{Z}{W} \tag{21}$$

in which $\Delta q_i$ is the small center-to-end change in slot flow rate and $\Delta L$ the small change in slot length. Eq. (19) then gives, to lowest order,

$$\frac{\Delta q_i}{\langle q \rangle} \approx \frac{H_i^3 W^2}{12\lambda_i L_e A_e^2} - \frac{\Delta L}{L} \tag{22}$$

The first term on the right is the ratio of the pressure drop along the cavity to that along the slot and must be small for high uniformity.

$$\varepsilon = \frac{H_i^3 W^2}{12\lambda_i L_e A_e^2} \tag{23}$$

The second term on the right is the fractional reduction in slot length from center to end; if it is order $\varepsilon$, it improves flow distribution in the primary slot.

For a single-cavity die, Eq. (22) estimates the final flow distribution. For a two-cavity die, it estimates the uniformity of the feed to the secondary cavity.

The continuity equation for the secondary cavity includes the difference in flow rates between the primary and secondary slots.

$$\frac{dQ_o}{dZ} = q_i - q_o = \left(q_i - \langle q \rangle\right) - \left(q_o - \langle q \rangle\right) \tag{24}$$

For estimation purposes, the flow nonuniformity in the secondary slot is presumed much smaller than that in the primary slot due to the evening effect of the cavity, and the final bracketed term in Eq. (24) is neglected. The flow distribution in the secondary slot is approximated as linear

$$q_o \approx \langle q \rangle + \Delta q_o \left(\frac{1}{2} - \frac{Z}{W}\right) \tag{25}$$

Here $\Delta q_o$ is the center-to-end variation in flow rate. Using Eq. (25) to integrate the approximate form of Eq. (24) leads to

$$Q_o \approx \Delta q_i \left(\frac{Z}{2} - \frac{Z^2}{2W}\right) \tag{26}$$

This result is consistent with zero values for $Q_o$ at the cavity center and ends.

The pressure drop across the secondary slot is given by

$$P_o = \frac{12 \mu q_o L_o}{H_o^3} \tag{27}$$

For purely viscous flow only, the main flow through the secondary cavity is uncoupled from flow along the cavity, and Eq. (11) applies

$$\frac{dP_o}{dZ} = -\frac{\mu Q_o}{\lambda_o A_o^2} \tag{28}$$

Here $\lambda_o$ is the shape factor for the secondary cavity. For highly uniform flow distribution, Eq. (27) implies that the pressure in the secondary cavity must be nearly constant. So, for evening of the flow distribution to occur, the flow resistance along the cavity must be small. Differentiating Eq. (27) with respect to $Z$ and combining the result with Eq. (28) gives

$$\frac{dq_o}{dZ} = -\frac{H_o^3}{12\lambda_o L_o A_o^2} Q_o \tag{29}$$

Substituting for $Q_o$ with Eq. (26) and integrating with respect to $Z$ from $0$ to $W$ gives the final result

$$\frac{\Delta q_o}{\Delta q_i} \approx \frac{s}{12} \tag{30}$$

The ratio of the resistance along the secondary cavity to that along the secondary slot appears here.

$$s = \frac{W^2 H_o^3}{12\lambda_o A_o^2 L_o} \tag{31}$$

For the approximate form of Eq. (24) used to be correct, $s$ must be a small number. Combining Eqs. (22) and (30) then gives an estimate for the complete die

$$\frac{\Delta q_i}{\langle q \rangle} \approx \left[ \varepsilon - \frac{\Delta L}{L} \right] \frac{s}{12} \tag{32}$$

Small values of $\varepsilon$ and $s$ promote uniform flow distribution. The nonuniformity resulting from pressure loss in the primary cavity is given by $\varepsilon$. This loss can be mitigated by the term $\Delta L / L_e$ representing the slot taper. For small values of $s$, the secondary cavity significantly evens the distribution.

## 4. INERTIAL EFFECTS IN THE PRIMARY CAVITY

Significant inertial effects require adjustment of the equation for developed flow and give rise to an entry region near the inlet requiring separate analysis.

When inertia is important, the pressure gradient in the primary cavity is not given by Eq. (11). There is no consensus in the literature on the proper equation, but the following form[11] is recommended here

$$\frac{dP_i}{dZ} = -\frac{\mu Q_i}{\lambda_i A_i^2} - \frac{\rho \vartheta}{A_i} \frac{d}{dZ} \left( \frac{Q_i^2}{A_i} \right) \tag{33}$$

The momentum shape factor, $\vartheta$, is defined by Eq. (8).

If the viscous term is neglected, the effect of inertia on the pressure can be considered. For the special case of constant flow rate but variable area, the integration of Eq. (33) gives a form suggestive of Bernoulli's equation

$$P + \frac{\rho \vartheta Q^2}{2A^2} = \text{constant} \qquad (Q \text{ constant}) \qquad (34)$$

For constant flow rate, pressure falls as area decreases. In the primary cavity the flow rate is not constant, but Eq. (33) can be integrated using Eqs. (14) and (16)

$$P + \frac{\rho \vartheta Q_e^2}{2A_e^2} \frac{(2-\gamma)}{(1-\gamma)} \left(1 - \frac{Z}{W}\right)^{2(1-\gamma)} = \text{constant} \qquad (35)$$

According to this estimate, the pressure increases from center to end of the primary cavity as liquid is lost to the slot. So, the effect of inertia on developed flow is expected to diminish the total pressure drop in the cavity.

Because of inertia, flow requires some distance to develop in the inner cavity, and the velocity profile near the inlet is not that of rectilinear flow. The velocity distribution near the inlet becomes important, but this depends on the detailed geometry of the inlet and flow conditions upstream of the inlet [8]. There is an entry region over which the velocity profile evolves to that for rectilinear flow.

From work on developing flow in conduits[12,15], high wall friction and a pressure gradient exceeding that of fully developed flow are expected in the entry region. Developing flow in the cavity of a die is complicated by variable cross-sectional area and by the loss of liquid to the slot. Nevertheless, for modest Reynolds numbers at least, developing flow in the cavity is similar to developing flow in a conduit.

The length and pressure drop of the developing flow region have been estimated[11]. The most conservative use for the estimates is predicting the Reynolds number for the onset of significant inertial effects, as for a die designed for viscous flow. Flow near the wall of the cavity is modeled as a viscous boundary layer. Outside the boundary layer, flow is presumed inviscid, and the velocity profile plug. As the liquid near the wall is slowed by viscous drag, that in the inviscid core speeds up to conserve mass. By Bernoulli's equation, pressure in the core falls. The core pressure is imposed on the boundary layer.

For purposes of estimation, the flow rate in the primary cavity is taken to be the ideal distribution, Eq. (16). For a cross-sectional area given by Eq. (14), the perimeter $C_i$ varies as

$$C_i = C_e \left(1 - \frac{Z}{W}\right)^{\gamma/2} \qquad (36)$$

where $C_e$ is the perimeter at the center of the cavity.

According to the analysis of Ruschak and Weinstein[11], the velocity of the inviscid core, $V$, is given approximately by

$$V = \frac{Q_e}{A_e} \frac{(1-\xi)^{1-\gamma}}{\left(1 - \frac{1}{3}\sqrt{\frac{\xi}{\xi_d}}\right)}, \qquad \xi = \frac{Z}{W}, \qquad \xi_d = \frac{Z_d}{W} \qquad (37)$$

Here, $Z_d$ is the length of the developing flow region given by

$$\frac{Z_d}{W} = \xi_d = \frac{K_0 Re}{\left[\frac{C_e^2}{A_e} + (K_1 + K_2\gamma)Re\right]} \qquad (38)$$

The three constants appearing in Eq. (38) are

$$K_0 = \frac{41}{20} - \frac{24}{5}\ln(1.5) \approx 0.10377 \qquad (39)$$

$$K_1 = \frac{1067}{60} - \frac{216}{5}\ln(1.5) \approx 0.26724 \qquad (40)$$

$$K_2 = \frac{52}{3} - \frac{216}{5}\ln(1.5) \approx -0.18276 \qquad (41)$$

Finally, the pressure in the entry region, determined by the inviscid flow, is

$$P_e - P_i = \frac{\rho}{2}\left(\frac{Q_e}{A_e}\right)^2 \left[\frac{(1-\xi)^{2-2\gamma}}{\left(1 - \frac{1}{3}\sqrt{\frac{\xi}{\xi_d}}\right)^2} - 1\right] \qquad (42)$$

A convenient though imprecise way of accounting for the additional pressure drop in the entry region is to apply the equation for developed flow, Eq.(33), along the entire cavity with an adjusted shape factor . For the purpose of correcting the shape factor, the ideal flow rate for the inner cavity, Eq. (16), is imposed.

$$\frac{dP_i}{dZ} = -\frac{\mu Q_e}{A_e^2}\left[\frac{1}{\lambda_i} - \vartheta Re(2-\gamma)\right]\left(1 - \frac{Z}{W}\right)^{1-2\gamma} \qquad (43)$$

This equation is integrated over the developed flow region, from $Z_d$ to the end of the cavity. It is also integrated over the entire cavity with the shape factor for developed flow replaced by a corrected shape factor, $\hat{\lambda}_i$. The value of $\hat{\lambda}_i$ is chosen such that the total pressure drop from center to end equals the sum of the drops for the entry and developed flow regions.

## 5. INERTIAL EFFECTS IN THE SECONDARY CAVITY

Inertia plays a role in the secondary cavity that has not been previously identified in the literature. The inertia of the main flow across the cavity inhibits secondary flow along the cavity. The results of a detailed analysis[11] are summarized here.

The three-dimensional flow in the secondary cavity is governed by the momentum and continuity equations. The primary flow direction, $X$, is across the cavity (Figure 2). A secondary flow occurs along the cavity in the $Z$ direction. The characteristic length in the $X$ and $Y$ directions is $\sqrt{A_o}$, and that in the $Z$ direction $W$. The characteristic velocity in the $X$ and $Y$ directions is $\langle q \rangle / \sqrt{A_o}$, and in the $Z$ direction $Q_e / A_o$. The characteristic pressure is determined by rectilinear flow in the secondary slot and is estimated as $\mu L_o \langle q \rangle / H_o^3$. These estimates of magnitude are used to define dimensionless coordinates $x$, $y$, $z$; slot heights $h_i$ and $h_o$; velocity components $v_x$, $v_y$, $v_z$; and pressure p.

First, a two-dimensional base flow is computed. This flow has no component along the secondary cavity, and therefore the flow distribution in the primary slot is perfect. An uneven distribution and the resulting flow along the cavity are subsequently considered as a perturbation.

The base flow is determined by the two-dimensional momentum and continuity equations solved in the domain shown in Figure 2. Appropriate boundary conditions are applied at the solid wall and in the metering slots just beyond the cavity . Fully developed flow in the primary slot extends to within about one slot height of the secondary cavity. The entry length in the secondary slot is proportional to $Re$. For Reynolds numbers of order 10, the entry length is a few slot heights[13]. The base flow can be computed using commercial software. If the aspect ratio of the cavity is small, an approximate solution can be obtained through the lubrication approximation to the momentum equation. The lubrication approximation has been used in die design before[14].

Nonuniform flow in the primary slot induces flow along the secondary cavity. The flow distribution in the primary slot is written as a perturbation of perfect flow distribution

$$\frac{q_i(z)}{\langle q \rangle} = 1 + \frac{\tilde{q}_i(z)}{\langle q \rangle} \tag{44}$$

Here $\tilde{q}_i(z) / \langle q \rangle$ is the small, fractional variation in flow supplied to the secondary cavity. Similarly, perturbations of the velocity components and pressures are written

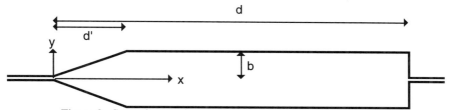

Figure 2. Secondary cavity consisting of an expansion and a channel.

$$v_x = \bar{v}_x + \tilde{v}_x, \quad v_y = \bar{v}_y + \tilde{v}_y, \quad v_z = \tilde{v}_z, \quad p = \bar{p} + \tilde{p} \tag{45}$$

Only first order terms in the perturbed quantities are retained. In addition, the dimensions groups $\delta$ and $\zeta$ are presumed small.

$$\delta = \frac{\sqrt{A_o}}{W} \tag{46}$$

$$\zeta = \frac{h_o^3}{L_o A_o} \tag{47}$$

Formally, the limit $\zeta \to 0$ and $\delta \to 0$ with $\delta^2 / \zeta$ fixed is taken. The resulting equations for flow along the cavity are

$$Re\left[\bar{v}_x \frac{\partial \tilde{v}_z}{\partial x} + \bar{v}_y \frac{\partial \tilde{v}_z}{\partial y}\right] = -\frac{\delta^2}{\zeta} \frac{\partial \tilde{p}}{\partial z} + \frac{\partial^2 \tilde{v}_z}{\partial x^2} + \frac{\partial^2 \tilde{v}_z}{\partial y^2} \tag{48}$$

$$\frac{\partial \tilde{v}_x}{\partial x} + \frac{\partial \tilde{v}_y}{\partial y} + \frac{\partial \tilde{v}_z}{\partial z} = 0 \tag{49}$$

By the $x$- and $y$-components of the momentum equation, not shown, the pressure perturbation is constant at a cross section, and so $\tilde{p}$ depends only on $z$. The solution of Eq. (48), the $z$-component of the momentum equation, has the form

$$\tilde{v}_z = -\frac{\delta^2}{\zeta} \frac{d\tilde{p}}{dz} T(x, y) \tag{50}$$

Making this substitution leads to the equation

$$Re\left[\bar{v}_x \frac{\partial T}{\partial x} + \bar{v}_y \frac{\partial T}{\partial y}\right] = 1 + \frac{\partial^2 T}{\partial y^2} + \frac{\partial^2 T}{\partial x^2} \tag{51}$$

As the notation suggests, if the Reynolds number is considered analogous to the Prandtl number, then the function $T$ is analogous to temperature in the equation for convective heat transfer with an internal heat source. The boundary condition is

$$T = 0 \qquad \text{(entire boundary)} \tag{52}$$

The solution to this problem can also be computed using commercial software, and an approximate solution can be obtained for small aspect ratio by neglecting the last term in Eq. (51). Equation (50) is then integrated over a cross section

$$\iint_{A_o} \bar{v}_z \, dx dy = -\frac{\delta^2}{\zeta} \frac{d\tilde{p}}{dz} \iint_{A_o} T(x,y) \, dx dy \tag{53}$$

The integral on the left gives the dimensionless flow rate along the cavity, $Q_o/Q_e$, and that on the right a shape factor corrected for inertial effects

$$\hat{\lambda}_o = \iint T(x,y) \, dx \, dy \tag{54}$$

The corrected shape factor $\hat{\lambda}_o$ depends on $Re$ through $T$ and is the usual viscous shape factor, $\lambda_o$, only for $Re = 0$. Eq. (53) leads to the final result

$$\frac{dP_o}{dZ} = -\frac{\mu Q_o}{\hat{\lambda}_o A_o^2} \tag{55}$$

Except for the shape factor, this equation is the same as that for viscous flow, Eq. (28).

Equation (49), the continuity equation, needs to be considered only in integral form. The end result is the mass balance for the secondary cavity found in the literature and given previously by Eq. (24).

As a particularly simple example, the result for a linear expansion of area $bd=1$ and aspect ratio $\beta = 2b/d \ll 1$ is

$$\hat{\lambda}_o = \frac{\beta}{12} \frac{1}{\left(1 + \frac{\beta Re}{5}\right)} \qquad \text{(expansion only)} \tag{56}$$

It is clear that the shape factor decreases, and hence flow resistance increases, as $Re$ increases.

For the cavity of Figure 2, the slope of the inclined wall is $k = b/d' << 1$, and the aspect ratio is $\beta = 2b/d << 1$. In terms of $\beta$ and $k$,

$$d = \sqrt{\frac{4k}{\beta(4k-\beta)}}$$

(57)

Results for the corrected shape factor are plotted in Figure 3. Here, the slope of the expansion, $k$, is 0.15, and the aspect ratio, $\beta$, ranges from 0.05 to 0.2. For sufficiently small aspect ratios, the expansion is a negligible portion of the cavity, and the shape factor is essentially that for a rectangle alone. Because Re appears only as a product with $\beta$, the effects of inertia are diminished by small aspect ratio. However, the shape factor is small and the resistance of the secondary cavity high. At the maximum possible value of $\beta$, the cavity consists of an expansion alone, and the shape factor is given by Eq. (56). According to these results, the shape factor increases with aspect ratio, but the adverse effect of increasing Re becomes more prominent.

## 6. SOLUTION FOR A TWO-CAVITY DIE

Flow in the secondary cavity is governed by Eq. (55) and that in the primary cavity by Eq. (33). These are augmented by the continuity equation for each cavity, Eqs. (15) and (24), and the equations for the pressure drop across the slots,

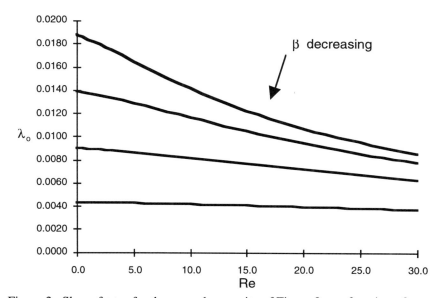

Figure 3. Shape factor for the secondary cavity of Figure 2 as a function of Reynolds number at aspect ratios of 0.05, 0.1, 0.2 and 0.4.

Eqs. (17) and (27). The boundary conditions are

$$Q_o = 0 \qquad (z = 0, \ W) \tag{58}$$

$$Q_i = 0 \qquad (z = W) \tag{59}$$

$$Q_i = Q_e \qquad (z = 0) \tag{60}$$

    The cavity pressures are eliminated from the equations, and the following dimensionless variables are defined

$$f = \frac{Q_i}{Q_e}, \quad g = \frac{Q_o}{Q_e}, \quad \xi = \frac{z}{W}, \quad a = \frac{A_i}{A_e}, \quad m = \frac{L_i}{L_e} \tag{61}$$

Here $f$ is the dimensionless flow rate in the primary cavity, $g$ the dimensionless flow rate in the secondary cavity, $\xi$ the dimensionless coordinate along the cavities, $a$ the normalized cross-sectional area of the primary cavity, and $m$ the normalized length of the primary slot. In terms of these variables the problem is

$$m\frac{d^2f}{d\xi^2} + \frac{dm}{d\xi}\frac{df}{d\xi} + rsg - \varepsilon\frac{f}{a^2}\left\{1 + \vartheta\,\hat{\lambda}_i\,\mathrm{Re}\left[2\frac{df}{d\xi} - \frac{f}{a}\frac{da}{d\xi}\right]\right\} = 0 \tag{62}$$

$$\frac{d^2f}{d\xi^2} + \frac{d^2g}{d\xi^2} - sg = 0 \tag{63}$$

$$f = 1 \qquad (\xi = 0) \tag{64}$$

$$f = 0 \qquad (\xi = 1) \tag{65}$$

$$g = 0 \qquad (\xi = 0, 1) \tag{66}$$

$$m = a = 1 \qquad (\xi = 0) \tag{67}$$

The dimensionless group of parameters $r$ appearing here is the ratio of the resistance of the secondary slot to that of the primary slot.

$$r = \frac{L_o H_i^3}{L_i H_o^3} \tag{68}$$

The equations are general to this point, but to simplify the mathematics, the area of the primary cavity is taken to taper according to Eq. (14)

$$a = (1-\xi)^\gamma \tag{69}$$

The only nonlinear term in the equations is preceded by $\varepsilon$, a small parameter. Therefore, a solution strategy is expanding the dependent variables in powers of $\varepsilon$

$$f \approx (1-\xi) + \varepsilon f_1 + O(\varepsilon^2) \tag{70}$$

$$g \approx \varepsilon g_1 + O(\varepsilon^2) \tag{71}$$

$$m \approx 1 + \varepsilon m_1 + O(\varepsilon^2) \tag{72}$$

The problem at first order in $\varepsilon$ is

$$\frac{d^2 f_1}{d\xi^2} - \frac{dm_1}{d\xi} + rsg_1 = (1-\xi)^{1-2\gamma}\left\{1 - \vartheta\,\hat{\lambda}_i\,\mathrm{Re}(2-\gamma)\right\} \tag{73}$$

$$\frac{d^2 f_1}{d\xi^2} + \frac{d^2 g_1}{d\xi^2} - sg_1 = 0 \tag{74}$$

$$f_1 = g_1 = 0 \qquad (\xi = 0,1) \tag{75}$$

$$m_1 = 0 \qquad (\xi = 0) \tag{76}$$

To solve this problem, equations (73) and (74) are combined to eliminate $f_1$.

$$\frac{d^2 g_1}{d\xi^2} - s(1+r)g_1 = -\frac{dm_1}{d\xi} - \left[1 - \vartheta\,\hat{\lambda}_i\,\mathrm{Re}(2-\gamma)\right](1-\xi)^{1-2\gamma} \tag{77}$$

From the right-hand-side of this equation, the form for the slot taper that compensates perfectly for the pressure loss in the primary cavity can be inferred

$$m = 1 - (1-m_t)\left[1 - (1-\xi)^{2-2\gamma}\right] \tag{78}$$

Here, $m_t$ is the value of the normalized slot length at the end of the cavity, that is, the length at the end divided by that at the center; $(1-m_t)$ is order $\varepsilon$.

Eq. (77) is straightforward to solve once a particular solution has been constructed. The particular solution is especially simple for a linear coat-hanger die, $\gamma = 1/2$. The result for peak-to-peak nonuniformity is given here; complete results are available[11].

$$\frac{q_o(0)-q_o(1)}{\langle q\rangle}=\frac{\left[1-1.5\vartheta\,\hat{\lambda}_i\,Re\right]\varepsilon-(1-m_t)}{1+r}\left\{1+\frac{2[1-\cosh(\alpha)]}{\alpha\sinh(\alpha)}\right\}\qquad(79)$$

$$\frac{q_o(0)-q_o(1)}{\langle q\rangle}\approx\left\{\left[1-\frac{3}{2}\vartheta\,\hat{\lambda}_i\,Re\right]\varepsilon-(1-m_t)\right\}\frac{s}{12}\qquad(\alpha<0.1)\qquad(80)$$

A combination of dimensionless groups appearing here is

$$\alpha=\sqrt{s(1+r)}\qquad(81)$$

Eq. (79) gives the difference between the flow at the center, $q_o(0)$, and that at the end, $q_o(1)$. The final bracketed term of Eq. (79) is plotted in Figure 4 as a function of $\alpha$. The graph shows that the uniformity of the flow distribution degrades as $\alpha$ increases until a plateau is reached. Eq. (80) is identical to Eq. (32) except for the term involving $Re$.

## 7. INTERPRETATION OF THE EFFECT OF THE SECONDARY CAVITY

The flows in the primary and secondary cavities are coupled; a change to the flow in one cavity induces change in the other. Special cases are considered here to explain this coupling.

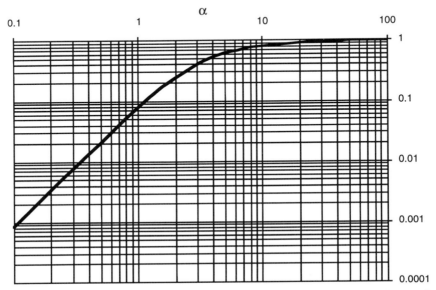

Figure 4. Plot of the final bracketed term in Eq. (77) showing the dependence of the flow distribution on the parameter $\alpha$.

(a) Spatial frequency response of the secondary cavity

Flow distribution in the primary slot is imposed, and the alteration of the distribution by the secondary cavity is computed. The applicable differential equation is

$$\frac{d^2g}{d\xi^2} - sg = \frac{1}{\langle q \rangle} \frac{dq_i}{d\xi} \qquad (82)$$

For the case where the imposed flow distribution has just one spatial frequency component, the amplitude of that component is diminished by the factor $s/(s+n^2\pi^2)$ in which $n$ is an integer. Thus, high spatial frequencies are efficiently damped. Small values of $s$ increase the effectiveness of the secondary cavity.

(b) Uncoupled cavities

For uncoupled cavities, the flow distribution in the primary slot can be estimated by imposing constant pressure at the end of the slot. The resulting flow distribution is then imposed on the secondary slot. The governing differential equation has the same general form as Eq. (77)

$$\frac{d^2g_1}{d\xi^2} - sg_1 = -\left\{\left[1 - \vartheta\hat{\lambda}_i(2-\gamma)\mathrm{Re}\right]\varepsilon - (2-2\gamma)(1-m_t)\right\}(1-\xi)^{1-2\gamma} \quad (83)$$

Comparing Eq. (83) with the coupled analysis, Eqs. (77) and (78), shows that the only difference is that $s(1+r)$ in the coupled analysis is just $s$ here. So, coupling between the cavities effectively increases the value of $s$ and makes the secondary cavity less effective at evening the flow distribution. The group $r$ measures the degree of coupling between the cavities. If the resistance to flow of the primary slot is small compared to that of the secondary slot, the degree of coupling is high, and the outer cavity is less efficient than would be the case if the cavities did not interact.

From Eq. (80), $r$ has no effect if $s$ is sufficiently small. The flows in the cavities uncouple in limit $s \rightarrow 0$, $r$ fixed.

(c) No flow along the secondary cavity

If the flow resistance along the secondary cavity is high, the flow distribution in the primary slot is unchanged by the secondary cavity. The primary and secondary slots can be considered as a single slot of equivalent flow resistance. Final flow distribution and its peak-to-peak variation can be shown to be

$$\frac{q_o(0) - q_o(1)}{\langle q \rangle} = \frac{\left[1 - 1.5\vartheta\lambda_i\mathrm{Re}\right]\varepsilon - (2-2\gamma)(1-m_t)}{(1+r)(2-2\gamma)} \qquad (84)$$

These results are identical to those obtained in the full analysis in the limit $s \rightarrow \infty$, $r$ fixed. In this limit a large value for $r$ is beneficial. The term $(1+r)$ represents the total flow resistance of the primary and secondary slots. An increase in r is favorable because the ratio of metering-slot resistance to cavity resistance increases.

## 8. CONCLUDING REMARK

There is little published work on several aspects of die design. Experimental data for die performance is lacking, perhaps because dies are expensive, or because flow distribution is difficult to measure precisely. The complications of elastic liquids have not been well explored. Finally, the optimization problem for die design has only been introduced here.

References

1. Carley, J. F., "Flow of melts in 'crosshead'-slit dies; criteria for die design," *J. Appl. Phys.*, 25, 1118 (1954)
2. Liu, T., C. Hong, and K. Chen, "Computer-aided analysis of a linearly tapered coat-hanger die," *Polym. Eng. Sci.*, **28**, 1517 (1988).
3. Leonard, W. K., "Effects of secondary cavities, inertia and gravity on extrusion dies," SPE *ANTEC Tech. Papers* , **31**, 144 (1985).
4. Lee, K. and T. Liu, "Design and analysis of a dual-cavity coat-hanger die," *Polym. Eng. Sci.*, **29**, 1066 (1989).
5. Sartor, L., Slot Coating: fluid mechanics and die design. PhD Thesis, University of Minnesota, Minneapolis (1990).
6. Yuan, S., "A flow model for non-Newtonian liquids inside a slot die," *Polym. Eng. Sci.*, **35**, 577 (1995).
7. Durst, F., U. Lange, and H. Raszillier, "Optimization of distribution chambers of coating facilities, *Chem. Eng. Sci.* **49**, 161 (1994).
8. Wen, S. and T. Liu, "Three-dimensional finite element analysis of polymeric fluid flow in an extrusion die. Part I: Entrance effect," *Polym. Eng. Sci.*, **34** (1994).
9. Weinstein, S. J., and K. J. Ruschak, "Asymptotic analysis of die flow for shear-thinning fluids, submitted to AIChE J., (1995a)
10. Weinstein, S. J., and K. J. Ruschak, "One-dimensional equations governing single-cavity die design," submitted to AIChE J., (1995b).
11. Ruschak, K. J. and S. J. Weinstein, "Perturbation model with inertia for flow in two-cavity dies," submitted to *AIChE J.* (1995).
12. Lundgren, T. S., E. M. Sparrow, and J. B. Starr, "Pressure drop due to the entrance region in ducts of arbitrary cross section," *J. Basic Engineering*, Trans. ASME Ser. D, **86**, 620 (1964).
13. Atkinson B., M. P. Brocklebank , and C. C. H. Card, "Low Reynolds Number Developing Flows," *AIChE J*, **15**, 548 (1969)
14. Vergnes, B., P. Saillard, and J. F. Agassant "Non-isothermal flow of a molten polymer in a coat-hanger die," *Polym. Eng. Sci.,* **24**, 980 (1984).
15. Schlichting, H. *Boundary-Layer Theory*, 7th ed., McGraw-Hill, New York (1970).

# GENERALISED JEFFERY-HAMEL FLOW :
# APPLICATION TO HIGH VELOCITY COATING

P. Bourgin, N. Tahiri
Université Louis Pasteur, Institut de Mécanique des Fluides
2, rue Boussingault-67000 Strasbourg (FRANCE)

## SUMMARY

A theoretical study of air recirculations involved in high velocity curtain coating processes is proposed. Assuming that the liquid free surface is known (for example as the result of a computation in which the third phase, namely air, is ignored), the flow configuration is locally assimilated to a wedge. The velocity of the upper wall spatially varies, which mimics the stretching of the liquid sheet and there is a leak at the apex, which corresponds to the amount of entrained air. This modified Jeffery-Hamel flow has been solved following Moffatt's approach. The general aspect of the streamlines is in good agreement with the experimental data available in the literature, as shown in the first example (flow between a steady needle and a translating plate). The second illustrative example resorts of plastic film quenching, where the existence of a stagnation point is predicted. The possible role of this point on air entrainment is discussed.

## 1. INTRODUCTION

One of the limiting factors in high velocity coating flows is the onset of air entrainment into the wedge between the moving solid substrate (generally flexible such as for instance paper or plastic web) and the liquid sheet being continuously coated on it. The effects of this entrapped air depend on the type of contacting process beeing carried out and can lead to bulk effects of froth and foam formation as well as imperfectly coated surfaces due to entrained collapsing bubbles. Experimental studies concerned with the mechanics of air entrainment in dynamic wetting have been reported by several authors since 1964 (1,2,3). Many studies summarised in (4) have shown that above a prescribed coating speed, the coating layer becomes unstable and discontinuous as air becomes entrained in the wedge. These phenomena which are very complex have been described in detail by O'Connell (5). The concept of a "velocity of air entrainment" has been introduced by Burley (6) who proposed various correlations for this critical speed value expressed in terms of non-dimensional groups based on the fluid properties and process conditions. From a theoretical point of view, there is a lack for studies devoted to the micro flows of air near the contacting zone between the liquid and the solid. This is due to the fact that in most theoretical work on dynamic wetting, the third phase (namely : air) is ignored, which is probably the reason why there is a "singularity" at the triple line.

The key idea is based on the fact that the pressure forces generated in the wedge of air are large enough to lift up the portion of liquid in the immediate vicinity of the apparent contact line. Therefore, it is necessary to investigate this region more thoroughly.

## 2. PRESENTATION OF THE THEORETICAL MODEL

### 2.1.Motivation

As a first step, the complex phenomena involved in the three-phase region are left aside. Only the overall amount of air likely to be entrained through the "triple line" is evaluated. The thickness of the equivalent continuous air layer results of the dynamic balance of the forces acting on the liquid layer portion near the contacting region, namely : the lifting force and the external forces of magnetic or electrostatic origins. The use of a computer code in the whole domain is to be avoided due to very large scale ratios between inlet (a few millimetres) and outlet (a few microns). The domain is therefore split into two parts: the upstream zone, where complex 3D flows take place, but without any significant pressure increase ; the contact zone, where the pressure is governed by a Reynolds type equation : see (7).

The present study is devoted to the upstream zone, which is a plane creeping flow if inertial and side-leakage effects are neglected. The configuration is analogous to Jeffery-Hamel classical problem (intersecting planes), but presents two major differences : (i) there is a leak flow at the wedge apex and (ii) the surface velocity of one of the planes may depend on the position (to mimic the stretching of the liquid sheet).

### 2.2. Extension of Moffatt's solution

The basic configuration is sketched in figure 1, where it is seen that the liquid sheet has been locally assimilated to a plane of variable velocity profile.

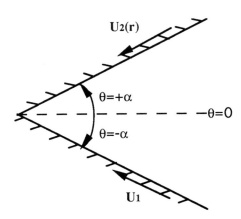

Fig.1 Flow domain

Following Moffatt (8), an analytical solution has been developed by expressing the stream function under the form of a truncated series in planar polar co-ordinates (r, θ) :

$$\psi(r,\theta)= \sum_{i=-M}^{+N} r^i f_i(\theta) \tag{1}$$

and by assuming that the velocity of the upper surface is a generalised polynomial of variable r. The numbers of terms of these two truncated series must be compatible to one another. For example if the following series is assumed for the upper wall velocity profile :

$$U_2(r,\theta)=a+\frac{b}{r}+cr \tag{2}$$

where a, b and c are parameters (see § 3.2), then the stream function must be of the form :

$$\Psi(r,\theta)=rf(\theta)+g(\theta)+r^2h(\theta) \tag{3}$$

where f, g and g are unknown functions of variable θ. Since the stream function is a biharmonic function, an elementary calculation leads to the following ordinary differential equations for functions f, g and h respectively :

$$\frac{d^4f(\theta)}{d\theta^4}+2\frac{d^2f(\theta)}{d\theta^2}+f(\theta)=0 \tag{4}$$

$$\frac{d^4g(\theta)}{d\theta^4}+2\frac{d^2g(\theta)}{d\theta^2}=0 \tag{5}$$

$$\frac{d^4h(\theta)}{d\theta^4}+4\frac{d^2h(\theta)}{d\theta^2}=0 \tag{6}$$

The boundary conditions merely express that the fluid (air) sticks at the walls :

- lower wall :
$$u(r,-\alpha)=-U_1$$
$$v(r,-\alpha)=0 \tag{7}$$

- upper wall :
$$u(r,+\alpha)=-U_2(r)$$
$$v(r,+\alpha)=0 \tag{8}$$

The velocity components are obtained in a straightforward way, and using equations (7,8), the following boundary conditions are derived for functions f, g and h respectively :

$$f'(-\alpha)=-U_1$$
$$f'(+\alpha)=-a$$
$$f(-\alpha)=0$$
$$f(+\alpha)=0$$

(9)

$$g'(-\alpha)=0$$
$$g'(+\alpha)=-b$$

(10)

$$h'(-\alpha)=0$$
$$h'(+\alpha)=-c$$
$$h(-\alpha)=0$$
$$h(+\alpha)=0$$

(11)

The problem is a "well posed problem" for functions f and h for which closed form solutions are obtained :

$$f(\theta)=-(a-U_1)\frac{\cos(\alpha)}{\sin(2\alpha)+2\alpha}\{\theta\sin(\theta)-\alpha\tan(\alpha)\cos(\theta)\}$$
$$-(a+U_1)\frac{\sin(\alpha)}{\sin(2\alpha)-2\alpha}\left\{\theta\cos(\theta)-\alpha\frac{\sin(\theta)}{\tan(\alpha)}\right\}$$

(12)

$$h(\theta)=\frac{c}{4}\left\{\frac{\cos(2\theta)-\cos(2\alpha)}{\sin(2\alpha)}-\frac{2\alpha\sin(2\theta)-2\theta\sin(2\alpha)}{2\alpha\cos(2\alpha)-\sin(2\alpha)}\right\}$$

(13)

As far as g is concerned, there are only two boundary conditions for a fourth order differential equation. Two additionnal B.C. are thus needed. The first one is derived by expressing that there is a leak whose flow rate is prescribed at the apex of the wedge :

$$Q=\int_{-\alpha}^{+\alpha}u(r,\theta)d\theta=g(\alpha)-g(-\alpha)$$

(14)

The stream function being defined within a constant, function g can be written as follows :

$$g(\theta)=\frac{b}{4}\frac{\cos(2\theta)-\cos(2\theta_0)}{\sin(2\alpha)}-\frac{b}{2}(\theta-\theta_0)$$
$$+\frac{Q+b\alpha}{2}\left\{\frac{\sin(2\theta)-\sin(2\theta_0)-2(\theta-\theta_0)\cos(2\alpha)}{\sin(2\alpha)-2\alpha\cos(2\alpha)}\right\}$$

(15)

where $\theta_0$ is a constant which can be interpreted as the value of some angle which cancels g.

Now it can be easily shown that there exists some value of $\theta$, say $\theta_1$, for which function f is null. In order to make it easier the drawing of the streamlines, and particularly the one associated to the stagnation point, we imposed the arbitrary constant $\theta_0$ be equal to $\theta_1$ (the detailed reasons for this choice are not given here).

## 3. APPLICATIONS TO COATING FLOWS

### 3.1. Validation of solution

To our best knowledge, there are no experimental data available in the literature, showing the air flow structure in the vicinity of the three phase boundary in coating flows. We had therefore to search for analoguous configurations letting it possible to make comparisons with our theoretical predictions. Dorémus and Piau (9) used a visualisation experimental set-up to study the flow between a steady small diameter needle and a rotating large diameter drum, seen as a translating plane. They observed that there is a stagnation line at the cylinder surface and they found its location in a particular case. In order to validate our model, at least qualitatively, the circle (projection of cylinder) was assimilated to its tangent drawn from point O (location of minimum gap). In such conditions, the computed location of the stagnation point is 1.6 mm, whereas its experimental value is 1.3 mm : see figure 2. The agreement is pretty good, regarding the poorness of the geometrical approximation.

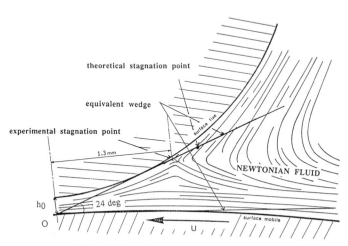

Fig. 2 Experimental streamlines (ref. 9)

### 3.2. Illustrative example

This example comes from polymer processing. A molten resin extruded through a rectangular die is rapidly cooled (quenched) on a rotating cylinder : see figure 3.

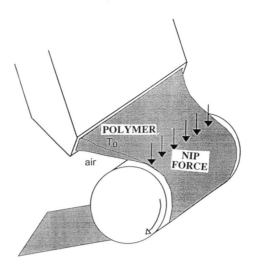

Fig. 3 Configuration of cast film

The challenge in this sort of coating flow is to achieve a good thermal transfer, and then to avoid air entrainment. In a previous study, carried out by Barq (10), the shape of the polymer sheet free surface and its velocity increase between the die outlet and the cylinder surface have been investigated, both experimentally and theoretically. The sheet curvature was shown to be small. Close to the contact, the liquid sheet can be readily assimilated to a plane. As far as the velocity profile is concerned, a typical example is presented in figure 4.

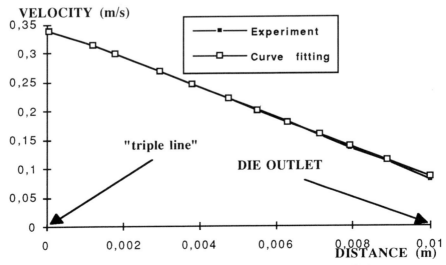

Fig. 4 Velocity profile : experimental data (10) and curve fitting

Using formula (2) to fit the curve in the least square sense, coefficients a, b and c were calculated and the corresponding streamlines were drawn. As shown in Figure 5, there is a large recirculating zone in the vicinity of the stretched plane. A magnification of the region close to the three-phase boundary allows the stagnation point location to be accurately determined : see figure 6.

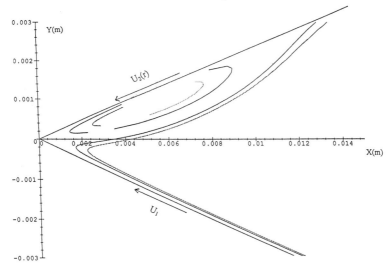

Fg. 5 Sketch of the streamlines

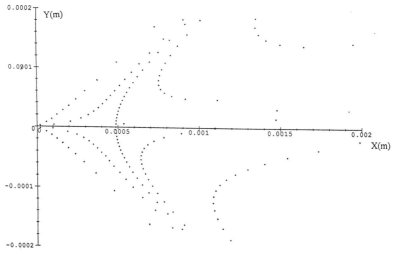

Fig.6  Streamlines in the vicinity of the stagnation point

## 4. CONCLUDING REMARKS

A simplified model has been proposed to describe the air flow kinematics upstream of the three-phase boundary typical of high velocity dynamic wetting. Neglecting inertia and side leakage effects, the stream function is governed by the bi-laplacian equation. The boundary conditions account for two important features : (i) the apex is not "locked", but there is a leak flow, (ii) the upper wall (which corresponds to the liquid sheet) is stretched. Following Moffatt's approach, this modified corner flow problem has been solved under the form of a series development for the stream function. In the particular cases considered here, a large eddy and one stagnation point have been predicted.

The future developments of the analysis will consist in studying the "contacting region" where the pressure generated in the air wedge deforms the liquid sheet shape. The stagnation point is considered as a good candidate to match the "upstream region", where no pressure rise occurs, with the "contacting region". Therefore, a preliminary study showing the influence of the stretching rate of the upper wall on the existence, unicity and location of the stagnation point is to be carried out.

## 5. REFERENCES

(1)     Burley, R. and Kennedy, B.S. (in) *Wetting, Spreading and Adhesion*, ed. J. Padday, Academic Press, London & New York, (1978).

(2)     Burley, R. and Jolly, R.P.S. Entrainment of Air into Liquids by a High    Speed Continuous Solid Surface, Chem. Eng. Sci., **39**, 9, 1357-1372   (1984).

(3)     Deryagin, B.M. and Levi S.M. *Film Coating theory* Focal Press, London, (1964).

(4)     Burley, R. Air Entrainment and the Limits of Coatability, JOCCA, **77**, 5, 192-202 (1992).

(5)     O'Connell, A. *Air Entrainment and the Limits of Coatability*, PhD Thesis, Heriot-Watt University, Edinburgh, (1989).

(6)     Burley, R. Fluid Behaviour between Horizontal Pad Rollers, J. Text. Inst. & Industry, **9**, 8, 164-188 (1971).

(7)     Emonot, J. *Contribution à l'Etude du Contact Dynamique entre une Nappe de Liquide et un Substrat Solide* , Thèse de Doctorat, Université de Lyon, (1994).

(8)     Moffatt, H. K. Viscous and Resistive Eddies near a Sharp Corner, J. Fluid Mech., **18**, 1-18 (1964).

(9)     Dorémus, P. and Piau, J.M. Experimental Study of Viscoelastic Effects in a Cylinder-Plane Lubricated Contact, J. Non-Newt. Fluid Mech., **13**, 79-91 (1983).

(10)   Barq, P. *Etude Théorique et Expérimentale de l 'Extrusion de Film à Plat. Application au Poly(éthylène)téréphtalate*, Thèse de Doctorat, l'Ecole Nationale Supérieure des Mines de Paris, (1992)

## 6. ACKNOWLEDGEMENTS

The authors wish to thank the French Ministry of Research and Education and BASF Company for their financial support.
They also acknowledge Professor G. M. Homsy for fruitful discussions during the Symposium .

# Recirculating Flows in Curtain Coating

A.Clarke

Kodak Limited, Research Labs.

Headstone Drive, Harrow, UK HA1 4TY

Recirculating viscous eddies are well known in liquid coating processes including curtain coating where they have been seen at low substrate speeds and high flow rates [1]. Recently, Gaskell et al [2] introduced a dynamical systems description of these flow structures in roll coating. This description topologically classifies the components (centres, saddle points, separatrices and snakes) of these structures and, in addition, systematically describes their evolution as a function of capillary number and flow rate. Flow visualisation techniques [3] using aqueous glycerol solutions have revealed that these components also exist in curtain coating. Experimental examples of each component will be presented together with a systematic study of their evolution as a function of both substrate speed and curtain flow rate.

## Introduction

Recirculating viscous eddies are endemic in all coating methods. Within a certain range of the controlling parameters of flow rate, substrate speed, viscosity, etc. there will exist one or more stable eddies within the flow. Flow visualisation techniques have graphically shown the existence of these structures, for example in curtain coating [1,4], forward roll coating [2], bead coating [5] and slot coating [6]. To date however, there has been no systematic experimental investigation of the way in which these eddies evolve as a function of the key process parameters. Recently, Gaskell, Savage and Summers [2] have described the recirculating eddies seen in their roll coating experiments in terms of four topological elements; centres, saddles, seperatrices and snakes. They have found the same stuctures in simple model calculations which show a wide variation of response to the imposed flow parameters. We have found similar behaviour in curtain coating flows; examples of each of the four topological elements have been observed.

Below we present flow visualisation measurements of the recirculating viscous eddies seen in curtain coating. The data show how the observed eddies evolve with curtain flow rate, substrate velocity and viscosity.

## Experimental

The experimental arrangement is based upon observing hydrogen bubble tracers travelling within the liquid flow. A detailed description of the techniques and apparatus may be found elsewhere [1,3], however for reference, figure 1 shows the experimental arrangement. A falling curtain of liquid is extruded from a hopper and impinges on a moving substrate. Near the top of the curtain, electrodes touch the two free liquid surfaces forming streams of hydrogen bubbles which index each surface. There is also a thin wire through the curtain that provides random bubble generation across the thickness of the curtain. A light "sheet" provides illumination perpendicular to the viewing direction so that the bubbles appear bright against a dark field.

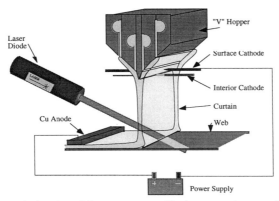

Fig.1. Schematic drawing of the apperatus used in these experiments. See text for a detailed discussion.

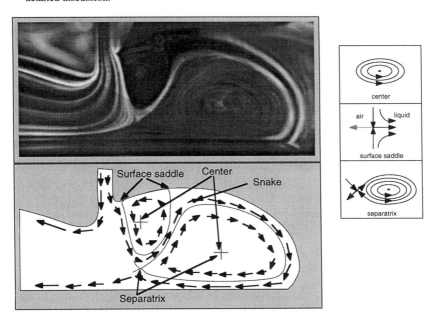

Fig.2. A flow visualisation image exhibiting the various flow patterns discussed in the text.

Fig.2. shows the result of summing several video frames together; both the flow profile and recirculation structure are clearly seen. In the lower panel of the figure, the flow profile together with the key points and surfaces have been abstracted from the image. In this example, the four topological structures can all be seen. The substrate is moving from right to left and the bank of liquid on the upstream (right hand) side of the curtain shows two recirculating eddies; one internal to the flow, and one attached to the surface of the flow. The former consists of a center bounded at the left with a saddle, this combination is termed a separatrix. The latter consists of a center and two surface saddle points. A narrow band of streamlines travel below the upper eddy and arround the separatrix forming a "snake".

In the current work we are interested in following the flow within recirculation structures at the base of the curtain. Since we are using hydrogen bubbles as our tracer particles and are interested in structures that necessarily have long residence times, we must be concerned about possible floatation of the bubbles leading to erroneous results. In order to calculate the flotation distance we use Stokes Law;

$$V = \frac{d^2 g \Delta \rho}{18 \eta} \qquad\qquad - (1)$$

taking the particle diameter, d, as 10μm, and a viscosity of 1mPas as the worst case, then V, the velocity of flotation, is 55 μms$^{-1}$. Assuming that the recirculation region rotates at S/πD revs s$^{-1}$, where D is the diameter of the region, and S is the substrate velocity, then taking S as 25cm s$^{-1}$ and D as 2mm the maximum deviation from the correct location is estimated as 0.8μm. Note that within the recirculation region the sense of flotation reverses every half revolution, so that a bubble trapped within the region should return to the correct position every half revolution.

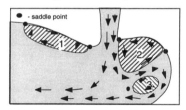

Fig.3. Schematic drawing of the location of the three observed recirculation regions.

In the following experiments, we have used aqueous glycerol solutions of varying concentration. Each solution has a small addition of salt ($NaHCO_3$) in order to enhance the generation of hydrogen bubbles via electrolysis. We chose four viscosities and investigated a range of flow rates and web speeds for a curtain height of 3cm.

We observed three recirculation regions labeled 1,2 and 3 in the schematic diagram of fig. 3. Region 1, located on the downstream side of the curtain, contains a centre bounded by two surface saddle points. Region 2 is the equivalent region on the upstream side of the curtain and again contains a centre bounded by two surface saddle points. Region 3 is internal to the flow and is a centre bounded on the downstream side by a saddle point. From the viewpoint of the diagram, regions 1 and 3 rotate clockwise, whereas region 2 rotates counterclockwise. Intuitively one might expect these three regions to exist. Regions 1 and 2 being simply the symmetric recirculation regions observed when a jet of liquid impacts a stationary tank of liquid, while region 3 is expected to be induced by the motion of the substrate. Table 1 shows that all three recirculation regions can be easily observed with all but the highest viscosities.

**Table 1**. For each viscosity used and each possible location of recirculation eddy the observed eddies are noted. "not observed" means that the eddy was not seen within the parameter range investigated, i.e. the solutions were investigated to a maximum flow rate of $3.81 \text{cm}^2/\text{s}$ at web speeds above 10.6cm/s for the 74.5mPas solution and 7.1cm/s for the 74.5mPas solution (see fig.6).

| viscosity (mPas) | zone 1 | zone 2 | zone 3 |
|---|---|---|---|
| 13.5 | √ | √ | √ |
| 25 | √ | √ | √ |
| 74.5 | not observed | √ | √ |
| 135 | not observed | √ | √ |

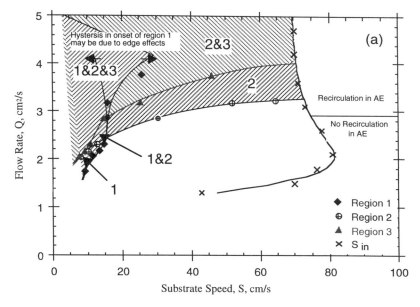

Fig.4. Plot showing regions of flow rate and substrate speed where each recirculation zone may exist. This data was obtained using 25mPas aqueous glycerol and a 3cm curtain.

The graph in figure 4 shows the range of flow rate per unit width (Q) and substrate speed (S) over which each of the three recirculation regions exist for a solution of 25mPas viscosity. Also shown in the figure is a curve describing the onset of air entrainment on increasing substrate speed. Some general trends can be seen within the plot;

- As the substrate speed increases, the flow rate for the onset of each recirculation zone also increases.

- The flow rate for the onset of region 3 is greater than the flow rate for the onset of region 2. It should be noted, however, that this sequence is reversed at higher viscosities.

- The behaviour of region 1 is significantly different to that of regions 2 and 3. Hysteresis was observed in the onset of region 1, i.e. on reducing substrate speed the region appeared at a lower speed than it cleared when increasing substrate speed.

Figure 5 shows the same data as figure 4 plotted together with outlines taken from image data showing in each case the recirculation regions. At moderate to high flow rates the region least stable to speed increase is the downstream region, region 1. The next least stable region is the internal upstream region, region 3, with region 2 being the most stable region. At low flow rates this order is altered, with the region most stable to speed increase being region 1.

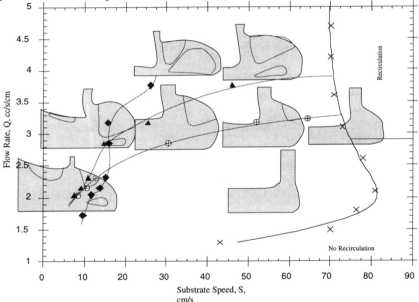

Fig. 6. The same data as shown in fig. 5, but here with outlines of the liquid flow taken from flow visualisation images. Note that the individual outlines are not drawn to the same scale.

Figure 6. shows, for solutions of viscosity (a) 74.5 mPas and (b) 125 mPas, the progression of flow patterns with $Q = 3.81 \text{cm}^2/\text{s}$ as the substrate speed is increased. Each image in the figure was generated by simply summing several sequential frames from a video tape of the experiment. In each case, as the substrate speed is increased the wetting line moves closer to the curtain and the size of the upstream bank decreases. As the bank diminishes in size, the recirculations also diminish and finally dissapear. Note however that the recirculation ceases well before the bank is fully removed. Comparing images at similar substrate speeds of 10.6cm/s (a5) and 10.8cm/s (b2), it can be seen that the higher viscosity liquid has a considerably smaller bank. This might be expected from a force balance along the substrate surface as given in [1], although the parameters in this case are well outside the range of validity of that analysis. Comparing the same two images, it can also be seen that whereas a5 shows both region 2 and 3, image b2 shows only region 3. It may be interesting to note that the saddle point defining the downstream side of the separatrix of region 3 appears to be approximately one curtain width behind the back of the curtain in all ten images.

The raw images such as those shown in figure 6 have been analysed in two ways. Firstly, the changes in position of the various saddle points and centres associated with each recirculation region have been plotted as a function of both flow rate and substrate speed. Secondly, the substrate speed at which each recirculation region vanishes has been measured as a function of both flow rate and solution viscosity.

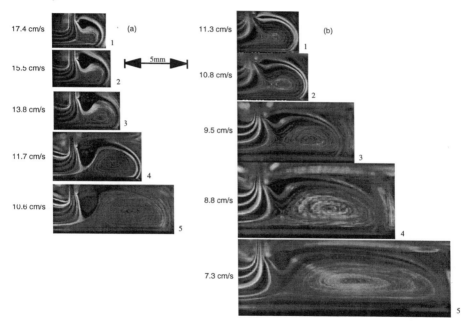

Figure 6. Images showing recirculation patterns for (a) 74.5mPas and (b) 125mPas aqueous glycerol at a flow rate of 3.81 cm2/s. All images are to the same scale. Note that as the substrate speed increases, the heel diminishes in size and the recirculation regions progressively dissapear. In (a) region 3 dissapears before region 2, whereas in (b) the sequence is reversed.

Figure 7 shows an outline of the flow profile for an aqueous glycerol solution of viscosity 36mPas. The substrate speed is 24.7 cm/s and the flow rate is 3.22 cm$^2$/s. Overlaid on the outline is a plot of the location of centres and saddles describing recirculation regions 2 and 3. The axes are scaled in μm and are plotted such that the origin is located on the substrate surface directly under the front surface of the curtain. The two surface saddle points (A & C) together with the associated centre (B) describe region 2 whilst the saddle point (E) and centre (D) describe region

3. Data are also plotted to show the locus of each saddle and center as the substrate speed is changed. As the substrate speed is increased, points A, B and C move together until at approximately 64cm/s they coalesce and region 2 is destroyed. Similarly, points D and E also move together and eventually coalesce, in this case at approximately 25cm/s. Figure 8 shows plots of similar data presented in a slightly different form. In this case, substrate speed is plotted as a function of the distance between a saddle and associated centre. Figure 8a shows how the distances BA and BC vary with substrate speed for three seperate solutions at three flow rates. Similarly, figure 8b shows how the distance ED varies with substrate speed, viscosity and flow rate. In both cases (figs 8a and b) for a given substrate speed, the recirculation region increases in size as flow rate increases. Also, as the viscosity increases, the size of the recirculation region at a given substrate velocity decreases. These two observations are true for both regions, however the size of region 3 is considerably more sensitive to substrate speed as indicated by the slopes of the respective graphs.

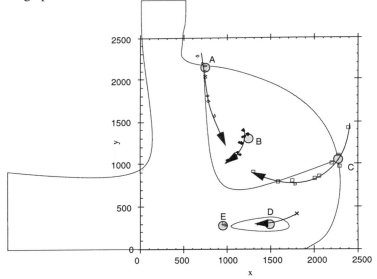

Fig. 7. Plot of saddle points and centers as a function of web speed. The other coating parameters are 3cm, Q=3.22 cm /s and viscosity of 36mPas. The flow outline is for 24.7cm/s. The large circles show where the five points (3 saddles and 2 centers) are for this flow. The x and y spatial axes have scales in μm. To interpret this plot imagine the 'substrate speed' axis pointing out of the paper - the curves for each type of saddle / center then trace out a line in 3D space. The direction of the arrows shows how the points move as speed is increaseed. Note that the saddle point labeled (E) apparently stays fixed relative to the front of the curtain and the web. Points D and E collide and anihilate at a speed of ~25cm/s . Points A,B and C collide and anihilate at a web speed of ~64cm/s.

Figure 9a shows a plot of flow rate against substrate speed. The data points correspond to the maximum speed at which,  for a given flow rate, recirculation is observed. The set of points for each recirculation region consists of data obtained with several viscosities.

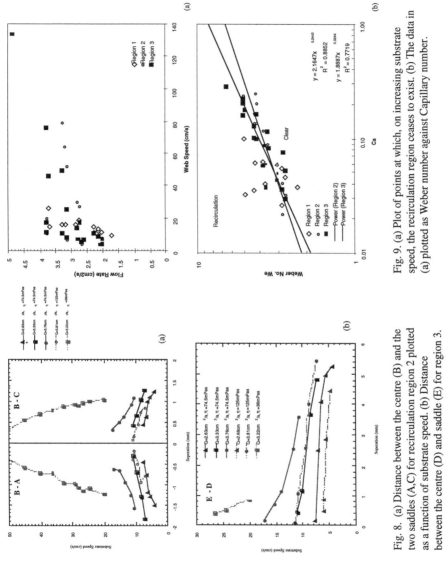

Fig. 8. (a) Distance between the centre (B) and the two saddles (A,C) for recirculation region 2 plotted as a function of substrate speed. (b) Distance between the centre (D) and saddle (E) for region 3.

Fig. 9. (a) Plot of points at which, on increasing substrate speed, the recirculation region ceases to exist. (b) The data in (a) plotted as Weber number against Capillary number.

It can be seen that there is a considerable spread of the data with no discernible trends. In figure 9b the same data have been plotted in such a way that there is a minimal spread of the points on the graph. It was found empirically that Weber number plotted against Capillary number fulfils this criteria. Although the data for recirculation regions 2 and 3 do not overlap completely, the curves have differing slopes, the plot is effective in collapsing the data to reveal a reasonably distinct boundary between regions where recirculation exists and where it does not. Recirculation region 1 can be seen to have a significantly different boundary.

## Discussion

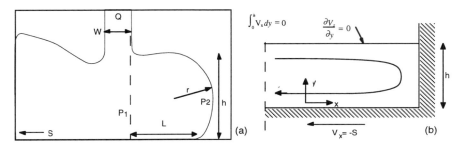

Fig. 10. (a) Shematic of the coating flow showing symbols used in the text. (b) The analogous flow used for the argument developed in the text.

The observed relation between Weber number and Capillary number at the onset of recirculation (fig. 9) may be explained fairly simply. We may construct a model that provides a qualitative understanding as follows; we suppose that it is the stagnation pressure developed at the base of the curtain which provides the driving force to move liquid into the recirculating bank. This stagnation pressure is balanced by the capillary pressure due to the curvature at the back surface of the bank, together with the viscous pressure gradient developed by the drag of the moving support. With reference to figure 10a, we can estimate the pressure close to the saddle point (E in fig 8.) by considering the Bernoulli equation, this leads to

$$P_1 + 0 = P_A + \frac{1}{2}\rho\frac{Q^2}{W^2}$$

- (2)

where $P_A$ is atmospheric pressure, $Q$ is the curtain flow rate, $W$ is the thickness of the curtain and $\rho$ is the density. The pressure $P_2$ just inside the back surface of the bank will be simply

$$P_2 \approx P_A + \frac{\sigma}{r}$$

- (3)

where $\sigma$ is the surface tension and r is the characteristic radius of curvature. The pressure gradient across the recirculating bank will therefore be,

$$\frac{P_2 - P_1}{L} \approx \frac{\sigma}{rL} - \frac{\rho Q^2}{2LW^2}$$

- (4)

where L is the length of the recirculating bank of liquid. The pressure gradient may also be estimated by considering the equivalent flow shown in figure 10b. Assuming the $V_y$ components are unimportant for the qualitative approach taken here, the Navier-Stokes equation gives

$$\frac{dP}{dx} = \eta \frac{\partial^2 V_x}{\partial y^2} \qquad - (5)$$

Equation (5) may be solved subject to the following boundary conditions,

$$V_x = -S, \text{ at } y = 0$$

$$\frac{\partial V_x}{\partial y} = 0, \text{ at } y = h \qquad - (6)$$

$$\int_0^h V_x dy = 0$$

where the first condition is substrate speed, the second condition is zero stress at the free surface and the third condition is that there is no net flow into or out of the recirculating bank. Equating the pressure gradients and rearranging terms then gives,

$$\left( \frac{\rho Q U}{\sigma} \right) \approx \left( \frac{2W}{r} \right) + \left( \frac{6WL}{h^2} \right) \left( \frac{\eta S}{\sigma} \right) \qquad - (7)$$

where U is the curtain velocity evaluated as Q=UW. Examining this result, we see that the Weber number, We $\approx f_1(\text{Ca},W/r,WL/h^2)$, where Ca is the capillary number. If there is then some critical values of the geometric ratios W/r and $WL/h^2$ for which recirculation will start, then this argument suggests that we should see We = $f_2(\text{Ca})$, at the onset of recirculation.

Equation (7) also demonstrates that the geometric ratios of the liquid bank depend on We and Ca. In particular, we would expect that as inertial forces ($\rho Q U$) increase at constant h, the length of the bank (L) would also increase, whereas as the viscous forces increase ($\eta S$), the length (L) should decrease. In figure 8b the distance E-D is plotted, this is roughly proportional to the length L. We see that for a given substrate speed, as Q increases (i.e. increasing inertia) so too does the distance E-D. Also, again for a given substrate speed, as viscosity increases the distance E-D shrinks as expected.

## Acknowledgements

The author would like to thank Dr. S.J.Weinstein for many invaluable transatlantic discussions and in particular for providing the physical explanation presented in the discussion. He is also indebted to Dr. T.D.Blake for careful reading for this manuscript.

## References

[1]    T.D.Blake, A.Clarke, K.J.Ruschak *AIChE Journal* **40**, 2, 229 (1994).
[2]    P.H.Gaskell, M.D.Savage and J.L.Summers, *AIChE 7th International Coating Process Science and Technology Symposium*, Atlanta, April 1994.
[3]    A.Clarke, *Chem. Eng. Sci.* **50**,15, 2397-2407 (1995).
[4]    L.E.Scriven, W.J.Suszynski, *Chem.Eng.Prog.* Sept. (1990).
[5]    P.M.Schweizer, *J. Fluid Mech.* **193**, 285-302 (1988).
[6]    D.Cohen, W.Suszynski, L.E.Scriven, *AIChE 6th International Symposium on Coating Science and Technology*, March 1992.

# NUMERICAL MODELING OF MULTILAYER SLOT COATING FLOWS

K. Kumar, S. Subbiah

*Fluent Inc.,*
*10 Cavendish Court,*
*Lebanon, NH 03766 USA*

and

P. Murtagh

*Fluent Europe Ltd.,*
*Holmwood House, Cortworth Road,*
*Sheffield, S11 9LP, England*

## SUMMARY

The spectral element method is successfully applied to the modeling of single- and two-layer slot coating flows. The free surface tracking and mesh deformation strategy are described first. The numerical model is then validated by comparing solutions for the flow structure and stability of single-layer slot coating flows with published experimental results. Finally, a two-layer slot coater is modeled where the fluid/fluid layer is treated as a distinct free surface, the position and shape of which is determined as part of the solution. The flow structure for this flow is then described.

## 1. INTRODUCTION

The objective of the present study is to analyze various slot coating flows, including multilayer flows, using the spectral element method. A description of the free surface tracking algorithm and mesh deformation strategy is given. For single-layer slot coating flows, we validate the computational model by comparison with published experimental observations of the flow structure and stability limits. For a two-layer slot coating flow, we verify the prediction of the flow structure and comment on the tracking of the additional free surface (the internal fluid/fluid interface).

Pioneering work in the analysis of slot caters was conducted by L. Sartor and L. E. Scriven at the University of Minnesota[1]. This work included both experimental measurements and flow visualization of single-layer slot coating, as well as numerical modeling using the finite element method (FEM). Experimental investigations into two-layer slot coating were performed by D. Cohen[2] and L. E. Scriven. FEM results for a two-layer coating were presented by M. Stevanovic and A. Hrymak[3]. In this work, the fluids were treated as miscible (that is, there was no internal free boundary).

In the present work, we use the spectral element method to model both single- and two-layer slot coating flows. For the two-layer problem, the fluids are treated as immiscible, therefore the model must predict the position and shape of the internal and external free boundaries. The free surfaces are tracked using an elastic mesh deformation strategy developed by L.-W. Ho[4]. The commercially-available software package NEKTON from Fluent Inc. was used to generate all of the results presented in this paper.

## 2. NUMERICAL MODEL

The governing equations for the fluid flow problem, namely the conservation of mass and momentum, are solved using the spectral element method. The spectral element method is a high-order discretization scheme utilizing unstructured meshes consisting of large elements. The basis functions are Legendre polynomials ranging from 4th-order through 14th-order. The polynomial order used to expand the pressure field ($P_{N-2}$) is actually two orders less than that used for the velocity field ($P_N$). A more complete description of our implementation of the spectral element method is given in references 5 and 6.

For the solution of coating flows, the location of one or more free surfaces must also be determined. We use an Arbitrary Legrangian Eulerian (ALE) algorithm to decouple the mesh velocity from the fluid velocity. Our deforming mesh strategy is summarized in the following steps:

1. On the current mesh, solve the Navier-Stokes equations (for the velocity and pressure fields). For 2-D coating flows, a direct Newton iteration is used.
2. Based on the new fluid flow solution, compute the new free surface position based on a stress balance or kinematic boundary condition.
3. Deform the remainder of the mesh so that it conforms to the new free surface position by treating it as an elastic body.
4. Update all flow coefficients based on the new geometry.
5. Go to step 1 and repeat until convergence.

Details regarding the surface intrinsic representation of the free surface are available in references 7 and 8, and more information on our steady state free surface solution algorithms is given in reference 9.

## 3. RESULTS

### 3.1. Single-layer Slot Coater Validation — Flow Structure

The first case to be considered is a single-layer slot coater for which experimental flow visualization and computational results are available[1]. The geometry and operating parameters are summarized in Table 1. The spectral element mesh used to solve this problem is shown in Figure 1. Sixth-order basis functions were used.

Table 1: Parameters for Single-layer Slot Coater

| Slot gap height | 250 microns | | Web speed | 13.3 cm/s |
|---|---|---|---|---|
| Slot lip length | 0.1 cm | | Flow rate | 0.10906 cm$^3$/s-cm |
| Lip to web distance | 500 microns | | Wet film thickness | 82 microns |
| Applied vacuum | 1.7 in. water | | Dyn. contact angle | 100 degrees |

The boundary conditions for this case are illustrated in Figure 2. The inlet velocity is specified as a fully-developed parabolic profile corresponding to the desired flow rate of 0.10906 cm$^3$/s-cm. The upstream meniscus is assumed to be pinned at the lower extremity of the die face (a contact angle of "0" implies that the static contact angle will be calculated as part of the solution). Where the free surface meets the web, the dynamic contact angle is specified as 100 degrees and the dynamic contact line position will be calculated as part of the solution. The upstream free surface is exposed to a vacuum pressure equivalent to 1.7 inches of water. The downstream meniscus is exposed to the ambient. The outflow is treated as a constant pressure boundary equal to the ambient.

The computed free surface positions and flow streamlines are shown in right side of Figure 3. Note that the upstream meniscus has become curved and the dynamic contact line has moved downward slightly relative to the initial mesh. The stream function contours indicate the primary flow path exits the inlet channel and turns downward towards the upstream meniscus where it turns and meets the web. There is a large recirculation zone adjacent to the downstream die lip.

The main flow features, including the position and shape of both free surfaces, the location of the dynamic contact line and the stream line patterns, compare quite well with the expermental flow visualization of Sartor. as shown in the side-by-side comparison in Figure 3. Sartor also performed finite element simulations of this flow that are also in good agreement with our results.

### 3.2. Single-layer Slot Coater Validation — Stability Analysis

As the flow rate into the slot die is reduced, the wet film thickness is correspondingly reduced until the flow becomes unstable. If the position of the dynamic contact point is tracked as the flow rate is reduced, Sartor1 has shown that this locus exhibits a "turning point" as the flow moves from a stable regime to an unstable one (or vice versa). Thus, tracking the dynamic contact line position serves as a useful tool to predict the limits of stable coating operation.

Lee, Liu and Liu[1]0 have published experimental measurements of the minimum stable wet film thickness for a slot coater operating over a range of Capil-

Table 2: Parameters for Minimum Wet Thickness Analysis

| Slot width | 0.025 cm | Density | 0.95 g/cm$^3$ |
|---|---|---|---|
| Lip to web distance | 0.02 cm | Surface tension | 21 dynes/cm |
| Web speed | 10 cm/s | Dyn. contact angle | 94 degrees |
| Viscosity | 0.1 poise | Stat. contact angle | 60 degrees |

lary numbers. By choosing a particular web speed, the NEKTON software was used to model this coater and to perform a parametric sweep by varying the flow rate. For each of the converged steady state solutions, the location of the dynamic contact line was plotted versus flow rate. The operating conditions simulated are listed in Table 2.

As shown in Figure 4, the fluid does not occupy the entire slot gap but rather the dynamic contact line is located adjacent to the slot channel exit even for moderate flow rates (h is the wet film thickness and V is the constant inlet velocity in this figure). As the flow rate is decreased (i.e., smaller inlet velocity, V), the wet film thickness (h) is correspondingly reduced and the dynamic contact line moves upward on the web considerably. Figure 5 shows a plot of the dynamic contact (or wetting) line location as a function of wet coating (or film) thickness (which is directly proportional to the flow rate). The dynamic contact line position moves upward significantly as the minimum wet film thickness is approached. Below a wet film thickness of 0.0062 cm, no steady states were found.

The minimum wet thickness predicted by NEKTON for a web speed of 10 cm/s is compared with data of Lee, Liu and Liu in Figure 6. The agreement is good, with the numerical prediction being slightly lower than the experimental data.

## 3.3. Single-layer Slot Coater Validation — High Vacuum Limit

Using a procedure similar to that described above, we have also studied the behavior of the upstream meniscus with respect to different levels of applied vacuum. For a given web speed, the maximum vacuum pressure for stable coating can be determined. A comparison was done with the experimental measurements and numerical predictions of Sartor[1] for a web speed of 100 feet/minute. The other operating conditions are listed in Table 3. For the critical vacuum pressure, Sartor gives an experimental value of approximately 9.34 inches of water ($\pm 4$ percent) and a numerical (FEM) value of 9.84, certainly in good agreement with the present study in which we obtained a value of 9.03.

Table 3: Parameters for High Vacuum Limit Case

| Slot width | 0.0271 cm | | Density | 1.19 g/cm$^3$ |
|---|---|---|---|---|
| Lip to web distance | 0.025 cm | | Surface tension | 61 dynes/cm |
| Viscosity | 25 cP | | Dyn. contact angle | 100 degrees |

### 3.4. Two-layer Slot Coater Results — Free Surface Positions and Flow Structure

The fluid dynamics of a two-layer slot coater are now considered. In this problem, a third free surface boundary exists, namely the fluid/fluid interface, for which the position must be determined. We apply the same free surface tracking algorithm described earlier to this boundary also.

The operating conditions and fluid properties for the two layer slot coater are described in Table 4. Note that there is a 10:1 viscosity ratio and a 4:1 flow rate ratio between the upper and lower fluid layers.

The initial spectral element mesh is shown in Figure 7. The liquid/liquid interface is tracked as the boundary between adjacent fluid elements. The initial position and shape of the interface were specified based on the known flow rate ratio and on the results of a preliminary simulation of the flow in this geometry using only one fluid. The interface is assumed to be pinned at the corner of the exit from the inlet channel for the top layer. The static contact line for the upstream meniscus is also pinned at the location shown. The dynamic contact line position was free to slide along the web at the specified contact angle of 125 degrees.

The final shape of all three free surfaces predicted by NEKTON is shown in Figure 8. The upper layer has "bowed out" some as it exits the slot and actually pushes upstream into the lower layer. The dynamic contact line has moved downstream significantly.

Contours of stream function are shown in Figure 9. This plot shows the existence of a rather large recirculation zone below the inlet for the lower layer. Thus, the primary flow path for the lower layer turns towards the upstream meniscus as it exits the slot die, then turns and meets the web.

### 4. Conclusions

The spectral element method has been successfully applied to several slot coater geometries, including both single-layer and two-layer flows. The predictions of flow structure, free surface locations and stability limits compare well with published experimental observations. The ALE algorithm, coupled with an elastic mesh deformation strategy, has been shown to be a convenient and

Table 4: Parameters for Two-Layer Slot Coater Analysis

|  | Upper layer | Lower Layer |
|---|---|---|
| Viscosity | 15 poise | 1.5 poise |
| Density | 1 g/cm$^3$ | 1 gm/cm$^3$ |
| Surface tension | 30 dynes/cm | 30 dynes/cm |
| Inlet velocity | 7.6 cm/s | 1.9 cm/s |
| Vacuum |  | 6 in. water |
| Web speed |  | 19 cm/s |
| Dynamic cotnact angle |  | 125 degrees |

robust way to track the positions of both internal and external free boundaries.

## 5. REFERENCES

1. L. Sartor, "Slot Coating: Fluid Mechanics and Die Design", Ph.D. Dissertation, University of Minnesota, 1990.

2. D. Cohen, "Two-Layer Slot Coating: Flow Visualization and Modeling", M.S. Dissertation, University of Minnesota, 1993.

3. M. Stevanovic and A. Hrymak, "Computer Simulation of Multilayer Slot Coating Flows", AIChE Meeting, Atlanta, GA, April 1994.

4. L.-W. Ho, "A Legendre Spectral Element Method for the Simulation of Incompressible Unsteady Viscous Free Surface Flows", Ph.D. Dissertation, Massachusetts Institute of Technology, 1989.

5. *NEKTON Version 2.9 User's Guide, Vol. II*, Fluent Inc., 1994.

6. E. M. Ronquist and A. T. Patera, "A Legendre Spectral Element Method for the Stefan Problem", *Int. J. Num. Meth. Eng.*, Vol. 24, pp. 2273-2299, 1987.

7. L.-W. Ho and A. T. Patera, "Variational Formulation of Three Dimensional Viscous Free Surface Flows: Natural Imposition of Surface Tension Boundary Conditions", *Int. J. Num. Meth. Fluids*, Vol. 13(6), pp. 691-698, 1991.

8. L.-W. Ho and A. T. Patera, "A Legendre Spectral Element Method for Simulation of Unsteady Incompressible Viscous Free Surface Flows", in *Spectral and High Order Methods for Partial Differential Equations*, Proc. of the ICOSAHOM '89 Conference, Como, Italy, Eds. C. Canuto and A. Quarteroni, North-Holland, pp. 355-366, 1990.

9. L.-W. Ho and E. M. Ronquist, "Spectral Element Solution of Steady Incompressible Viscous Free-Surface Flows", from Proc. of the Inter-

national Conference on Spectral and High Order Methods for Partial Differential Equations, Montpellier, France, 1992.

10. K.-Y. Lee, L.-D. Liu and T.-J. Liu, "Minimum Wet Thickness in Extrusion Slot Coating", *Chem. Eng. Sci.*, Vol. 47, No. 7, pp. 1703-1713, 1992.

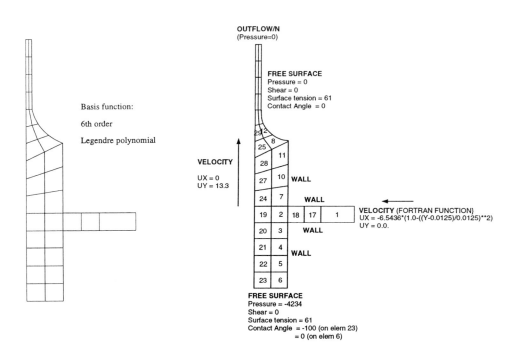

Figure 1.

Figure 2.

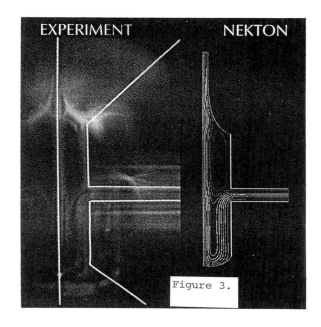

Figure 3.

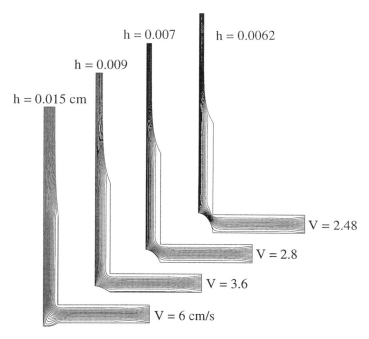

h = 0.0062

h = 0.007

h = 0.009

h = 0.015 cm

V = 2.48

V = 2.8

V = 3.6

V = 6 cm/s

Figure 4.

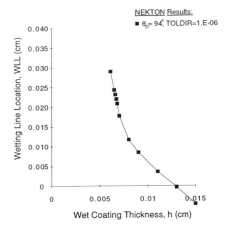

Figure 5.

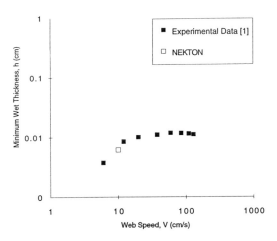

Figure 6.

Figure 7.

Figure 8.

Figure 9.

# TOWARDS A NUMERICAL SIMULATION OF THIN LIQUID DROPS

Andreas Münch

*Institut für Angewandte Mathematik, Technische Universität München*
*Postfach 20 10 32, 80010 München*
*email:* `amue@appl-math.tu-muenchen.de`

### SUMMARY

In the case of viscous fluid films with a free surface, standard lubrication theory yields one- and two-dimensional, degenerate, fourth order parabolic equations describing the evolution of the film profile $h(x,t)$. In spite of their widespread use in practical numerical simulations, investigators have only very recently begun to acknowledge some of the unique analytical properties which makes these equations specially adapted to modeling thin liquid films. A survey of recent analytical results in one space dimension is given in this article, and a numerical example is included to indicate that some of these results can be expected to apply in two dimensions as well.

## 1. INTRODUCTION

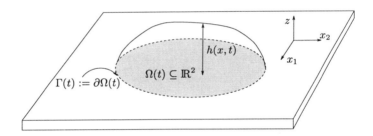

Figure 1: A spreading droplet on a horizontal wall.

Although *lubrication theory* was originally introduced to describe the behaviour of closed thin fluid films, as in journal bearings, widespread use has been made for a considerable time in the treatment of films which feature a free surface, as is for example the case for the thin droplet in fig. 1. If the film is predominantly driven by surface tension, one usually obtains a fourth order degenerate parabolic equation for the profile $h(x,t)$ of the kind

$$\left. \begin{aligned} h_t &= -\nabla \cdot (f(h)\nabla \Delta h) \qquad \text{on} \quad \Omega \times ]0, T[ \\ \text{with} \quad f(h) &= h^n, \quad T > 0 \quad \text{and} \quad \Omega \subset \mathbb{R}^N \quad \text{with } N = 1, 2. \end{aligned} \right\} \tag{1}$$

An inital profile $h_0(x)$ and boundary values must be added to turn (1) into a complete initial boundary value problem (IBVP).

More general forms for $f$, inclusion of lower-derivative terms, different values for $n > 0$ and a variety of boundary conditions are possible depending on the specific physical problem being studied. Some typical situations are:

- $n = 3$ for films with a non-slip condition imposed at the wall.

- $1 \leq n = 3 - s \leq 2$ if a slip condition of the type $v|_{z=0} = C \cdot h^{-s} \partial_z v|_{z=0}$, $0 \leq s \leq 2$, is imposed at the wall.

- $n = 1$ in one space dimension (i.e. $\Omega = ]-1,1[$, $\Gamma = \{\pm 1\}$) and one of the two pairs of boundary conditions

$$h|_\Gamma = 1, \qquad h_{xx}|_\Gamma = p \tag{2}$$
$$\text{or} \quad h|_\Gamma = 1, \qquad h_{xxx}|_\Gamma = \pm q. \tag{3}$$

These conditions are henceforth referred to as *pressure* or *current* boundary conditions, respectively[3].

- As another example we give the complete problem formulation for the motion of surface tension driven droplet according to Greenspan[11] It uses a modified form for $f$ and features a dynamical contact line $\Gamma(t)$. At points near the contact line where $h \to 0$ we have an effective $n = 1$ as a result of the special slip condition embedded into our model. The full set of equations is:

$$-\gamma h_t = \nabla \cdot \left( \left( h^3 + \beta^2 h \right) \nabla \Delta h \right)$$

$$h(x,0) = h_0(x) \qquad x \in \Omega(0)$$

$$h = 0 \qquad \text{on } \Gamma(t)$$

$$\frac{\beta^2}{\gamma} \frac{\partial \Delta h}{\partial n} + \frac{\partial h}{\partial n} + 1 = 0 \qquad \text{on } \Gamma(t)$$

$$\frac{\partial \Gamma}{\partial t} = \frac{\partial h}{\partial n} + 1 \qquad \text{on } \Gamma(t)$$

In all these cases, serious doubts arise about the validity of the lubrication theory for free films alone or in connection with the formulation of explicit free boundary problems for $\Gamma(t) = \partial\{x; h(x,t) > 0\}$. The approximations made to derive the above equation from the full Navier-Stokes formulation for Newtonian fluids are applied under the assumption that the Reynolds number is small and the gradient of the film's surface is everywhere of the same small scale $\epsilon \ll 1$.

Especially the last assumption may be violated near *singularities* or *contact lines* where $h \to 0$. Lubrication theory might break down in these cases, so a thorough analysis of (1) is sensible and can help to clarify whether this equation retains essential features of the full fluid-model from which it was derived. Such a rigorous discussion of the analytical properties of (1) has only recently been published by several authors[1,3,5]. Their results focus on the one-dimensional case and treat the following questions:

**Existence and non-negativity.** Existence and non-negativity are minimum requirements for the equations and their solutions to be physically relevant. In contrast to the second-order analogues of (1), we have no maximum principle at hand to settle the question of the conservation of the sign of $h$ easily.

**Singularities and uniqueness.** If the initial data is positive, we would like to know whether the solutions obtained remain strictly positive or whether they can develop singularities, i.e., whether $h$ can go to zero, reflecting a spontaneous rupture of the film.

**Solutions for initial data with compact support.** These reflect situations in which a finite droplet of a viscous liquid is allowed to evolve. How does the support behave, does it spread or rather shrink, is the velocity of propagation finite and are the dynamics of the support completely determined or must additional conditions be imposed?

A survey of the results published by these authors in recent and very recent articles will be given in the following section. All of the theorems are strictly one-dimensional, and little is known about whether they carry over to higher dimensions. A numerical example for two dimensions at the end of the article, however, indicates that some of the assertions may hold in higher dimensions as well.

## 2. ANALYTICAL RESULTS IN ONE DIMENSION

For simplicity, most authors focus on the following boundary conditions:

$$h_x(\pm 1, t) = 0, \quad h_{xxx}(\pm 1, t) = 0. \tag{4}$$

Here, $\Omega = ]-1, +1[$, $\Gamma = \{\pm 1\}$, and the PDE is slightly modified

$$h_t = -(|h|^n h_{xxx})_x \quad \text{on} \quad \Omega \times [0, T], \quad T > 0. \tag{5}$$

The value $n \geq 0$ is treated as a real parameter. Wherever possible, we shall outline how the results generalize to IBVPs with pressure and the current boundary conditions (2) and (3).

The analysis hinges on two basic tools, the application of one of two regularisation schemes which approximate the original problem with a family of uniformly parabolic or higher degenerate initial boundary value problems and, secondly, on a set of three conservation laws. The basic ideas presented here and theorems 1–3 can be found in Bernis[1], together with a weaker formulation of theorem 4, theorems 4 and 5 were published by Beretta et al.[5]

The regularisation schemes introduce the following families of initial boundary value problems, $\epsilon > 0$:

$$\left.\begin{aligned} h_t &= (-f_\epsilon(h)h_{xxx})_x \\ h_\epsilon(x,0) &= h_{\epsilon 0}(x) \end{aligned}\right\} \tag{6}$$

with

$$\left.\begin{aligned} f_\epsilon(h) &= |h|^n + \epsilon \\ h_{\epsilon 0} &\in C^{4+\alpha}(\Omega), \qquad \alpha > 0, \\ \text{and} \quad h_{\epsilon 0} &\to h_0 \qquad \text{in } H^1(\Omega) \text{ for } \epsilon \to 0 \end{aligned}\right\} \tag{7}$$

for the first family and

$$\left.\begin{aligned} f_\epsilon(h) &= \frac{|h|^{n+4}}{\epsilon|h|^n + h^4} \\ h_{\epsilon 0}(x) &= h_0(x) + \epsilon^\theta, \qquad \text{with} \quad 0 < \theta < 1/2 \end{aligned}\right\} \tag{8}$$

for the second. According to standard theory, the first family of problems is uniformly parabolic and hence possesses unique local solutions on $\Omega \times [0, \sigma[$ with a $\sigma \in ]0, T]$. The other family of problems is again degenerate, but of higher order of degeneracy if $n < 4$. Again, unique local solutions can be shown to exist. As we will see, higher degeneracy of (5) surprisingly often yields better properties, so this second family can be used to push results down to lower $n$ once they have been obtained for higher degenerate equations.

With the help of two conservation laws for the regularised problems, namely *volume conservation*

$$\int_\Omega h_\epsilon(x,t)dx = \int_\Omega h_{0\epsilon}(x)dx, \qquad 0 < t < \sigma \tag{9}$$

and *energy conservation*

$$\frac{1}{2}\int_\Omega h_{\epsilon,x}^2(x,t)dx + \int_0^t \int_\Omega f_\epsilon(h_\epsilon)h_{\epsilon,xxx}^2(x,t)dx = \frac{1}{2}\int_\Omega h_{0\epsilon,x}^2(x)dx. \qquad 0 < t < \sigma, \tag{10}$$

Bernis obtains a-priori bounds for the solution which are independent of $\sigma$ and $\epsilon$. They allow the solutions to be extended to a family of global solutions, i.e. solutions on $\Omega \times [0, T]$, of which a subsequence converges uniformly. For the limit function $h$, one can prove that it solves the original initial value problem (5) in the special sense of the following theorem:

**Theorem 1 (Bernis: Existence of very weak solutions)**
*The limit function $h$ of the first family of regularised problems (6),(7) satisfies the following properties:*

$$h \in C^{1/2,1/8}(\overline{Q}_T), \qquad Q_T := \Omega \times ]0, T[,$$

*i.e., $h$ is Hölder continuous on $\overline{Q}_T$ with Hölder exponent $1/2$ in $x$ and $1/8$ in $t$,*

$$h_t, h_x, h_{xx}, h_{xxx}, h_{xxxx} \in C(P), \qquad P := \overline{Q}_T \setminus (\{h = 0\} \cup \{t = 0\})$$

*$h$ fulfills (4), (5) in the following sense:*

$$f(h)h_{xxx} \in L^2(P) \qquad and \qquad \int \int_{Q_T} h\Phi_t + \int \int_P f(h)h_{xxx}\Phi_x = 0$$

*$\forall \Phi \in Lip(\overline{Q}_T)$ with $\Phi = 0$ near $t = 0$ and $t = T$, $h(x,0) = h_0(x)$, and $h$ satisfies (4) whenever $h > 0$ on the boundary.*

**Remark.** Note that $h$ depends on the regularisation scheme and the subsequence selected and cannot be expected to be a unique solution.

In addition to the two conservation laws already mentioned, another entity introduced by Bernis and denominated *entropy* by Bertozzi[3,4], is conserved. This term denotes the first integral in the conservation law

$$\int_\Omega G_\epsilon(h_\epsilon(x,t))dx + \int_0^t \int_\Omega h_{\epsilon,xx}^2 dx dt = \int_\Omega G_\epsilon(h_{0\epsilon}(x))dx, \qquad 0 < t < T. \quad (11)$$

where $A$ is a constant $> \max|u_\epsilon|$ and $G_\epsilon$ is given by

$$G_\epsilon(y) := \int_y^A \int_s^A \frac{dr ds}{f_\epsilon(r)}. \quad (12)$$

In fact, there is a certain degree of freedom in the choice of $G_\epsilon$. Clever exploitation of this freedom and combination with the different regularisation schemes can significantly extend the scope of the theorems.

As a first application of (12) and (11) **Bernis** obtains the following result:

**Theorem 2 (Regularity and existence of distributional solutions)**
*Assume that $h_0 \in H^1(\Omega)$ is entropy bounded and that $n > 1$. Then the solution $h$ obtained through the first regularisation scheme satisfies*

$$h_x \in L^2((0,T); H_0^1(\Omega))$$

*and (4), (5) holds in the following sense:*

$$\int \int_{Q_T} h\Phi_t dx dt = \int \int_{Q_T} f(h)h_{xx}\Phi_x dx dt + \int \int_{Q_T} f'(h)h_x h_{xx}\Phi_x dx dt$$

*$\forall \Phi \in C^2(\overline{Q}_T)$ with $\Phi = 0$ near $t = 0$ and $t = T$ and $\Phi_x = 0$ on $T \times (0,T)$.*

**Remark.** Bertozzi et al.[4] extend Bernis' methods by generalizing his concept of distributional solutions and furthermore derive additional regularity for $h$, if $h$ is obtained through the second regularisation scheme (6), (8).

The full merit of the entropy methods is revealed in the discussion of non-negativity/positivity and uniqueness of the solutions obtained. To exhibit the main idea, we note that an entropy conservation law for the degenerate IBVP itself can be *formally* obtained by setting $\epsilon = 0$ in (12) and (11). Using this law, we can summarize Bernis' argumentation in the the following *formal* way: If the initial data is entropy bounded – because it is strictly positive, for example – the conservation law guarantees that it will remain bounded for all times $t \in [0, T]$. As the function $G_\epsilon(y)$ which appears in the definition of the entropy (12) becomes singular for $\epsilon = 0$ and $y \to 0$, and $h$ is known to behave like $|x - x_0|^{1/2}$, one can show that this a-priori bound on the entropy limits the possible set of zeros of $h$. The limitation increases in severity if $n$ increases, due to the increasing singularity of the the integral (12). A rigorous proof based on the regularised problems yields the following theorem:

**Theorem 3 (Bernis: Non-negativity and uniqueness)**
*For non-negative initial data $h_0 \in H^1(\Omega)$ the following holds for the limit function $h$ obtained by the first method of regularisation:*

   (i) *If $1 \leq n < 2$, $h \geq 0$ throughout $\overline{Q_T}$.*

   (ii) *If $2 \leq n < 4$ and $h_0$ is entropy bounded, then again $h \geq 0$ and the set $\{h = 0\}$ has zero measure.*

   (iii) *If $n \geq 4$ and $h_0$ is strictly positive in $\overline{\Omega}$, then $h > 0$ throughout $\overline{Q_T}$. Such a solution is unique.*

**Remarks.** In (iii), the bound for $n$ kann be replaced by 3.5 as was shown by Bertozzi et al. and Beretta et al.[4,5], and by 2, if $h^2_{xx}$ is a-priori bounded[3], if $h$ is obtained as a limiting function of the second family of regularised problems (6), (8). In addition, for this type of regularisation, non-negativity can be proved for all values of $n > 1$ and for any non-negative (but not necessarily entropy-bounded or strictly positive) initial data.

Furthermore, appropriately modified versions of theorem 3 have been obtained by Bertozzi and her co-authors[3] for the pressure and current boundary conditions (2), (3) as well. In the first case, $n \geq 4$ or $n \geq 2$ and a-priori-boundedness of $h^2_{xx}$ excludes the appearance of *finite time* singularities. In the second, note that the draining of liquid at a constant rate as in (3) always forces a singularity to appear in finite time, so this cannot be the objective of a theorem. Instead, Bertozzi shows that one can distinguish *where* this singularity appears: On the boundary or in the interior of the domain. The latter cannot occur for $n \geq 3.5$ or $n \geq 2$ and second derivatives $h_{xx}$ which are bounded on every compact subset of the support of $h$.

Results about the asymptotic long-time behaviour of solutions for the principal boundary condition (4) show that for $0 < n < 3$ and non-negative $h_0$ all singularities will eventually vanish after a finite time $T^*$. The upper limit can be dropped if $h_0$ is required to be entropy-bounded instead.

Finally, Bernis showed that by using results for entropy bounded data and another, especially simple regularisation, one can always construct a non-negative solution provided $n \geq 1$ and one starts with nonnegative inital data $h_0$. This need not be, of course, the solution treated in theorem 3.

The above theorem is useless for initial data with compact support if $n \geq 2$. However, a refinement of the entropy method using cut-off functions allows Bernis to make statements about the behaviour of the support in time. The following theorem is an improvement of his result given by Beretta et al.[5]:

**Theorem 4 (Initial data with compact support)**
*Consider the limit function obtained through the second regularisation scheme for initial data $h_0 \in H^1(\Omega)$ with compact support. Then the following holds:*

*(i) If $n \geq 3/2$, then the support of $h$ is non-shrinking:*

$$\operatorname{supp} h(\cdot, t_1) \subseteq \operatorname{supp} h(\cdot, t_2) \qquad \forall t_1, t_2 \in [0, T], \quad t_2 \geq t_1.$$

*(ii) If $n \geq 4$, then the support of $h$ is constant:*

$$\operatorname{supp} h(\cdot, t) = \operatorname{supp} h_0 \qquad \forall t \in [0, T].$$

The last theorem fits in nicely with several facts known for dynamical contact lines. The increasing immobility of the support for increasing $n$ demonstrates that the lubrication equation retains an essential feature of the full fluidmechanical model from which it was derived. It is a widely known fact that the combination of the Navier-Stokes equations with a non-slip condition are incompatible with moving contact lines[9,10]. Typical remedies relax the non-slip condition at the wall, which effectively lowers the value of $n$ the lubrication approximated model as can be seen from the examples in the introduction.

As a kind of complement to the theorems mentioned so far, Beretta et al. prove results which exhibit non-uniqueness or the spontaneous appearance of singularities $h \to 0$ below certain values of $n$.

**Theorem 5 (Beretta, Bertsch, dal Passo)**
1. *If $0 < n < 3$ and $h(x_0, t_0) = 0$ for some $x_0 \in \Omega$, $t_0 \in [0, T]$, then there exist at least two distinct solutions. The first approaches its mean value $\frac{1}{2} \int_{-1}^{1} h_0(x) dx$ uniformly for $t \to \infty$ and thus becomes positive after a finite time, the second remains zero at $x_0$ for all $t \in [t_0, T]$. Hence both solutions must necessarily differ.*

2. *If we replace (4) with the pressure boundary condition (2) with $p > 2$, we can show that for $n < 1/2$ there exists a finite time $T_0 < \infty$ such that this IBVP has a unique classical solution which touches down to $0$ as $t$ approaches $T_0$, i.e.*

$$h > 0 \qquad in \quad \Omega \times [0, T_0[$$
$$\min_{x \in \Omega} h(x, t) \longrightarrow 0 \qquad as \ t \nearrow T_0.$$

Ideally, one would want these lower bounds to coincide with the bounds obtained in the theorems stated earlier. However, especially in the question of the appearance of singularities for $n$ in the range of $0.5 \ldots 3.5$ or $4$ respectively, few rigorous results are known. To date, mainly numerical experiments backed with more or less rigorous asymptotic analysis for special types of scaling solutions, namely source type and traveling wave solutions, have shed some light on this gap and have lead to tentative conjectures[2,3,4,6,7,8].

At this point, the reader should recall that the non-uniqueness of solutions arises only because $h$ cannot be uniformly bounded from below on $\Omega \times [0, T]$. A strictly positive a-priori lower bound for $h$ would make the equation uniformly parabolic. Standard theory would then easily supply us with unique solutions. Thus, the uncertainty about the behaviour of the set of zeros of $h$ appears to be the main source for non-uniqueness. It is therefore natural to ask whether and which additional conditions at the boundary can be introduced to make the solution unique.

Beretta et al.[5] suggest several explicit conditions which they deem adequate to define the dynamics of the free boundary. Rigorous results confirming their heuristic argumentation could ultimately serve to guarantee the validity of models like Greenspan's.

The same question can be analysed from a second point of view. According to Bernis et al.[2], so-called source-type solutions of (5) are unique if the one with the highest regularity is selected. The regularity obtained agrees nicely with the assertions proved for general solutions of (4),(5) in theorem 2. This prompts Bertozzi et al.[4] and Beretta et al. to conjecture that uniqueness holds for general solutions in a sufficiently high regularity class. Note that higher regularity ultimately implies horizontal tangents at the boundary of the support, so that the lubrication equation may turn out to be be especially well suited for the case of complete wetting, where contact-angles are zero.

## 3. A NUMERICAL EXAMPLE IN TWO DIMENSIONS

As was emphasized throughout this article, all rigorous analytical results only apply to one dimensional problems. The methods fail at several points, starting with the absence of Hölder estimates for $h$. A brief numerical exam-

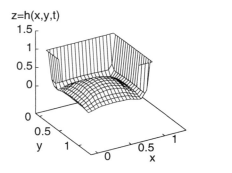
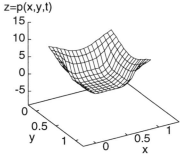

Figure 2: A cell drained at a constant rate $q|_\Gamma = 1$, with $n = 2$.
Left plot: $h(x, y, t)$ at $t = 0.013$, right: $p = \triangle h(x, y, t)$ at $t = 0.011$.

ple may serve as an indication that, nevertheless, the qualitative behaviour predicted and proved for the one dimensional case may be valid for the two dimensional case as well.

In order to make sure we would observe a finite time singularity, we used the current boundary conditions (3) and calculated a solution for a cell of fluid drained at a constant rate $q|_\Gamma = 1$. Our goal was to compare the results obtained with the predictions made for the one-dimensional case in the remarks following theorem 3. The calculations where done using a mixed finite element scheme based on linear elements for space and Crank-Nicholson for time discretization. In detail, the underlying weak formulation was

$$
\begin{aligned}
(\overset{*}{h_t}, \phi) &= (h^n \nabla p, \nabla \phi) - \int_\Gamma q \cdot n\phi \, ds && \forall \phi \in H^1(\Omega) \\
0 &= (p, \psi) + (\nabla h, \nabla \psi) && \forall \psi \in H^1_0(\Omega).
\end{aligned}
$$

with $h(\cdot, t), p(\cdot, t) \in H^1(\Omega)$ and an essential boundary condition for $h$: $h|_\Gamma = 1$. The parentheses denote the $L^2$-scalar product, and $n$ is the normal to $\Gamma = \partial\Omega$. Note that space discretization of this combination of a parabolic with an elliptic equation leads to a system of differential-algebraic equations, so that special care has to be taken to provide conforming initial data for $h_0$ and $p_0$.

For $\Omega$ we took the unit square split into $25 \times 25$ squares which themselves were halved along their diagonals to provide a uniformly triangulized mesh. The time step remained fixed at 0.001. Non-linearities where resolved to a high precision – forcing the residual to be less than $3.0 \times 10^{-11}$ when using double-precision arithmetics – with the help of a simple fixed-point iteration.

Starting with conforming initial data $h = 1$, $p = \triangle h = 0$, the integration routine terminated after 13 time steps; in figure 2, we can see the final $h$-profile on the left. Obviously, $h$ approached zero fastest near the boundary of $\Omega$, especially in the corners, whereas in the interior a certain bulge was retained. In consistency with the behaviour of $h$, the second derivative $p = \triangle h$ 'exploded' at the boundary with an emphasis at the corners. The effect was so strong that the final values resulted in a rather confusing graphic, so that we resolved to take $p$ at an earlier time step, namely $t = 0.011$.

## 4. REFERENCES

1. Bernis, F. and Friedman, A. Higher Order Nonlinear Equations, *Journal of Differential Equations*, **83**, 179–206 (1990).
2. Bernis, F., Peletier, L. A., Williams, S. M. Source Type Solutions of a Fourth Order Nonlinear Degenerate Parabolic Equation, *Nonlinear Analysis, Theory, Methods & Applications*, **18**(3), 217–234 (1992).
3. Bertozzi, A. L. and Pugh, M. The Lubrication Approximation for Thin Viscous Films: Regularity and Long Time Behaviour, *Preprint, submitted to: C.P.A.M.*
4. Bertozzi, A. L., Brenner, M. P., Dupont, T. F. and Kadanoff, L. P. Singularities and Similarities in Interface Flows, Ch. 6 in: *Trends and Perspectives in Applied Mathematics, Vol. 100*. L. Sirovich, Ed., Springer-Verlag New York, (1994).
5. Beretta, E., Bertsch, M., dal Passo, R. Nonnegative Solutions of a Fourth-Order Nonlinear Degenerate Parabolic Equation, *Archive for Rational Mechanics and Analysis*, **129**, 175–200 (1995).
6. Boatto, S., Kadanoff, L. P., Olla, P. Traveling-Wave Solutions to Thin-Film Equations, *Physical Review E*, **48**(6), 4423–4431 (1993).
7. Constantin, P., Goldstein, R. W., Zhou, S. Droplet Breakup in a Model of the Hele-Shaw Cell, *Physical Review E*, **47**(6), 4169–4181 (1993).
8. Dupont, T. F., Goldstein, R.E., Kadanoff, L. P., Zhou, S. Finite-Time Singularity Formation in Hele-Shaw Systems, *Physical Review E*, **48**(6), 4182–4196 (1993).
9. Dussan V, E. B. and Davis, S. On the motion of a fluid-fluid interface along a solid surface, *Journal of Fluid Mechanics*, **65**, 71–95 (1974).
10. de Gennes, P. G. Wetting: Statics and Dynamics, *Review of Modern Physics*, **57**, 827–863 (1985).
11. Greenspan, H. P. On the Motion of a Small Viscous Droplet that Wets a Surface, *Journal of Fluid Mechanics*, **84**, 125–143 (1978).

# NEW RESULTS IN REVERSE ROLL COATING

C. A. Richardson[†], P. H. Gaskell[‡], M. D. Savage[†]

†*Deptartment of Applied Mathematical Studies,*
‡*Department of Mechanical Engineering,*
*University of Leeds, Leeds, LS2 9JT, UK.*

## SUMMARY

The flow in a fixed gap reverse roll coater is examined by Finite Element solutions of the Navier-Stokes equations as the feed condition varies. Attention is focussed on the two free surface problem in which the locations of both free surfaces are parameterised by a method of inter-dependent spines. As flow rate, $\lambda$, is increased the metered downstream film thickness is found to asymptote towards a limit whilst that upstream increases with $\lambda$. The structure of the associated pressure and velocity fields are found to be influenced strongly by the degree of starvation. Three distinct flow structures are identified, their onset being dependent upon the degree of starvation.

## 1. INTRODUCTION

Since its introduction over 60 years ago[1], reverse roll coating has become a widely used and effective coating process. Due to its versatility, speed and precision it is a process used for the manufacture of coated products over a diverse range of industries - electronic, photographic, food, paper etc.

The inlet-flooded case, shown in Figure 1, has been investigated extensively by several authors[2,3,4]. This coating geometry consists of two co-rotating rigid rolls a fixed distance, $2H_0$, apart, which rotate with peripheral speeds $U_1$ and $U_2$. Fluid is dragged through the nip by the lower roller and metered by the upper one to produce a downstream film, thickness $H_2$. The objective is to control $H_2$ by determining its dependence upon various physical parameters, e.g. $U_1$, $U_2$, $U_1/U_2$, fluid properties and process geometry, $H_0/R$. Note that the thickness of the film leaving the fluid reservoir attached to the upper roll is that determined by viscous lifting.

As the supply of fluid is reduced the inlet becomes starved and a second meniscus is produced upstream of the nip with a film of thickness $H_1$ attached to the upper roll, as shown in Figure 2. The thickness of both outlet films ($H_1$ and $H_2$) and the film split mechanism are highly dependent upon the degree of starvation. This process, termed inlet-starved reverse roll coating, seems to have escaped the interest of the coating community at large and hence this paper is concerned with examining numerically the effect of inlet flux on the associated velocity and pressure fields and the outlet film thicknesses.

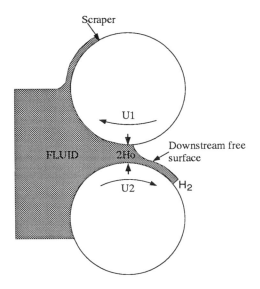

Figure 1: Reverse roll coater with flooded inlet

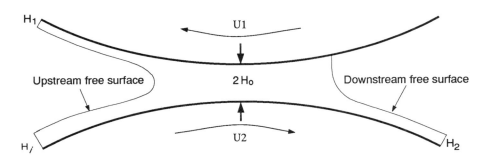

Figure 2: Nip region of a reverse roll coater with starved inlet

It is convenient to define a non-dimensional flux per axial length by

$$\lambda = \frac{U_2 H_i}{2 U_2 H_0},$$

where $H_i$ is the thickness of the inlet film.

## 2. GOVERNING EQUATIONS AND GEOMETRY

The governing equations for an incompressible, Newtonian fluid of density $\rho$, viscosity $\mu$ and surface tension $T$, in the absence of gravitational and inertial effects, can be cast in the following non-dimensional form

$$\nabla.\underline{\underline{\sigma}} = 0, \tag{1}$$
$$\nabla.\underline{u} = 0, \tag{2}$$

where $\underline{\underline{\sigma}} = -p\underline{\underline{I}} + (\nabla\underline{u}) + (\nabla\underline{u})^T$, $\underline{x} = (x,z) = \underline{X}/H_0$, $\underline{u} = (u,w) = \underline{U}/U_2$, pressure, $p = H_0 P/\mu U_2$ and $\underline{\underline{\sigma}} = H_0\underline{\underline{\Sigma}}/\mu U_2$ ($\underline{\underline{\Sigma}}$ being the dimensional stress tensor). The attendant boundary conditions are shown in Figure 3.

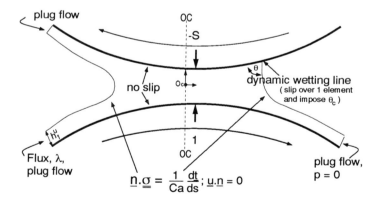

Figure 3: Stokes flow boundary conditions

The roll surfaces, $z_1(x)$ and $z_2(x)$, are modelled as cylinders with the origin of the co-ordinates located midway between them at the nip.

## 3. METHOD OF SOLUTION

### 3.1. Finite Element (F.E.) Formulation

A Galerkin weighted residual F.E. approach is employed to solve Eq.'s (1) and (2) in terms of V6/P3 elements. For optimal convergence and no 'locking'[8] the pressure interpolation is one order of magnitude less than that for the velocity and the $div - stability$[9] condition for conforming elements is satisfied. At the element level, the F.E. calculations are performed in a local co-ordinate system, namely $(\xi, \eta)$, by means of an isoparametric mapping from global $(x, y)$ to local co-ordinates. The partial differential Eq.'s (1) and (2) may then be reduced to a large set of non-linear algebraic equations which can be solved by an iterative technique. All computations were performed on a grid containing 1371 elements, 3004 nodes and 6398 degrees of freedom which was found to be sufficient to guarantee grid independent solutions. Adopting the frontal method and Newton iteration allowed convergence of all solutions to be found within a maximum of five iterations[6]. Finally the Poisson equation

$$\nabla^2 \psi = \frac{\partial u}{\partial y} - \frac{\partial v}{\partial x}, \tag{3}$$

may be solved subject to appropriate boundary conditions to calculate the streamfunction $\psi(x, y)$. Stagnation points and their classification may be then calculated by seeking local maxima and minima of $\psi$[10].

### 3.2. The Computational Mesh

The solution domain, tessellated by a series of contiguous triangular elements, divides naturally into five distict regions, as shown in Figure 4. It is convenient to parameterise regions 2 and 3 by a vertical spine, $O_c$, whose position typically lies midway between the two free surfaces. Region 2, defined as $x \in [x_m^u, O_c]$, $y \in [z_1(x), z_2(x)]$, contains $n_d^u \times n_{nip}$ nodes with co-ordinates

$$x^{(i)}(x_m^u) = O_c + (x_m^u - O_c)\frac{i}{n_d^u}, \qquad i \in [0, n_d^u],$$

$$z^{(i,j)}(x_m^u) = z_1(x(i)) + \frac{j}{n_{nip}}(z_2(x(i)) - z_1(x(i))), \qquad j \in [0, n_{nip}],$$

where $n_d^u$ is the number of nodes that lie along either roll surface between $x_m^u$

66

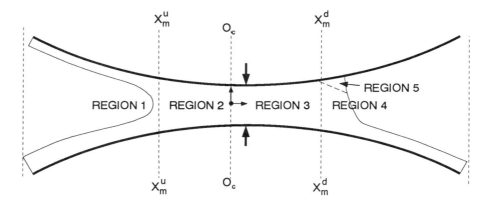

Figure 4: Reverse roll coater with starved inlet

and $O_c$ and $n_{nip}$ the number along the vertical spine at $O_c$. Region 3 is defined as $x \in [O_c, x_m^d]$, $z \in [z_1(x), z_2(x)]$ and contains $n_u^d \times n_{nip}$ nodes with co-ordinates

$$x^{(i)}(x_m^d) = O_c + (x_m^d - O_c)\frac{i}{n_u^d}, \qquad i \in [1, n_u^d],$$

$$z^{(i,j)}(x_m^d) = z_1(x(i)) + \frac{j}{n_{nip}}(z_2(x(i)) - z_1(x(i))), \qquad j \in [0, n_{nip}],$$

where $n_u^d$ is the number of nodes along either roll surface between $O_c$ and $x_m^d$. The free surface regions, 1 and 4, are dependent upon the location of the free surface which is unknown *a priori*. The free surfaces are parameterised using a spinal approach[5]
Region 1 has $n_{sp}^u \times n_{fs}$ nodes with co-ordinates:

$$\underline{x}^{(i,j)}(x_m^u, h_i^u) = \underline{x}_b^u(i) + \frac{j}{n_{sp}^u}h_i^u\hat{\underline{e}}_i^u; \qquad i \in [1, n_{sp}^u] \text{ and } j \in [0, n_{fs}],$$

whilst region 4 (minus the contact line region) has $n_{sp}^d \times n_{fs}$ nodes with co-ordinates:

$$\underline{x}^{(i,j)}(x_m^d, h_i^d) = \underline{x}_b^d(i) + \frac{j}{n_{sp}^d}h_i^d\hat{\underline{e}}_i^d; \qquad i \in [1, n_{sp}^d] \text{ and } j \in [0, n_{fs}],$$

where $h_i^u$ and $h_i^d$ are the lengths of the spines and $n_{sp}^u$, $n_{sp}^d$ the number of these spines in regions 1 and 4 respectively. In the above two regions, $\underline{x}_b^u$ and $\underline{x}_b^d$, corresponding to the spinal base points (j=0) and $\hat{\underline{e}}_i^u$ and $\hat{\underline{e}}_i^d$ to their direction vectors, which are dependent upon $x_u^m$ and $x_d^m$, respectively - thus all co-ordinates in these regions depend upon $x_m$'s and $h_i$'s[6]. The unit direction vectors in regions 1 and 4 are calculated using the base points in conjunction with a series of multiple origins, specifically tailored for the geometry.

By fixing the length of an upstream spine and employing the kinematic equation that remains from this unknown variable, $x_m^u$ may be calculated. As a consequence the whole of region 1 moves with the free surface. The downstream variable, $x_m^d$, is calculated from the contact angle equation

$$\underline{t}_2.\underline{t}_{fs} = -cos(\theta_c^d), \qquad (4)$$

where $\underline{t}_2$ is the tangent to the upper roll and $\underline{t}_{fs}$ is the tangent to the free surface. The contact angle $\theta_c^d$ is prescribed and set at $90°$. In region 5 a multiple-origin inter-dependent spine formulation is used to avoid any conflict between $\theta_c^d$ and a neighbouring spine to the dynamic wetting line[7]. Constructing the grid in such a way allows the elements between the two $x_m$ lines to move in a 'concertina' type fashion which helps alleviate distortion in regions 2 and 3.

## 4. RESULTS

We now examine the effect on the dimensionless outlet film thicknesses, $h_1 = H_1/H_0$ and $h_2 = H_2/H_0$, of varying the dimensionless inlet flux, $\lambda$, over the interval $0.235 \leq \lambda \leq 0.665$. Results are restricted to the particular parameter set $Ca = 0.1, S = 0.5$ and $R/H_0 = 100.0$. Figure 5 shows how $h_1$ and $h_2$ vary with $\lambda$. As $\lambda$ increases the net flux passing through the nip does so too resulting in a corresponding increase of $h_2$. Clearly there is a limit how much fluid can pass through the nip and thus $h_2$ asymptotes towards a limiting value of approximately $h_2 = 0.63$. The upstream film thickness, $h_1$, continues to increase with $\lambda$. Figure 6 shows the F.E. pressures inside the fluid bead measured along a line equidistant between the roll surfaces and bounded by the two free surfaces. For a small inlet flux, e.g. $\lambda = 0.235$, the curve is effectively linear and entirely sub-ambient with a positive gradient, typical of those generated in meniscus forward roll coating[11]. As $\lambda$ increases fluid pressures throughout the bead become less sub-ambient and at approximately $\lambda = 0.400$ the capillary pressures at either free surface are the same. The profile now exhibits two turning points which is more characteristic of a classical roll coating flow operating under conditions of a flooded or moderately starved inlet. Beyond $\lambda = 0.400$, the pressure curve 'flips' as the capillary pressure at the upstream meniscus exceeds that at the downstream. As the inlet flux approaches $\lambda = 0.665$, the net flux passing through the nip approaches its maximum value and the pressure curve tends towards that of the inlet-flooded case[4].

Figure 7 shows (a) a streamline plot and (b) a schematic representation of the

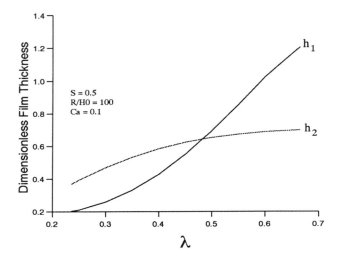

Figure 5: Predicted dimensionless film thicknesses

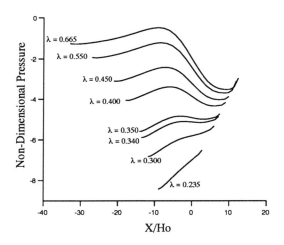

Figure 6: Predicted pressure contours

flow field for the case $\lambda = 0.235$. The flow structure consists of a recirculation region bounded by a heteroclinic orbit connecting stagnation points $P_6$ and $P_7$ at the upstream meniscus. Fluid entering the bead above the dividing streamline $P_4 P_5$ circumnavigates this closed eddy to depart the bead as the upstream outlet film. This fluid path is termed the *primary transfer − jet*. All remaining fluid entering the bead passes directly through the nip to exit the fluid domain as the downstream outlet film.

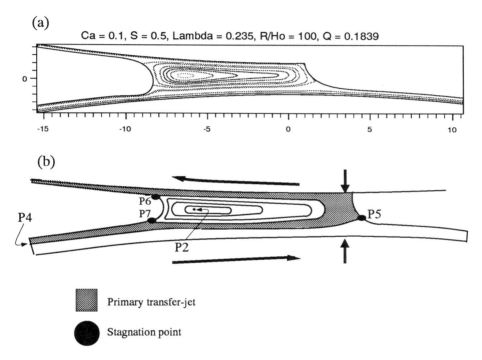

Figure 7: (a) Streamline plot and (b) schematic for $\lambda = 0.235$

Figure 8 shows (a) a streamline plot and (b) a schematic representation of the flow field for the case $\lambda = 0.300$. Both free surfaces have receded away from the nip and the extra flux has caused the recirculation region bounded by the streamline $P_6 P_7$ to extend downstream of the nip. The consequent pinching of the recirculating fluid inside the closed eddy results in the birth of a homoclinic orbit attached to a saddle point, $P_1$ at the nip.

Figure 9 shows (a) a streamline plot and (b) a schematic representation of the flow field for the case $\lambda = 0.500$ in which the separatrix is no longer a closed curve. The streamline passing through the saddle point, $P_1$, enters the fluid

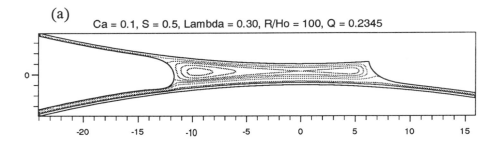

(a)

Ca = 0.1, S = 0.5, Lambda = 0.30, R/Ho = 100, Q = 0.2345

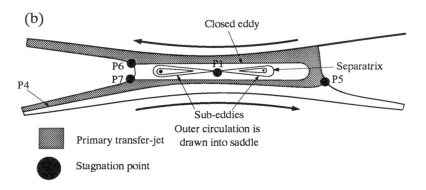

(b)

Closed eddy

Separatrix

P6
P7
P4
P1
P5

Sub-eddies

Outer circulation is
drawn into saddle

Primary transfer-jet

Stagnation point

Figure 8: (a) Streamline plot and (b) schematic for $\lambda = 0.300$

domain via the inlet film and exits via the upstream outlet film. Fluid entering the bead above this streamline is able to transfer to the upper roll without ever reaching the nip. It simply circumnavigates a small eddy attached to the upstream meniscus at $P_6P_7$ and is referred to as the *secondary transfer − jet*. Fluid entering the bead below the separatrix but above $P_4P_5$ is conveyed to the upper roll via the primary transfer-jet; all remaining fluid entering the bead passes directly through the nip to depart as the downstream outlet film.

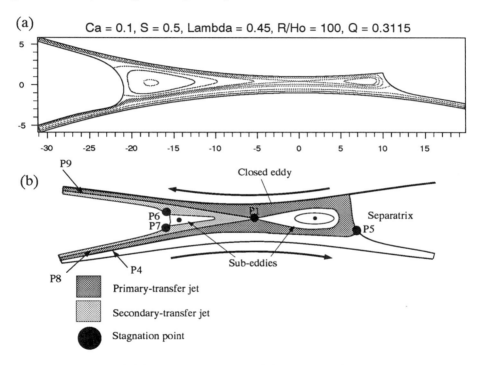

Figure 9: (a) Streamline plot and (b) schematic for $\lambda = 0.450$

## 5. CONCLUSION

For a fixed speed ratio, $S$, the solutions show that, as the supply of fluid entering the bead is increased both upstream and downstream outlet film thicknesses increase accordingly. However, whilst the thickness of the downstream film, $h_2$, approaches a limiting value, the upstream outlet film, $h_1$, continues to grow. The pressure contours are also highly dependent upon the degree of starvation. At low values of $\lambda$ the contours are effectively linear with a positive gradient - typical of those generated by a forward meniscus roll coating process[11]. As the inlet is slowly flooded the curve can be seen to 'flip' as capillary pressure at the upstream meniscus becomes less that at the downstream one. Two turning points are born and the curve becomes more typical of those characteristic of inlet-flooded reverse roll coating. The F.E. solutions reveal intricate flow patterns, the structure of which are highly dependent upon the level of starvation. There exist two mechanisms for conveying fluid from the lower to upper roll; a primary transfer-jet, which is always present and a secondary transfer-jet which is switched on only if the system is sufficiently flooded. Work is currently in progress to relate these observations to other physical parameters such as speed ratio, $S$ and capillary number.

## 6. REFERENCES

1. Munch, C. U. S. Pat. 1847065, (Feb 23, 1932).
2. Ho, W. S. and Holland, F. A. Between-Rolls Coating Technique: A Theoretical and Experimental Study, *Tappi*, **61**, (2) pp 53–56 (1978).
3. Greener, J. and Middleman, S. Reverse Roll Coating of Viscous and Viscoelastic Fluids, *Ind. Eng. Fund.*, **20**, pp 63–66 (1981).
4. Coyle, D. J., Makosko, C. W. and Scriven, L. E. The Fluid Dynamics of Reverse Roll Coating, *A.I.Ch.E.*, **36**, 161–173 (1990).
5. Kistler, S. F. and Scriven, L. E. Coating Flows *Computational Analysis of Polymer Processing*, Eds Pearson, J. R. A. and Richardson, SM, Publishers Apl. Sci. Computational Analysis of Polymer Processing, *A.I.Ch.E.*, pp 243-299 (1983).
6. Thompson, H. M. A Theoretical Investigation of Roll Coating Phenomena, *PhD Thesis, University of Leeds* (1992)
7. Gaskell, P. H, Savage, M. D, Summers, J. L. and Thompson, H. M. Flow Toplogy and Transformation in a Fixed Gap Forward Roll Coating System, *Numerical Methods in Laminar and Turbulent Flow*, Eds. Taylor, C. and Durbetaki, P, Publishers Pineridge Press **9 II** (1995)
8. Hughes, T. R. J, The Finite Element Method: Linear Static and Dynamic Finite Element Analysis, (1987).
9. Gunzburger, M. D. Finite Element Methods for Viscous Incompressible Flows, *Computer Science and Scientific Computing, Academic Press*, (1989).
10. Summers, J. L. A Private Communication, [1995]
11. Gaskell, P. H, Savage, M. D, Summers, J. L. and Thompson, H.M. Modelling and Analysis of Meniscus Roll Coating, *J.F.M.*, **298**, 113–137

# SECTION 2

# Instability and

# Coating Defects

# Deformable Roll Coating:
# Analysis of Ribbing Instability and Its Delay

M. S. Carvalho and L. E. Scriven

Coating Process Fundamentals Program
Center for Interfacial Engineering and
Department of Chemical Engineering & Material Science
University of Minnesota, Minneapolis, MN 55455

*Abstract*

'Ribs', 'pinstripes', or 'corduroy' are names of a common patterning that arises at the film-split meniscus in roll coating flows. Specifically, ribs or ribbing refers to nonuniform coating whose thickness profile is wavy in the transverse direction and more-or-less steady in the coating direction. The onset of ribbing is a flow instability that arises when the roll speed is too high or the liquid is too viscous. If ribbing does not level rapidly enough it can be deleterious or even an unacceptable defect. Understanding this phenomenon can lead to better control of roll coating processes and thus to better quality and/or faster production rate (line speed). The mechanism of ribbing can be understood in terms of the competition between surface tension forces that stabilize the flow, and viscous flow-induced pressure gradients that destabilize the flow.

This type of instability has received a lot of attention in the literature. However, all previous work has addressed the flow between two rigid rolls. Often, in practice, one of the rolls of a pair is covered by a layer of elastomer. The deformation of the roll cover alters the conformation of the gap and thus the pressure gradient at the film-split meniscus. Consequently the critical parameters at the onset of ribbing change.

The Navier-Stokes system for the two-dimensional flow coupled with linear elastic elements to represent the resilient roll cover was solved by the Galerkin/finite element method, as described in the companion paper. Pseudo-arc-length continuation was employed not only to explore the parameter space, but also to find steady-state solutions at turning points, where the stability of flow to two-dimensional disturbances changes.

Stability of the flow with respect to three-dimensional disturbances was analyzed by examining the linearized time-dependent response to infinitesimal disturbances and identifying those that grow fastest. This approach leads to an asymmetric generalized eigenvalue problem, which was solved for the leading eigenvalues and corresponding eigenfunctions by Arnoldi's method.

The results indicate how a deformable cover can be used to delay the onset of ribbing in forward-roll coating. Some of the results can be compared with observations of deformable roll coating flows made in our laboratory.

## 1. Introduction

When a substrate is coated with a thin layer of liquid by the action of rotating rolls, spreaders or brushes, the layer as coated is often not uniform. If the substrate speed is too fast, or the liquid viscosity too high, the thickness profile is wavy in the transverse direction. This type of nonuniformity of the flow pattern is commonly called *ribbing*, sometimes *corduroy* or *pin-striping*. It is a result of a flow instability: above a critical value of one or another parameter, the two-dimensional flow is unstable and therefore cannot be obtained in reality. The stable flow state is then a three-dimensional flow characterized by a thickness profile that is periodic in the transverse direction.

The instability of the splitting of a liquid layer, or film, as it exits from between rotating rigid rolls or spreaders has been extensively studied. Using a lubrication approximation,

Pearson (1960) was the first to analyze why a flow that otherwise leads to a uniform film can turn unstable. He showed that the viscous forces destabilize the film-split meniscus whereas surface tension tends to stabilize it. With that, he recognized that the capillary number, a measure of the ratio of viscous to capillary forces, is the critical parameter in the stability of this class of flows. He was unable to derive a stability criterion because he did not have available an appropriate boundary condition on Reynolds' equation at the film-split.

Pitts & Greiller (1961) went further and developed a stability criterion based on the pressure gradient in the vicinity of the meniscus of the splitting film. Their criterion was derived by using a force balance at the perturbed meniscus. With several further simplifications, they estimated that the critical capillary number at onset of ribbing ought to be about

$$Ca = 28 \frac{H_0}{R}$$

$H_0$ is half of the distance between the rolls, and $R$ is the roll radius. This first stability criterion was plainly an estimate.

Savage (1977 and 1984) also used the lubrication approximation, but employed the boundary condition derived by Coyne & Elrod (1971) to account roughly for the capillary pressure at the film-split and developing film flows downstream of it. The pressure gradient at the film split they got was more accurate than what Pitts & Greiller estimated. Savage went on to use linear stability analysis to predict the condition of marginal stability. Fall (1985) followed the same approach but examined the time-dependent response to infinitesimal perturbations in order to identify those that grow fastest.

Coyle et al. (1990) went beyond the lubrication approximation and examined the time-dependent response of the flow between counter-rotating rigid rolls to three-dimensional disturbances by applying linear stability theory to base flows described by the Navier-Stokes system for viscous free-surface flows. They solved for both the base flow and its stability by Galerkin's method and employed finite element basis functions. They found that the stability analysis based on the lubrication approximation developed by previous researchers underpredicted the onset of ribbing.

All previous studies of stability of roll-coating flows dealt with flows between two *rigid* rolls. In practice, a pair of rigid rolls is seldom used in forward-roll coating. The deformation of the roll cover affects the flow rate, meniscus positions, and pressure distribution close to the film-split meniscus (see Carvalho & Scriven 1995). Therefore, it also affects the stability of the flow. Here, the stability of the film-splitting flows in deformable gaps is analyzed. The method used is an improvement over previous stability analyses of free surface flows by the Galerkin / finite element approach. This improved method reduces substantianlly the size of the algebraic eigenvalue problem to which the Galerkin / finite element method leads. And it does so at no cost in accuracy, as compared to the previously available method for analyzing this class of flows (Christodoulou & Scriven 1990). The improved method is described and discussed in the following section.

## 2. Stability Analysis of Viscous Free-Surface Flows

In order to obtain accurate predictions of the onset of ribbing, or of any other flow instability, it is fundamental to develop accurate theoretical analysis of both the base flow and the response of that flow to all physically admissible infinitesimal disturbances.

The Navier-Stokes system of equation for two-dimensional, steady-state viscous flows with free surfaces can be solved by Galerkin's method with appropriate basis functions, of which finite element basis functions appear to be most effective. Methods for solving this class of problems have evolved over twenty years (see Orr 1976, Silliman 1979, Kistler 1984, Coyle 1984). The stability of the base flow with respect to three-dimensional infinitesimal disturbances also can be determined by using Galerkin's method and the very same basis functions. Methods for doing this have been evolving over fifteen years (see Bixler 1982, Coyle 1984, Christodoulou 1988, and Chen 1992). The amplitude of the disturbances and their rate of growth are governed by a generalized asymmetric eigenvalue problem:

$$\mathbf{J u}' = \beta \mathbf{M u}'$$

$\mathbf{u}'$ is a vector that contains the amplitudes of the combination of velocity, pressure and free surface (when applicable) perturbations that constitute a mode; $\beta$ is the exponential growth factor of that mode of disturbance. If the real part of $\beta$ is positive, i.e. $\Re(\beta) > 0$, the disturbance mode grows with time and the flow is unstable. If $\Re(\beta) < 0$, it decays and the flow is stable. $\mathbf{M}$ is the mass matrix, which is singular; and $\mathbf{J}$ is the Jacobian matrix, the elements of which are the sensitivities of the weighted residuals of local momentum balances, mass balances, mesh generation equations, and boundary conditions, to the velocity, pressure, and nodal position values all over the domain.

To determine the stability of the flow, the generalized eigenvalue problem has to be solved. Practically, a complete solution is not required. Only the eigenvalues (and their corresponding eigenvectors) with largest real part need to be computed.

Because the size of the algebraic eigenproblem generated by Galerkin's method with an adequately large subset of finite element basis functions to represent significant disturbances is typically quite large, standard algorithms, which compute all eigenvalues, cannot be used. Ruschak (1983) and later Coyle (1984) (see also Coyle et al. 1990) dodged this difficulty by neglecting the transient terms in the momentum equations. With this approximation they could reduce considerably the size of the eigenproblem by using static condensation and then computing all the eigenvalue-eigenvector pairs of the reduced problem with a utility program, EISPACK in particular. However, this approximation is valid only at vanishing Reynolds number or when the marginal stability (onset of instability) of non-periodic disturbances suffices.

Like Bixler (1982), Christodoulou (1988) (see also Christodoulou & Scriven 1990) did not resort to this approximation. Christodoulou solved the complete algebraic eigenproblem by Arnoldi's method but computed only the leading modes, i.e. the eigenvalue-eigenvector pairs whose eigenvalues have the largest real parts. However, owing to the mesh generation schemes they employed, the infinitesimal disturbances included perturbations of positions of all the finite element nodes, including those interior to the domain. Although these perturbations are consistent with the domain-perturbation formulation of the linear stability theory of free boundary flows, it is now clear that perturbations of location of nodes in the interior of the domain are not relevant to the stability of the flow. Moreover, the mass and Jacobian matrices obtained with the method adopted by Christodoulou (1988) include entries that describe the sensitivity of the solution of the mesh generation to the locations of the mesh nodes.

This section describes a method that is still consistent with the domain-perturbation formulation, but that avoids the perturbation of the finite element interior nodes. This approach reduces the size of the eigenproblem and thereby reduces the computational cost substantially. Moreover, it is satisfyingly and completely independent of the mesh generation scheme used to deal with the base flow.

The idea behind linear stability analysis is to perturb a two-dimensional steady 'base' flow with infinitesimal disturbances, insert all in the equation system that models the time-dependent situation and solve for the amplitude of the perturbation and its rate of growth. Because the main goal of this study is to analyze the onset of ribbing, the disturbances have to be three-dimensional. Because the disturbances are infinitesimal, all possible disturbances can be represented as linear combination of a complete basis set of linearly independent normal modes, and so it suffices to examine the fate of a generic normal mode. Most convenient in the present case is a set of Fourier modes in one direction, the coefficients of which contain the dependences of those modes in the other directions. Here, the variation of disturbances in the transverse direction is represented by Fourier modes, i.e. by sines and cosines, of which the wave number is $N$. Accordingly, the perturbed fields are written as ($\Re$ denotes the real part of)

$$\mathbf{v}(\mathbf{x}, t) = \mathbf{v_0}(\mathbf{x_0}) + \epsilon \Re \{ \mathbf{v}'(\mathbf{x}) \cdot \mathbf{D}(Nz) e^{\beta t} \} \qquad (1-a)$$

$$p(\mathbf{x}, t) = p_0(\mathbf{x_0}) + \epsilon \Re\{p'(\mathbf{x})cos(Nz)\ e^{\beta t}\} \qquad (1-b)$$

If the flow domain is bounded by free surfaces, the positions of those boundaries also have to be perturbed, i.e.

$$\mathbf{x} = \mathbf{x_0} + \epsilon \Re\{\mathbf{x'} \cdot \mathbf{E}(Nz)\ e^{\beta t}\} \qquad (1-c)$$

$N$ is the wavenumber of the perturbation ($N = 2\pi/\lambda$, where $\lambda$ is the wavelength of the perturbation); $\beta$ is the growth factor. The matrices $\mathbf{D}$ and $\mathbf{E}$ are

$$\mathbf{D} = \begin{pmatrix} cos(Nz) & 0 & 0 \\ 0 & cos(Nz) & 0 \\ 0 & 0 & sin(Nz) \end{pmatrix} \quad \text{and} \quad \mathbf{E} = \begin{pmatrix} cos(Nz) & 0 & 0 \\ 0 & cos(Nz) & 0 \\ 0 & 0 & 1 \end{pmatrix}$$

$\mathbf{v_0}$, $p_0$ and $\mathbf{x_0}$ are the two-dimensional, steady-state solution, which is customarily called the base flow.

The velocity, pressure, and free surfaces of the disturbed flow are governed by the time-dependent Navier-Stokes system for three-dimensional flows:

$$Re(\frac{\partial \mathbf{v}}{\partial t} + \mathbf{v} \cdot \nabla \mathbf{v}) - \nabla \cdot \boldsymbol{\sigma} - St\ \mathbf{F} = 0$$

$$\nabla \cdot \mathbf{v} = 0 \qquad (2)$$

$$\mathbf{n} \cdot \frac{\partial \mathbf{x}}{\partial t} - \mathbf{n} \cdot \mathbf{v} = 0$$

$Re \equiv \rho V L/\mu$ and $St \equiv \rho g L^2/\mu V$ are, respectively, the *Reynolds number* characterizing the ratio between inertial and viscous forces and the *Stokes number* characterizing the ratio between gravity·and viscous forces. $V$ and $L$ are characteristic flow velocity and length. $\mathbf{v}$ is the velocity vector, $\mathbf{F}$ is the unit vector in which gravity acts, and $\boldsymbol{\sigma}$ represents the stress tensor.

In order to evaluate the perturbations $\mathbf{x'}$, $\mathbf{v'}$, and $p'$, and their rate of growth $\beta$, Galerkin's weighted residual method is applied to eqs. (2). The weighting function used for the momentum equation is $\mathbf{D}(Nz) \cdot \boldsymbol{\phi_i}$; and for the continuity equation, $\chi_i cos(Nz)$. $\boldsymbol{\phi_i}$ is the same vector basis function used to compute the base flow velocity and $\chi_i$ is the same scalar basis functions used to compute the pressure field of the base flow. The domain of integration $V^*$ is a volume obtained by extending the perturbed two-dimensional flow domain ($\Omega^*$) over one wavelength in the transverse direction, as indicated in Fig. 1.

The perturbed physical domain $V^*$ (now a three-dimensional domain) is mapped into a known reference domain $V_0$, as sketched in Fig. 1. The mapping $\mathbf{x} = \mathbf{x}(\boldsymbol{\xi})$ can be written as a perturbation of the mapping $\mathbf{x_0} = \mathbf{x_0}(\boldsymbol{\xi})$ between the base flow domain $V$ and the reference domain $V_0$, which is part of the base flow solution and is known.

The perturbed fields (eqs. 1) are inserted into the weighted residual equations, and terms of order $O(\epsilon^2)$ or larger are neglected — this analysis is restricted to infinitesimal disturbances.

The same finite element basis function used to represent the base flow are used to represent the infinitesimal disturbances $\mathbf{v'}$, $p'$ and $\mathbf{x'}$ in the equations that govern them. Therefore, these variables are written as linear combinations of the basis functions $\phi_i$ and $\chi_i$:

$$u' = \sum_{i=1}^{N} U_i' \phi_i \ , \quad v' = \sum_{i=1}^{N} V_i' \phi_i \ , \text{and} \quad w' = \sum_{i=1}^{N} W_i' \phi_i$$

$$p' = \sum_{i=1}^{M} P_i' \chi_i$$

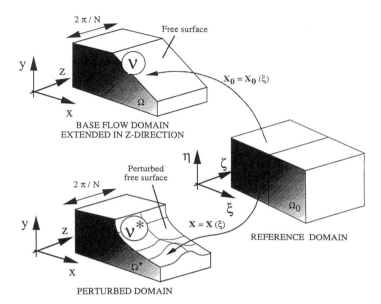

**Figure 1.** *Sketch of perturbed three-dimensional domain*

Christodoulou (1988) (see also Christodoulou & Scriven 1990) allowed all of the nodal locations to be perturbed and represented the perturbations in the same finite element basis that had been used to obtain the base flow:

$$x' = \sum_{i=1}^{N} X_i' \phi_i \quad , \quad y' = \sum_{i=1}^{N} Y_i' \phi_i$$

In that analysis, the weighted residuals associated with the perturbations of nodal position were those of the elliptic partial differential equation system used to create the mesh of the two-dimensional, steady-state flow. The equations of the algebraic eigenproblem included the perturbations of the positions of all the interior nodes. However, only the position of the free-surface is physically relevant to the problem. All the extra unknowns related with the mesh position inside the physical domain are non-physical. This is the main difference between the method developed in this work and that used first by Bixler (1982) and later by Coyle et al. (1990) and Christodoulou & Scriven (1990). In the next sub-section, details of how to restrict a mesh perturbation to free surfaces are presented.

## 2.1. Restriction of Domain Perturbation to Free Surfaces

The main idea of this formulation is to restrict the perturbation of nodal position $\mathbf{x}'$ to where it is physically relevant, i.e. free-surfaces or deformable walls. This is accomplished by setting the domain perturbation to

$$\mathbf{x}' = H^{(0)}(\mathbf{x_0}) \, h' \, \mathbf{n} \tag{3}$$

where

$$H^{(0)} = \lim_{\delta \to 0} H^{(\delta)}$$

$h'$ is a scalar function that gives the amplitude of the perturbation, and $\mathbf{n}$ is the unit normal vector to the unperturbed domain boundary.

$H^{(\delta)}$ is a smooth function defined by

$$H^{(\delta)}(\mathbf{x_0}) = \begin{cases} 1, & \text{if } \mathbf{x_0} \in \Gamma; \\ 0, & \text{if } \mathbf{x_0} \in \Omega \text{ and } |\mathbf{x_0} - \mathbf{x_b}| > |\delta \, \mathbf{n}|, \, \forall \mathbf{x_b} \in \Gamma; \\ \text{smooth function}, & \text{if } \mathbf{x_0} \in \Omega \text{ and } |\mathbf{x_0} - \mathbf{x_b}| < |\delta \, \mathbf{n}|, \, \forall \mathbf{x_b} \in \Gamma \end{cases}$$

It vanishes inside the domain, it is equal to one at the boundaries, and close to the boundaries (inside a ring of thickness $\delta$), it smoothly decays from 1 to 0. It follows from eq. (3) that the nodal perturbation at the boundaries is in the normal direction. The amplitude of the boundary displacement $h'$ is only different from zero at positions where the domain perturbation is relevant (free-surfaces and deformable walls). It is written as a linear combination of basis functions $\phi_i$:

$$h' = \sum_{i=1}^{L} H'_i \phi_i$$

With this approach, the number of degrees of freedom related to the disturbance of the domain is reduced to one unknown per node located at free-surfaces (or deformable walls).

The set of algebraic equations that governs the disturbances and their growth is the following generalized eigenproblem:

$$\mathbf{Ju'} = \beta \mathbf{Mu'} \tag{4}$$

$\mathbf{M}$ is called the mass matrix and $\mathbf{J}$, the Jacobian matrix.

All entries of the Jacobian and mass matrices are presented by Carvalho (1995).

## 2.2. Method of Solving the Generalized Eigenproblem

The stability of the flow is dictated by the eigenvalues with largest real parts, which are called the *leading eigenvalues*. Therefore, there is no need to calculate all the eigenvalues of problem (4), only those that are, or are candidates to become, the leading ones.

When the mass matrix $\mathbf{M}$ is singular and the Jacobian $\mathbf{J}$ is nonsingular, as in the class of problems here, the eigenvalues that are associated with the equations that are not time dependent (continuity equation and essential boundary conditions) are indefinitely large (see Golub & Van Loan 1983). According to Christodoulou & Scriven (1990), they are related to sound waves that travel at infinite speed in incompressible material. Before the eigenproblem can be solved, these 'infinite eigenvalues' have to be removed from the equation system, for otherwise they will be the ones with largest real part. A way of accomplishing this is by using a transformation known as shift-and-invert. The eigenproblem is rewritten as

$$\mathbf{Au'} = \mu \mathbf{u'} \; ; \quad \mathbf{A} = (\mathbf{J} - \sigma \mathbf{M})^{-1}\mathbf{M} \quad \text{and} \quad \mu = \frac{1}{\beta - \sigma} \tag{5}$$

Here, the shift $\sigma$ is real. With the shift-and-invert transformation, the generalized eigenproblem (4) is modified to a simple eigenproblem — eq. (5). The leading eigenvalues of (5) are those eigenvalues of the original problem that are closest to the value of the shift $\sigma$. The infinite eigenvalues of the generalized problem are mapped into zero eigenvalues of the simple eigenvalue problem. The computation of $(\mathbf{J} - \sigma \mathbf{M})^{-1}$ was done by means of a frontal solver (de Almeida 1995).

In this work, the simple eigenvalue problem described by eq. (5) was solved by an iterative Arnoldi method with implicit deflation developed by Saad (1988) for asymmetric eigenproblems.

## 3. Stability of Film-Splitting Flows in Deformable Gaps

Linear stability analysis was also applied to the onset of ribbing in film-splitting flows in deformable gaps. The flow domain considered is sketched in Fig. 2. The top roll is rigid and the bottom roll is covered with a resilient rubber layer. The latter was modeled by an array of radially oriented independent Hookean springs. The base flows are those reported by Carvalho & Scriven (1995).

First, the stability of the flow between rigid rolls is compared with the stability of the flow in a deformable gap at the same operating conditions. Figure 3 illustrates how the real part of the leading eigenvalue for both rigid and a deformable gaps varies with wavenumber at $H_0/R = 2.6 \times 10^{-3}$, $Ca = 0.2$. The elasticity number of the deformable roll case is $Ne^* \equiv \mu \overline{V}/K R^2 = 10^{-6}$. For this set of conditions, the flow between rigid rolls is unstable — there is a small range of wavenumbers $N$ such that the growth factor $\beta$ is positive — and the flow in the deformable gap is stable. This result indicates that the roll deformation delays the onset of ribbing.

The critical capillary number $Ca^*$ was determined by searching at each gap $H_0/R$ for the growth factor curve that is tangent the horizontal axis. The critical capillary number is plotted against gap-to-roll radius in Fig. 4 at $Ne^* = 10^{-6}$. The calculations were extended all the way to negative gaps. The critical capillary numbers for a film-splitting flow between rigid rolls are also shown. When the distance between the rolls is large ($H_0/R > 10^{-1}$), the liquid traction is not large enough to deform the roll cover and the deformable roll behaves as if it were rigid. As the rolls are pushed against each other, the pressure in the liquid between them rises, and the rubber cover deforms in response. The critical capillary numbers in a deformable gap, i.e. $Ne^* = 10^{-6}$, are always larger than those in a rigid gap at the same center-to-center distance. At a fixed center-to-center distance between rolls, a deformable gap can be operated at higher speeds and produce a uniform film.

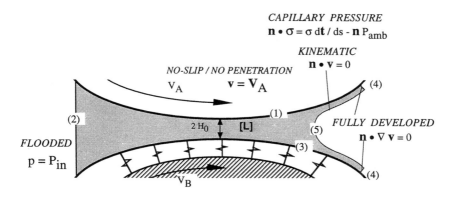

**Figure 2.** *Sketch of film-splitting flow in a deformable gap. The top roll is rigid and bottom roll covered with an elastic layer, modeled here by an array of independent springs.*

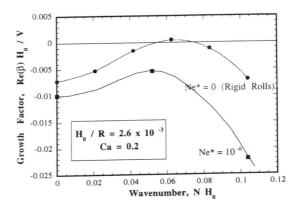

**Figure 3.** *Real part of the leading eigenvalue as a function of wavenumber at modified elasticity number* $Ne^* = 0$ *(rigid rolls) and* $Ne^* = 10^{-6}$, *and* $H_0/R = 2.6 \times 10^{-3}$.

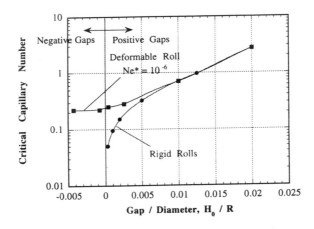

**Figure 4.** *Critical capillary number versus gap-to-roll radius ratio for rigid and deformable gaps.*

As reported by Carvalho & Scriven (1995), the roll deformation pushes the film-split meniscus away from the gap, reducing the pressure gradient at the free surface: Reducing the pressure gradient there tends to stabilize the flow. When the rubber cover is very soft (large elasticity number, viz. $Ne^* = 10^{-6}$), the sensitivity of the critical capillary number to gap is small. At negative gaps, i.e. when the center-to-center distance is smaller than the sum of the undeformed roll radii, the critical capillary number scarcely changes at all as the rolls are pushed together: it falls by less than 2% as the interference goes from $H_0/R = (-)7.5 \times 10^{-4}$ to $(-)4.5 \times 10^{-3}$.

## 4. Final Remarks

The operability limits of coating processes are often imposed by hydrodynamic instabilities that produce defects in the coated film. A way of predicting theoretically these limits is to study the stability of two-dimensional steady flows with respect to infinitesimal disturbances. The mathematical formulation of the problem leads to large, sparse and nonsymmetric generalized eigenvalue problems.

A more efficient formulation than has been in use recently was found. Its essence is to restrict domain perturbations to free surfaces, where they are physically relevant. This approach reduces the size of the eigenproblem. It is completely independent of the equation system used to generate the mesh for solving the equation system of the two-dimensional steady-state base flow.

The method was used to study the stability of film-splitting flows in rigid and deformable gaps. The predicted critical capillary numbers at the onset of ribbing in rigid gaps agree with previous results reported by Coyle et al. (1990). The main results show that the deformation of the roll cover delays the onset of ribbing, i.e. deformable gaps can be operated at higher speeds without producing nonuniform profiles in the cross-web direction.

## 5. References

[1] DE ALMEIDA 1995 *Gas-Liquid Counter-flow in Constricted Passages*. Ph.D. Thesis, University of Minnesota, Minneapolis.

[2] CARVALHO M. S. & SCRIVEN L. E. 1995 Deformable Roll Coating: Modeling of Steady Flow in Gaps and Nips. $1^{st}$ *European Coating Sysmposium, Leeds*.

[3] CHRISTODOULOU C. N. & SCRIVEN L. E. 1988 Finding leading modes of a viscous free surface flow: An asymmetric generalized eigenproblem. *Journal of Scientific Computing*. **3(4)**, 355.

[4] CHRISTODOULOU C. N. 1989 *Computational Physics of Slide Coating Flow*. Ph.D. Thesis, University of Minnesota, Minneapolis.

[5] COYLE D. J. 1984 *The Fluid Mechanics of Roll Coating: Steady Flows, Stability and Rheology*. Ph.D. Thesis, University of Minnesota, Minneapolis.

[6] COYLE D.J., SCRIVEN L.E. AND MACOSKO C.W. 1986 Film splitting flows in forward roll coating. *Journal of Fluid Mechanics*. **171**, 183.

[7] COYLE D. J. 1988 Forward roll coating with deformable rolls: A simple one-dimensional elastohydrodynamic model. *Chemical Engineering Science*. **43**, 2673.

[8] COYLE D. J., MACOSKO C. W. & SCRIVEN L. E. 1990 Stability of symmetric film-splitting between counter-rotating cylinders. *Journal of Fluid Mechanics*. **216**, 437.

[9] FALL C. 1982 A theoretical model of striated film-rupture. *Journal of Lubrication Technology*. **104**, 165.

[10] FALL C. 1985 A theoretical model of striated film-rupture applied to the cylinder-plane. *Journal of Tribology*. **107**, 419.

[11] HUGHES T. J. R. 1987 *The Finite Element Method. Linear Static and Dynamic Finite Element Analysis*, Prentice-Hall, New Jersey.

[12] KISTLER S. F. & SCRIVEN L. E. 1983 Coating Flows. *Computational Analysis of Polymer Processing* (Eds. J. R. A. Pearson and S. M. Richardson), Applied Science Publishers, London and New York, 243.

[13] KISTLER S. F. & SCRIVEN L. E. 1984 Coating flow theory by finite element and asymptotic analysis of the Navier-Stokes system. *Int. J. Numer. Methods Fluids.* **4**, 207.

[14] IOOSS G. & JOSEPH D. D. 1980 *Elementary Stability and Bifurcation Theory.* Springer-Verlag, New York.

[15] PEARSON J. R. A. 1960 The stability of uniform viscous flow under rollers and spreaders. *Journal of Fluid Mechanics.* **7**, 481.

[16] PITTS E. & GREILLER J. 1961 The flow of thin liquid films between rollers. *Journal of Fluid Mechanics.* **11**, 33.

[17] RUSCHAK K. J. 1983 A three-dimensional linear stability analysis for two-dimensional free boundary flows by the finite element method. *Computers & Fluids.* **11(4)**, 391.

[18] SAAD Y. & SCHULTZ M. H. 1986 GMRES: a generalized minimal residue algorithm for solving nonsymmetric linear systems. *SIAM J. Sci. Stat. Comput..* **7**, 856.

[19] SAAD Y. 1988 Numerical solution of large nonsymmetric eigenvalue problems *RIACS Thechnical Report. NASA Ames Research Center.* 88.39.

[20] SAAD Y. 1994 *Numerical methods for large eigenvalue problems.* Manchester University Press Series in Algorithms and Architetrures for Advanced Scientific Computing.

[21] SAVAGE M. D. 1977 Cavitation in lubrication. Part 1: On boundary conditions and cavity - fluid interfaces. *Journal of Fluid Mechanics.* **80**, 743.

[22] SAVAGE M. D. 1977 Cavitation in lubrication. Part 2: Analysis of wavy interfaces. *Journal of Fluid Mechanics.* **80**, 757.

[23] SAVAGE M. D. 1982 Mathematical models for coating processes. *Journal of Fluid Mechanics.* **117**, 443.

[24] SAVAGE M. D. 1984 Mathematical model for the onset of ribbing. *AIChE Journal.* **30**, 999.

## 6. Acknowledgments

M. S. Carvalho was supported by a fellowship from CAPES (Brazilian Federal Government). Further support came from cooperating corporations through the Center for Interfacial Engineering and was supplemented by the National Science Foundation.

# MINIMIZATION AND CONTROL OF RANDOM EFFECTS ON FILM THICKNESS UNIFORMITY BY OPTIMIZED DESIGN OF COATING DIE INTERNALS

F. Durst, U. Lange, H. Raszillier

*Lehrstuhl für Strömungsmechanik*
*Universität Erlangen-Nürnberg*
*Cauerstr. 4, D-91058 Erlangen (Germany)*

## SUMMARY

In this paper, model equations for the flow of arbitrary shear-dependent liquids in the distribution chambers inside a curtain coating die are presented. Assuming random manufacturing inaccuracies, a stochastic error analysis is carried out. It is illustrated that the film thickness uniformity produced by the die may depend strongly on these inaccuracies. Moreover, the consequences of stochastic effects on film thickness control strategies – usually based on adjusting parts of the internal geometry – are analyzed. Finally, a systematic optimization strategy for the internal design of a coating die with respect to film thickness uniformity is outlined. This strategy takes into account both the the effects of manufacturing inaccuracies and practical constraints on flow parameters such as wall shear stress, residence time, or maximum pressure.

## 1. INTRODUCTION

The quality of photographic films and of many other coated products is very sensitive to thickness variations of the applied functional layers. In many coating processes the required uniformity has to be achieved by designing the internal geometry of the coating die in such a way that the coating liquid is evenly distributed in the cross–web direction (cf. fig. 1). This internal design has to comply with a number of constraints, which relate certain flow parameters – for example minimum wall shear rates or maximum residence times in the die – to practical aspects of the production process, such as minimization of cleaning effort or aging of the liquid. Because of these constraints, the simple "infinite cavity"-design shown in fig. 1 is not suitable for most applications and more elaborate internal geometries have to be used, as for example the so-called "coat-hanger"-design (cf. e.g. [1], [4]), or dies with a secondary distribution chamber like the one shown in fig. 2 in the next section. The reliable and efficient prediction of such a die geometry, which complies with all the constraints of a given coating application and yields uniform film profiles for the whole range of the operating parameters is a challenging problem of flow optimization.

The various simplified models for the three–dimensional flow situation in the die, which have been discussed in the literature (cf. e.g. [2], [3], [4] and many others), can be considered as a prerequisite for the solution of this optimization problem, but a strategy to take into account constraints in a systematic way has not been published yet. Moreover, even if a good design is known for a given range of fluid and process parameters, the film uniformity can still be deteriorated significantly by random influences such as manufacturing inaccuracies (cf. [5]). Because of the high manufacturing expenses for coating dies, it is very important that these random effects are also minimized by an optimal design.

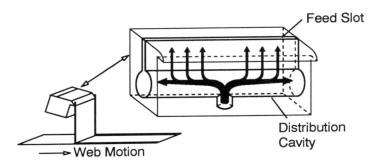

Fig. 1: Internal design of a curtain coating die.

In this paper, a stochastic error analysis of model equations, which were derived by the authors for the flow of arbitrary shear-dependent liquids through the die, is presented. Based on this analysis for random manufacturing inaccuracies in the internal geometry, an optimization strategy is outlined, which takes into consideration these random effects as well as the most important practical contraints.

## 2. MODEL EQUATIONS AND DETERMINISTIC PREDICTIONS

The most appropriate simplified models for die flows are those based on averaging the mass and momentum balances in the distribution cavities, since they allow for a consistent consideration of inertial effects and complex rheological behaviour of the fluids (cf. e.g. [2], [3]).

Averaging the balances for the two cavities of a dual cavity die (cf. fig. 2) yields four ordinary differential equations in the arc-length coordinates $s_1, s_2$, which relate the flow rates $\dot{Q}_i$ in the axial directions of the cavities to the local fluxes (per unit length) $\dot{q}_i$ in the feed slots:

$$\frac{d\dot{Q}_1}{ds_1} = \dot{q}_1 \quad , \quad \frac{d\dot{Q}_2}{ds_2} = \dot{q}_2 - \dot{q}_1 , \tag{1}$$

$$\frac{dp_i}{ds_i} = \alpha_i(\dot{Q}_i) \cdot \dot{Q}_i - \frac{d}{ds_i}\beta_i(\dot{Q}_i) \cdot (\dot{Q}_i)^2 \quad (i = 1, 2). \tag{2}$$

Here, $p_i$ denote the (averaged) pressures in the cavities. While the global mass balances (1) are exact, the accuracy of the momentum balances (2) depends strongly on how well the axial velocity profiles in the cavities can be approximated, since these "shape functions" determine the values of the coefficients $\alpha_i$ and $\beta_i$ in the viscous and the inertial terms on the right hand side of eq. (2). A suitable choice for the shape functions are (approximations to) fully developed flow profiles through the local cross sections of the cavities.

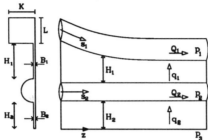

Fig. 2: Definition sketch of a dual-cavity coating die.

In principle, this approach is feasible for arbitrary shear–dependent liquids. The model calculations presented in this paper are based on the three–parameter Carreau model:

$$\eta(\dot\gamma) = \frac{\eta_0}{\left(1 + \left(\frac{\eta_0 \dot\gamma}{\tau_*}\right)^2\right)^{(1-n)/2}} \cdot \tag{3}$$

Using an asymptotic approximation to this model, the coefficients $\alpha_i, \beta_i$, can still be evaluated analytically with the help of a computer algebra system. Hence the model equations can, on one hand, be evaluated almost as fast as for simple "power law"–fluids. On the other hand the dependency of the profiles on the typical rheological behaviour of the coating liquids used in the photographic industry, is predicted with much higher accuracy as with the power law ansatz.

The system of ordinary differential equations (1), (2) can be closed by suitable relations $F_i$ between the local flow rates in the feed slots and the local pressure differences across the slots:

$$p_i - p_{i+1} = F_i(\dot q_i) \qquad \left(= -12\eta\dot q_i \frac{H_i}{B_i^3}\right) \quad (i = 1, 2). \tag{4}$$

Here, $p_3$ denotes the atmospheric pressure and the functions $F_i$ can be derived by assuming one-dimensional flow in the slots. In the newtonian case, the $F_i$ simplify to the Poiseuille law given by the bracketed term in eq. (4), in which $\eta$ denotes the newtonian viscosity, $H_i$ the slot lengths and $B_i$ the slot widths. (cf. fig. 2). In order to predict the local flux $-q_2(z)$, which yields the film thickness distribution, the equations are expressed in the cartesian coordinate $z$ (see fig. 2) an then numerically solved using Newton's method in order to treat the nonlinearities arising from inertia and shear–dependence and the shooting method in order to solve the linear boundary value problem in each Newton iteration. This procedure allows for a very fast prediction of the film thickness profiles, which agrees very well with numerical simulations of the full three–dimensional flow.

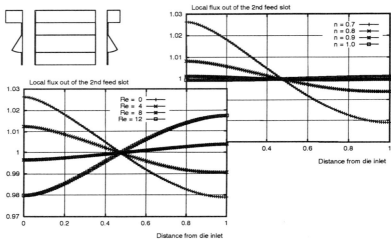

Fig. 3: Effect of inertia and shear–dependency on predicted profiles.

As an illustration of the most important mechanisms of the internal flow in a typical dual cavity die (cf. upper left corner of fig. 3), some predictions of the local flux $-q_2$ along the production width are shown in fig. 3 for different Carreau fluids and different operating conditions. (Throughout this paper, lengths are normalized with the production width $A$ and the local fluxes $-\dot{q}_2$ are normalized by the desired uniform flux $\dot{q}_u$. The operating state is then described by the Reynolds number $Re = \rho\dot{q}_u/\eta_{in}$, with $\rho$ being the density of the liquid and $\eta_{in}$ being a characteristic value of the viscosity.) With no inertial effects present, the film thickness decreases with the distance from the die inlet, due to the viscous pressure drop in the first cavity. However, with increasing Reynolds number, the profile is levelled and finally even increases with the distance from the die inlet. This is due to the fact that the principal inertial effect is a pressure recovery caused by the deceleration of the flow in the first cavity.

In the dimensionless notation, only the power law index $n$ and the normalized critical stress. $T = \tau_* A^2/\eta_0\dot{q}_u$ are required for the characterization of the properties of Carreau fluids. As the shear–thinning behaviour gets stronger, i.e. with decreasing $n$, the nonuniformity of the profiles increases (see back of fig. 3), because the viscosity in the narrow slots is then very low compared to the viscosity in the cavities.

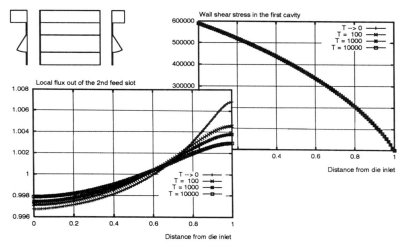

Fig. 4: Predicted profiles for different critical shear stresses.

Fig. 4 shows the importance of modeling the rheological behaviour properly by comparing the predicted profiles for Carreau–fluids with the same power law index $n$ and same value of $m = \eta_0^n \tau_*^{1-n}$ , but with different values of the critical stress $\tau_*$. All these fluids correspond to the same power law model. The predictions differ in particular far away from the inlet, where the shear rate in the first cavity goes to zero, due to the vanishing flow rate $\dot{Q}_1$. According to the power law model, the viscosity would therefore tend to infinity, which is of course not realistic and leads to erroneous profile predictions.

## 3. STOCHASTIC ERROR ANALYSIS

The practical relevance of the deterministic results outlined so far depends on how well their sensitivity to stochastic effects can be modeled. In particular, the effect of manufacturing inaccuracies $b_i(z)$ in the narrow slot widths $B_i(z)$ has to be estimated. They can be treated by a linear error analysis, but some care has to be taken, since the deviations $b_i$ from the deterministic slot widths are not just random variables, but random valued functions. Hence, the deviation from the determistic profile caused by the inaccuracies must be linearized in the proper functional sense, which yields

$$\dot{q}_2(B_1 + b_1, B_2 + b_2) - \dot{q}_2(B_1, B_2) \approx$$
$$D\dot{q}_2[B_1](B_1, B_2) \cdot b_1 + D\dot{q}_2[B_2](B_1, B_2) \cdot b_2 , \tag{5}$$

where $D\dot{q}_2[B_i](B_1, B_2) : C^1[0, A] \to C^0[0, A]$ denote Fréchet–derivatives of $\dot{q}_2$. They can be evaluated approximately using finite–dimensional approximations to the functions $b_i(z)$ and their standard deviations

$$b_i(z) \approx \sum_{k=0}^{N} \lambda_k^i \phi_k(z) \ , \qquad \sigma_{b_i}(z) \approx \sum_{k=0}^{N} \sigma_k^i \phi_k(z) \ . \tag{6}$$

Thus the local standard deviation of the film thickness $\sigma_{\dot{q}_2}(z)$ can be computed, in its dependence on the local standard deviations of the slot widths $\sigma_k^i$, by

$$\sigma_{\dot{q}_2}(z) = \left( \sum_{k=0}^{N} (\sigma_k^1)^2 (\psi_k^1(z))^2 + (\sigma_k^2)^2 (\psi_k^2(z))^2 \right)^{1/2} , \tag{7}$$

with $\psi_k^i = D\dot{q}_2[B_i](B_1, B_2) \cdot \phi_k$ being the directional derivatives with respect to the basis functions $\phi_k$.

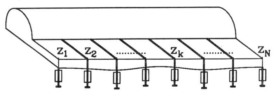

Fig. 5: Control of the thickness profile by adjusting the slot width.

In practice, uniform film profiles are often achieved by adjusting the slot widths locally in response to thickness measurements (cf. fig. 5). The effect of small errors in the tuning of the adjustment screws can be treated by an alternative interpretation of the error analysis. The deterministic slot width $B_i(z)$ can be described in this case by a cubic spline though the screw positions $z_i$, $(i = 1 \ldots N)$. The tuning inaccuracy for the $k$-th screw at the $i$-th slot is a random variable $b_i(z_k)$ with zero mean and a standard deviation $\sigma_k^i$. If $\phi_k$, $(k = 1 \ldots N)$ is a nodal basis of the cubic splines through the screw positions, the inaccuracies yield a deviation $b_i(z)$ of the slot width, which is described by

$$b_i(z) \approx \sum_{k=0}^{N} b_i(z_k) \phi_k(z) \ , \qquad \sigma_{b_i}(z) \approx \sum_{k=0}^{N} \sigma_k^i \phi_k(z) \ . \tag{8}$$

This is exactly the situation treated by the error analysis, which therefore can also be used to model the influence of tuning inaccuracies.

Fig. 6 shows the predicted standard deviation of a thickness profile, if manufacturing inaccuracies with standard deviations of one percent are assumed. Obviously the random effect can be of the same order of magnitude as the deterministic nonuniformity. Hence, as a reliable measure for nonuniformity, one should take the difference between the maximum and the minimum of this stochastic profile, given by

$$ UG_\sigma = \frac{1}{\dot{q}_{gl}} \left( \max_{0 \le z \le 1} \{ \dot{q}_2(z) + \sigma_{\dot{q}_2}(z) \} - \min_{0 \le z \le 1} \{ \dot{q}_2(z) - \sigma_{\dot{q}_2}(z) \} \right) . \qquad (9) $$

In the calculation shown in the back of fig. 6 it was assumed, that there are only inaccuracies in the first slot. The effect of those inaccuracies is almost damped out by tranversal flow in the secondary cavity.

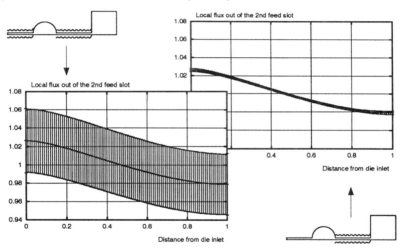

Fig. 6: Effect of manufacturing inaccuracies in the feed slots.

This function of a secondary cavity can also be illustrated by the example of the effect of tuning inaccuracies shown in fig. 7. Here, the slot width was adjusted by eleven screws, until the deterministic profile was perfectly uniform, and a tuning inaccuracy of 0.1 percent was assumed for all screws. Adjusting the second slot instead of the first one results in about five times bigger standard deviations of the profile. It should be pointed out that the wavy pattern of the standard deviation in the case of a tuned second slot is due to the fact that the slot width between the screws is correlated to the screw positions. If the first slot is tuned, this effect is also damped out by the flow in the secondary cavity.

## 4. OPTIMIZATION STRATEGY

Based on the model outlined so far, a systematic strategy to determine optimal die designs can be developed. To this end, it is useful to pose the die design problem as a problem of constrained optimization: The constraints for wall shear stress etc. can be written as inequalities, which have to be fulfilled in all operating states $b$ out of the operating range $B_V$:

$$\vec{N}_V(b, g_o) \leq 0 \text{ for all } b \in B_V \ . \tag{10}$$

Let $G_V$ denote the set of all geometries, which meet the constraints given by eq. (10). The task is then to find a die geometry $g_o$ out of $G_V$, which minimizes the maximum stochastic nonuniformity $UG_\sigma$:

$$\max_{b \in B_V}\{UG_\sigma(b, g_o)\} = \min_{g \in G_V}\left\{\max_{b \in B_V}\{UG_\sigma(b, g)\}\right\} \ . \tag{11}$$

Because of the many variations of die geometries that are possible, the complexity of this optimization problem is considerable. It can be reduced by the idea of posing so–called inverse problems, which will be briefly sketched below:

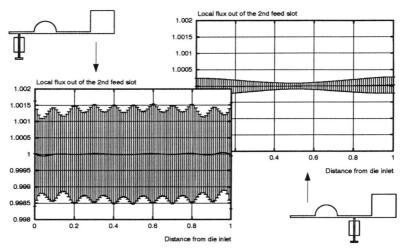

Fig. 7: Effect of tuning inaccuracies, if a feed slot is adjusted by screws.

For a newtonian fluid and a circular cavity cross section, for example, the wall shear stress in the first cavity $\tau_{w1}$ depends on the flow rate $\dot{Q}_1$ – which is almost linearly decreasing from the inlet value $\dot{Q}_{In}$ to zero – and on the radius $R_1$ of the cavity:

$$\bar{\tau}_{1w}(z) = -\frac{2\eta\,\dot{Q}_1(z)}{\pi R_1(z)^3} \approx -\frac{2\eta\dot{Q}_{In}\,(1 - z/A)}{\pi R_1(z)^3} \,. \tag{12}$$

If a constraint $\tau_{min}$ is prescribed, one can demand this value for the flow variable ($\tau_{w1} \equiv \tau_{min}$) and then invert eq. (12) in order to determine the geometry variable $R_1(z)$. In a practical realization, one would rather determine $R_1(z)$ from a suitable approximation to eq. (12). Using the same idea, several geometry variables can be parameterized via the constraints by a finite number of parameters. The resulting reduced set of geometries automatically complies with those constraints.

A plausible way to reduce the set of admissible geometries even further is to consider only those geometries which yield a perfect distribution in at least one operating state. This leads also to an inverse problem: If the desired uniform flow ($\dot{q}_1(z) \equiv \dot{q}_2(z) \equiv \dot{q}_u$) is inserted into the model equations, they can be interpreted as a determining relation for the slot length $H_1(z)$ of the first slot:

$$\frac{d}{dz}\left(\frac{H_1}{B_1^3}\right) = \alpha_{11}(A - z)\left(1 + \left(\frac{dH_1}{dz}\right)^2\right)^{1/2} + Re\frac{d}{dz}\left(\beta_{21}(A - z)^2\right). \tag{13}$$

The geometries remaining as solutions of this equation are parameterized by the operating parameters, e.g. the Reynolds number. Hence the reduced optimization problem is finite–dimensional and can be solved by standard methods.

Typical features of the die designs which result from the outlined optimization strategy are, for example, a tapering of the first cavity resulting from the inverse problem for the wall shear stress and a "coat hanger like" shape resulting from the inverse problem for the profile. If manufacturing inaccuracies are taken into account, the geometries resulting from the optimization have typically a wider second slot, since this reduces the influence of the manufacturing inaccuracies. It should be pointed out that the shape of the optimal design depends strongly on the size of the manufacturing inaccuracies. For a typical operating range for photographic applications, the design obtained by the optimization yields a reduction of about 30 percent in the stochastic nonuniformity compared to a good conventional system, if it is required that both systems should comply with same the constraints for wall shear stress, residence time, pressure level and available space.

## 5. CONCLUSION

In most industrial applications the internal design of coating dies, which is crucial for the uniformity of film thickness profiles, must comply with important constraints. Hence, it is useful to consider the die design problem as a problem of constrained optimization. An approximate solution to this problem can then be found efficiently by the idea of posing inverse problems, which on one hand are a systematic way to take into account constraints and on the other hand provide a careful reduction of the complexity of the problem.

As a basis for the presented optimization strategy, simplified model equations were discussed, which provide an appropriate treatment of inertial and rheological effects and therefore allow for reliable predictions of the flow inside coating dies. Carrying out a linear error analysis of these model predictions, it is possible to take into account stochastic effects on the internal die flow, e.g. the effect of manufacturing inaccuracies, which can be the most important effect for nonuniformities of the thickness profiles. The error analysis is also applicable to stochastic effects on film thickness control strategies, such as tuning inaccuracies of screws used for the adjustment of feed slot widths.

Because of the outlined features of the underlying flow model, the presented optimization strategy provides an efficient and comprehensive tool for the design and for the prediction of the performance of coating dies.

## 6. REFERENCES

1. Durst, F., Lange, U., Raszillier, H.: Optimization of Distribution Chambers of Coating Facilities, *Chem. Eng. Sci.* 48 (1994), pp. 161-170.

2. Leonard, W.K.: Effects of Secondary Cavities, Inertia and Gravity on Extrusion Dies, *43rd Annual Technical Conference of the Soc. of Plastics Eng.*, 1985, pp. 144-148.

3. Liu, T.-J., Hong, C.-N., Chen, K.-C.: Computer-Aided Analysis of a Linearly Tapered Coat-Hanger Die, *Polymer Eng. Sci.* 28, 23 (1988), pp. 1517-1526.

4. Sartor, L.: Slot Coating: Fluid Mechanics and Die Design, Ph.D. Thesis, University of Minnesota, 1990.

5. Schweizer, P.M.: Fluid Handling and Preparation, in: Cohen, E., Gutoff, E. (Hrsg.) *Modern Coating and Drying Technology*, VCH Publishers, New York, 1992.

## 7. ACKNOWLEDGEMENTS

This paper presents results of a research project, which was financially supported by Troller & Co. AG, Murgenthal, Switzerland. The authors would like to thank U. Troller and also P. Schweizer from Ilford AG, Switzerland for many suggestions and helpful discussions on the subject.

# CHEBYSHEV COLLOCATION METHOD ON SOLVING STABILITY OF A LIQUID LAYER FLOWING DOWN AN INCLINED PLANE

A.T.L. Horng

Department of Applied Mathematics
Feng Chia University, Taichung, Taiwan

## SUMMARY

The Orr–Sommerfeld equation governing the linear stability of a liquid layer flowing down an inclined plane appears as a differential eigenvalue problem with the eigenvalue involved nonlinearly in the free surface boundary conditions which causes the problem more difficult to solve than most other Orr–Sommerfeld problems. A numerical solver based on Chebyshev spectral collocation discretization auxiliary with the companion matrix method is employed to solve the problem. The superiority in the accuracy of the solver is shown by its validation with other papers.

## 1. INTRODUCTION

The linear stability Orr–Sommerfeld problem of a viscous fluid layer flowing down an inclined plane under the action of gravity is re–solved here by a novel numerical approach–spectral collocation method. Since the paper focuses on the details of numerical method, the derivation of the governing differential equation, that is Orr–Sommerfeld equation, and the associated boundary conditions can be found in Yih[9] and Lin[8]. As to its physics, a recent review by Chang[3] describing the linear and nonlinear instabilities of a free–falling film is worth referring to.

Following Lin[8], the geometric configuration of the flow is shown in Fig. 1. with $d$ denoting the Nusselt flat–film thickness, $\beta$ the angle of inclination of the plane. Using the maximum velocity of the base flow $U_m = g \sin \beta / 2\nu$ ($g$ is the gravitational acceleration, $\nu$ the kinematic viscosity) as the characteristic velocity and $d$ the characteristic length, the dimensionless governing differential equation is the famous Orr–Sommerfeld equation,

$$\phi^{iv} - 2\alpha^2 \phi'' + \alpha^4 \phi = i\alpha Re[(U - c)(\phi'' - \alpha^2 \phi) - U''\phi], \qquad (1)$$

where $\alpha$ is the wavenumber, $c$ the complex wave speed, $Re = U_m d / \nu$ the Reynolds number, $U$ the dimensionless velocity profile of the base flow:

$$U(y) = 1 - y^2, \qquad (2)$$

and $\phi$ a function of $y$ representing the $y$–direction distribution of the perturbed streamfunction $\psi$ by

$$\psi = \phi(y) \exp[i\alpha(x - ct)]. \qquad (3)$$

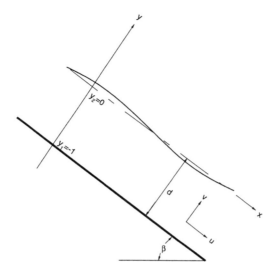

Figure 1: Definition sketch for a liquid layer flowing down an inclined plane.

The velocity perturbations $u'$ and $v'$ in the $x$ and $y$ directions can be expressed in terms of $\psi$ by

$$u' = \partial\psi/\partial y, \tag{4}$$
$$v' = -\partial\psi/\partial x. \tag{5}$$

The boundary conditions are nonslip conditions at the bottom wall ($y = -1$),

$$\phi(-1) = \phi'(-1) = 0, \tag{6}$$

and the continuity of shear and normal stresses at the free surface ($y = 0$),

$$\phi''(0) + \left(\alpha^2 - \frac{2}{c'}\right)\phi(0) = 0, \tag{7}$$

$$-\left(\alpha\frac{2\cot\beta + \alpha^2 S Re}{c'}\right)\phi(0) + \alpha(Re\, c' + 3i\alpha)\phi'(0) - i\phi'''(0) = 0, \tag{8}$$

where $S = We^{-1} = T/\rho dU_m^2$ ($We$ is the Weber number, $T$ the surface tension, $\rho$ the density) and $c' = c - 1$. Since $S$ is dependent on $U_m$ (and thus on flow rate), some papers prefer using the Kapitza number $\gamma = 3^{1/3}T/\rho\nu^{4/3}g^{1/3} = S Re^{5/3}(\frac{3}{2}\sin\beta)^{1/3}$ which is only a function of the physical properties of the liquid. Then, Eq.(8) is replaced by

$$-\left(\alpha\frac{2\cot\beta + \alpha^2\gamma Re^{-2/3}(\frac{3}{2}\sin\beta)^{-1/3}}{c'}\right)\phi(0) + \alpha(Re\, c' + 3i\alpha)\phi'(0)$$
$$-i\phi'''(0) = 0, \tag{8a}$$

Eq.(1), (6), (7) and (8) or (8a) are homogeneous and a nontrivial solution exists if there exists a relation between $\alpha, c, \beta, Re$, and $S$ (or $\gamma$):

$$f(\alpha, c, \beta, Re, S \text{ or } \gamma) = 0. \tag{9}$$

They construct a differential eigenvalue problem with the eigenvalue $c$ and the associated eigenfunction $\phi$ to be solved. Although, the dimensionless velocity profile of the base flow is same as plane Poiseuille flow, the current Orr–Sommerfeld problem is much more difficult to solve than that for plane Poiseuille flow. It is because the boundary conditions Eq.(7) and (8) at free surface involve the eigenvalue nonlinearly. For temporal instability, $c$ is complex ($c = c_r + ic_i$). Eq.(9) implies $c_r = c_r(\alpha, Re)$, $c_i = c_i(\alpha, Re)$ if $\beta$ and $S$ (or $\gamma$) are given. The flow is stable if $c_i < 0$ and unstable if $c_i > 0$, and $c_i(\alpha, Re) = 0$ gives a neutral stability curve on $\alpha$–$Re$ plane for given $\beta$ and $S$ (or $\gamma$).

There are two linear stability modes encountered in the current problem: surface mode (also named as soft mode) and shear mode (also named as hard mode). The surface mode is essentially surface wave driven by gravity–capillary effects slightly modified by viscosity, and the shear mode is basically the Tollmien–Schlichting wave modified by the presence of the free surface. The surface mode usually with long wavelength (compared with $d$) is the dominant unstable mode when Reynolds number is small to moderate ($1 < Re < 300$ suggested by Chang[3]), while the shear mode becomes dominant when Reynolds number is large ($Re > 1000$ suggested by Chang[3]) with wavelength comparable to or shorter than $d$.

Yih[9] solved the problem by perturbation expansion with the restriction to long waves (surface mode) and small Reynolds numbers, and obtained the famous result for the critical Reynolds number

$$Re_{cr} = \frac{5}{4} \cot \beta.$$

As to the shear mode, Lin[8] analytically solved the problem motivated by the Orr–Sommerfeld solutions of plane Poiseuille flow. However, Lin made a mistake on boundary condition Eq.(8) (the sign of the first term at the left–hand side of Eq.(8) is positive in Lin[8]), which is pointed out by De Bruin[5]. De Bruin uses Runge–Kutta integration to integrate Eq.(1) shooting for $c$ with orthogonalization procedure to solve the problem with the correct boundary condition. The orthogonalization procedure is chiefly to keep particular solutions from depending on each other (caused by the stiffness of Eq.(1)) during the integration. Floryan et al.[6] following De Bruin's numerical method solved the problem with Eq.(8a) with nonzero surface tension extensively for both surface and shear modes. Chin et al.[4] also calculates the problem with Eq.(8a) by finite difference method with an additional consideration on the from factor effect on the velocity profile of the base flow. Ho and Patera[7] computes the problem with Eq.(8) using Hermitian finite element method as a preliminary study in their spectral element Navier–Stokes simulation of a free falling film.

The numerical methods used in the papers mentioned above are either Runge–Kutta integration shooting method or finite difference/element methods. The former can not calculate the whole Orr–Sommerfeld spectrum and is sensitive to the initial guess on $c$. Also, during iteration, the search on $c$ complex plane can sometimes be tedious. The latter can calculate the whole Orr–Sommerfeld spectrum but the order of accuracy is low. In this paper, the novel spectral collocation method with much higher order of accuracy than conventional finite difference/element methods is employed to solve the whole Orr–Sommerfeld spectrum (for both surface and shear modes) from which the most unstable eigenvalue $c$ is identified. Based on spectral collocation method, the solver is easily developed and highly efficient. The numerical procedure is derived in section 2 (numerical method) and the results are validated with Lin[7], De Bruin[5] and Floryan et al.[6] in section 3 (results and discussion).

## 2. NUMERICAL METHOD

The physical domain $y \in [-1, 0]$ needs to be mapped into the computational domain $z \in [-1, 1]$ by

$$y = \frac{z - 1}{2}. \tag{10}$$

$\phi(y(z))$ is then expanded as a series of Chebyshev polynomials,

$$\phi(y(z)) = \sum_{k=0}^{N} a_k T_k(z). \tag{11}$$

The reason to choose Chebyshev polynomials among all kinds of orthogonal polynomials is that Chebyshev polynomials are orthogonal with respect to the weight function $1/(1 - z^2)^{1/2}$ which underlies its emphasis near the boundaries and is particularly suitable to describe the boundary layer phenomenon frequently encountered in flow instability. Eq.(1), (6), (7) and (8) or (8a) can then be rewritten in terms of differential operators with respect to $z$:

$$L_4\phi = cL_2\phi, \tag{12}$$
$$B_0\phi = B_1\phi = 0, \qquad \text{at} \quad z = -1, \tag{13}$$
$$E_2\phi = cF_2\phi, \qquad \text{at} \quad z = 1, \tag{14}$$
$$E_3\phi = cF_3\phi + c^2F_1\phi, \qquad \text{at} \quad z = 1, \tag{15}$$

where

$$L_4 = 16\,d^4/dz^4 - (8\alpha^2 + 4i\alpha Re\,U)\,d^2/dz^2$$
$$+ (\alpha^4 + i\alpha^3 Re\,U + i\alpha Re\,U''), \tag{16}$$

$$L_2 = -4i\alpha Re\,d^2/dz^2 + i\alpha^3 Re, \tag{17}$$

$$B_0 = 1, \tag{18}$$

$$B_1 = d/dz, \tag{19}$$

$$E_2 = 4\,d^2/dz^2 + (\alpha^2 + 2), \tag{20}$$

$$F_2 = 4\,d^2/dz^2 + \alpha^2, \tag{21}$$

$$E_3 = 8\,d^3/dz^3 - (6\alpha^2 + 2i\alpha Re)\,d/dz - i(2\alpha\cot\beta + \alpha^3 S Re), \tag{22}$$

$$F_3 = 8\,d^3/dz^3 - (6\alpha^2 + 4i\alpha Re)\,d/dz, \tag{23}$$

$$F_1 = 2i\alpha Re\,d/dz. \tag{24}$$

or

$$E_3 = 8\,d^3/dz^3 - (6\alpha^2 + 2i\alpha Re)\,d/dz - i(2\alpha\cot\beta$$
$$+\alpha^3\gamma Re^{-2/3}(\frac{3}{2}\sin\beta)^{-1/3}), \tag{22a}$$

for boundary condition (8a).

Eq.(12)–(15) are to be collocated at Chebyshev–Gauss–Lobatto quadrature points (Canuto et al.[2]), distributed as

$$z_j = \cos\left(\frac{j\pi}{N}\right) \qquad j = 0, 1, 2, \dots, N. \tag{25}$$

Let $\phi$ with the entry $\phi_j = \phi(y(z_j))$ denotes the value of $\phi(y(z))$ at the collocation points in Eq.(25). $d/dz$ is then approximated by the Chebyshev collocation derivative matrix $D_N$ (Canuto et al.[2]):

$$(D_N)_{ij} = \begin{cases} \frac{\bar{c}_i(-1)^{i+j}}{\bar{c}_j(z_i - z_j)}, & i \neq j, \\[2mm] \frac{-z_j}{2(1-z_j^2)}, & 1 \leq i = j \leq N-1, \\[2mm] \frac{2N^2+1}{6}, & i = j = 0, \\[2mm] -\frac{2N^2+1}{6}, & i = j = N, \end{cases} \tag{26}$$

and, likewise, $d^n/dz^n$ is approximated by $D_N{}^n$. The differential operators in Eq.(16)–(24) can then be transformed into matrix operators:

$$(L_4)_{ij} = 16\,(D_N{}^4)_{ij} - \left(8\alpha^2 + 4i\alpha Re\,U(z_i)\right)(D_N{}^2)_{ij}$$

$$+ \left( \alpha^4 + i\alpha^3 Re\, U(z_i) + i\alpha Re\, U''(z_i) \right) \delta_{ij}, \tag{27}$$

$$(\boldsymbol{L_2})_{ij} = -4i\alpha Re\, (\boldsymbol{D_N}^2)_{ij} + i\alpha^3 Re\, \delta_{ij}, \tag{28}$$

$$(\boldsymbol{B_0})_{ij} = \delta_{ij}, \tag{29}$$

$$(\boldsymbol{B_1})_{ij} = (\boldsymbol{D_N})_{ij}, \tag{30}$$

$$(\boldsymbol{E_2})_{ij} = 4\,(\boldsymbol{D_N}^2)_{ij} + (\alpha^2 + 2)\,\delta_{ij}, \tag{31}$$

$$(\boldsymbol{F_2})_{ij} = 4\,(\boldsymbol{D_N}^2)_{ij} + \alpha^2 \delta_{ij}, \tag{32}$$

$$(\boldsymbol{E_3})_{ij} = 8\,(\boldsymbol{D_N}^3)_{ij} - (6\alpha^2 + 2i\alpha Re)(\boldsymbol{D_N})_{ij} - i(2\alpha \cot\beta + \alpha^3 S Re)\,\delta_{ij}, \tag{33}$$

$$(\boldsymbol{F_3})_{ij} = 8\,(\boldsymbol{D_N}^3)_{ij} - (6\alpha^2 + 4i\alpha Re)(\boldsymbol{D_N})_{ij}, \tag{34}$$

$$(\boldsymbol{F_1})_{ij} = 2i\alpha Re\, (\boldsymbol{D_N})_{ij}, \tag{35}$$

or

$$(\boldsymbol{E_3})_{ij} = 8\,(\boldsymbol{D_N}^3)_{ij} - (6\alpha^2 + 2i\alpha Re)(\boldsymbol{D_N})_{ij} - i(2\alpha \cot\beta$$
$$+\alpha^3 \gamma Re^{-2/3}(\tfrac{3}{2}\sin\beta)^{-1/3})\,\delta_{ij}, \tag{33a}$$

for boundary condition (8a), where $\boldsymbol{L_4}, \ldots,$ etc. denote the matrix operators approximating $L_4, \ldots,$ etc. and $\delta_{ij}$ denotes Kronecker delta.

By Eq.(27)–(35), Eq.(12)–(15) are transformed into a nonlinear matrix generalized eigenvalue problem:

$$\boldsymbol{G}\phi = c\,\boldsymbol{H}\phi + c^2 \boldsymbol{K}\phi, \tag{36}$$

where

$$(\boldsymbol{G})_{ij} = \left\{ \begin{array}{ll} (\boldsymbol{E_3})_{0j}, & i = 0, \\ (\boldsymbol{E_2})_{0j}, & i = 1, \\ (\boldsymbol{L_4})_{ij}, & 2 \leq i \leq N-2, \\ (\boldsymbol{B_0})_{Nj}, & i = N-1, \\ (\boldsymbol{B_1})_{Nj}, & i = N, \end{array} \right\}, \quad 0 \leq j \leq N, \tag{37}$$

$$(\boldsymbol{H})_{ij} = \left\{ \begin{array}{ll} (\boldsymbol{F_3})_{0j}, & i = 0, \\ (\boldsymbol{F_2})_{0j}, & i = 1, \\ (\boldsymbol{L_2})_{ij}, & 2 \leq i \leq N-2, \\ 0, & i = N-1, N, \end{array} \right\}, \quad 0 \leq j \leq N, \tag{38}$$

$$(\boldsymbol{K})_{ij} = \left\{ \begin{array}{ll} (\boldsymbol{F_1})_{0j}, & i = 0, \\ 0, & 1 \leq i \leq N, \end{array} \right\}, \quad 0 \leq j \leq N. \tag{39}$$

Eq.(36) can be solved for the eigenvalue $c$ satisfying

$$\det(\boldsymbol{G} - c\boldsymbol{H} - c^2\boldsymbol{K}) = 0. \tag{40}$$

Eq.(40) is numerically approximated by minimizing $|\det(\boldsymbol{G} - c\boldsymbol{H} - c^2\boldsymbol{K})|$, which can be performed by an IMSL routine UMINF based on a quasi–Newton method with an initial guess on $c = c_r + ic_i$. However, when $N$ is large, the computer calculation of determinant overflows easily. A remedial alternative is to minimize the reciprocal of the condition number of the matrix rather than the determinant itself. The idea is that when the reciprocal of the condition number is close to 0, the matrix is extremely ill–conditioned and therefore nearly singular. Another problem in using IMSL routine UMINF is that a close initial guess on $c$, usually not available, is required for this minimization scheme to succeed. To cure this, Eq.(36) can be transformed into a linear matrix generalized eigenvalue problem by the companion matrix method (Bridges and Morris[1]):

$$\begin{bmatrix} \boldsymbol{G} & \boldsymbol{0} \\ \boldsymbol{0} & \boldsymbol{I} \end{bmatrix} \begin{bmatrix} \phi \\ c\phi \end{bmatrix} = c \begin{bmatrix} \boldsymbol{H} & \boldsymbol{K} \\ \boldsymbol{I} & \boldsymbol{0} \end{bmatrix} \begin{bmatrix} \phi \\ c\phi \end{bmatrix}, \tag{41}$$

where $\boldsymbol{0}$ and $\boldsymbol{I}$ are the identity and null matrices with the same ranks as $\boldsymbol{G}, \boldsymbol{H}$ and $\boldsymbol{K}$. Eq.(41) can be directly solved by the IMSL routine GVCCG based on QZ algorithm. However, the matrix size in Eq.(41) is twice as large as Eq.(40), and the computation is not economic when $N$ is large. Besides, its round–off error will accumulate to significant figures when $N$ is very large. Hence, $c$ is better first calculated by Eq.(41) with a smaller $N$, then this $c$ is used as a good initial guess for Eq.(40) for further improvement with a larger $N$.

## 3. RESULTS AND DISCUSSION

Although the boundary condition (8) is mistaken in Lin[8], Lin's result is still worth comparing with for numerical validation. Using the incorrect boundary condition, the current solver recomputes a series of neutral $c$ of shear mode in Lin[8], and the comparison with Lin's result is shown in TABLE I, in which the agreement is well.

The eigenvalue $c$ under $\alpha = 1.062553, Re = 8126.813538$, in TABLE I is arbitrarily chosen to demonstrate the spectral accuracy of the current method by computing $c$ against increasing $N$. The result is shown in TABLE II. The spectral accuracy can be more clearly observed from the diagram of the relative error vs. $N$ in Figure 2 based on TABLE II. From Figure 2, the order of accuracy can reach as high as 33 when $N$ is between 40 and 50.

The results in Floryan et al.[6] are also recalculated here for validation. The comparison is shown in TABLE III and IV. The agreement in TABLE III and IV is even better than TABLE I.

Table I: Current computation of shear mode eigenvalue $c$ compared with Lin[8] with $S = 0, \beta = 1°$.

| $\alpha$ | $Re$ | $c$ | |
|---|---|---|---|
| | | Lin [8] | current result, $N = 60$ |
| 0.371858 | 4.152194(5) | 0.0706 | $0.7055577(-1) + 0.3700788(-4)i$ |
| 0.539594 | 5.164135(4) | 0.1295 | $0.1294119 - 0.1004532(-3)i$ |
| 0.656387 | 1.976407(4) | 0.1725 | $0.1724954 - 0.3032945(-3)i$ |
| 0.849765 | 7327.351990 | 0.2367 | $0.2372869 - 0.6205254(-3)i$ |
| 0.966308 | 5791.709595 | 0.2619 | $0.2629749 - 0.3972811(-3)i$ |
| 1.039895 | 6476.701111 | 0.2640 | $0.2651037 + 0.1893931(-3)i$ |
| 1.062553 | 8126.813538 | 0.2550 | $0.2558536 + 0.5700734(-3)i$ |
| 1.060653 | 1.195320(4) | 0.2370 | $0.2374371 + 0.8192571(-3)i$ |
| 1.014058 | 2.406454(4) | 0.2044 | $0.2043330 + 0.7798818(-3)i$ |
| 0.946671 | 4.799068(4) | 0.1750 | $0.1747119 + 0.5498975(-3)i$ |
| 0.852935 | 1.138520(5) | 0.1430 | $0.1426459 + 0.2646038(-3)i$ |
| 0.736120 | 3.382576(5) | 0.1100 | $0.1097519 - 0.3297150(-4)i$ |
| 0.618260 | 1.141233(6) | 0.0815 | $0.8116823(-1) - 0.1218521(-3)i$ |

Table 2: eigenvalue $c$ vs. $N$

| $N$ | $c$ |
|---|---|
| 20 | $0.2560783 + 0.7455761(-3)i$ |
| 22 | $0.2541468 + 0.1469023(-4)i$ |
| 24 | $0.2565135 - 0.1089530(-2)i$ |
| 26 | $0.2568238 + 0.1155089(-2)i$ |
| 28 | $0.2555736 + 0.8622085(-3)i$ |
| 30 | $0.2558768 + 0.4605091(-3)i$ |
| 32 | $0.2558753 + 0.6403534(-3)i$ |
| 34 | $0.2558131 + 0.5617639(-3)i$ |
| 36 | $0.2558629 + 0.5570625(-3)i$ |
| 38 | $0.2558553 + 0.5742978(-3)i$ |
| 40 | $0.2558526 + 0.5694982(-3)i$ |
| 42 | $0.2558540 + 0.5700405(-3)i$ |
| 44 | $0.2558535 + 0.5701773(-3)i$ |
| 46 | $0.2558536 + 0.5700383(-3)i$ |
| 48 | $0.2558536 + 0.5700790(-3)i$ |
| 50 | $0.2558536 + 0.5700739(-3)i$ |

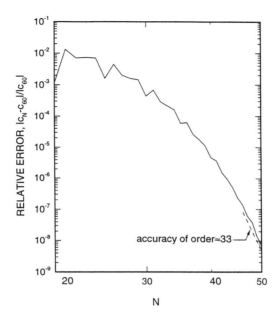

Figure 2: Fast decay of the relative error of $c$ as $N$ increases based on TABLE II.

Table 3: Current result compared with Floryan et al.[6]: surface modes with $\alpha = 0.27, \gamma = 4899.38, \beta = 4°$.

| $Re$ | $c$ | |
|---|---|---|
| | Floryan et al.[6] | current result, $N = 60$ |
| 1000 | $1.09780 + 0.260614(-1)i$ | $1.09978 + 0.260605(-1)i$ |
| 2000 | $1.06222 + 0.203353(-1)i$ | $1.06222 + 0.203355(-1)i$ |
| 5000 | $1.03441 + 0.141295(-1)i$ | $1.03441 + 0.141295(-1)i$ |
| 10000 | $1.02241 + 0.105362(-1)i$ | $1.02241 + 0.105362(-1)i$ |
| 40000 | $1.00984 + 0.569076(-2)i$ | $1.00984 + 0.569076(-2)i$ |
| 100000 | $1.00582 + 0.373391(-2)i$ | $1.00581 + 0.373666(-2)i$ |
| 1000000 | $1.00163 + 0.125600(-2)i$ | $1.00163 + 0.125521(-2)i$ |

Table 4: Current result compared with Floryan et al.[6]: neutral shear modes

| $\beta$ | $\gamma$ | $Re$ | $\alpha$ | $c$ | |
|---|---|---|---|---|---|
| | | | | Floryan et al.[6] | current result, $N = 60$ |
| 0.5' | 0 | 8369.69 | 2.893 | 0.176474 | $0.176474 - 0.224745(-6)i$ |
| 1' | 100 | 5375.60 | 2.588 | 0.205978 | $0.205977 - 0.258216(-6)i$ |
| 3' | 500 | 3707.23 | 1.898 | 0.248637 | $0.248637 - 0.104392(-7)i$ |
| 4' | 1000 | 3829.26 | 1.691 | 0.254429 | $0.254429 - 0.100337(-6)i$ |
| 1° | 10000 | 5414.59 | 1.091 | 0.263420 | $0.263420 - 0.163627(-7)i$ |
| 4° | 20000 | 5498.74 | 1.074 | 0.263643 | $0.263643 - 0.407821(-7)i$ |

Typical profiles of eigenfunction $\phi$ and its derivative $\phi'$ for shear and surface modes are shown respectively in Figure 3 and 4. The results agree well with De Bruin[4]. For shear mode, shown in Figure 3, $\phi$ and $\phi'$ resembles their counterparts of Tollmien–Schlichting wave for plane Poiseuille flow with slight modification at the free surface. For surface mode, shown in Figure 4, the magnitudes of $\phi$ and $\phi'$ are small near the wall but large as approaching the free surface, which indicates the general feature of gravity–driven free surface wave.

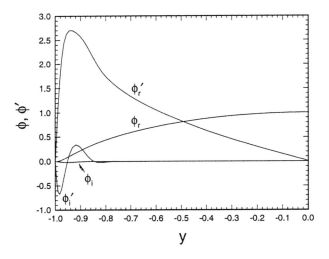

Figure 3: A shear mode eigenfunction $\phi$ and its derivative $\phi'$: $\alpha = 1.0354, c = 0.2044281 - 0.3341756(-4)i, Re = 24065, S = 0$, and $\beta = 1°$.

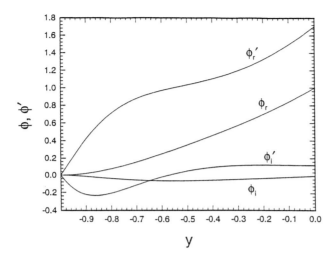

Figure 4: A surface mode eigenfunction $\phi$ and its derivative $\phi'$: $\alpha = 0.3597, c = 1.814062 + 0.3843755(-6)i, Re = 100, S = 0$, and $\beta = 1°$.

## 4. CONCLUSION

The current solver based on Chebyshev spectral collocation method together with the companion matrix method and the reciprocal of condition number minimization scheme shows high accuracy and computing efficiency in solving the whole Orr–Sommerfeld spectrum. Since the problem is physically better interpreted by the concept of convective instability which is actually observed in experiments, the future work is to solve the spatial version of the problem. It is more complicated and difficult to solve because the eigenvalue, that is the wavenumber $\alpha$, appears nonlinearly up to $\alpha^4$ in the governing equation. This also causes the rank of matrices involved in the companion matrix method twice as large as the current situation, which is definitely more cpu time consuming.

## REFERENCES

1. Bridges, T. J. and Morris, P. J. Differential Eigenvalue Problems in which the Parameter Appears Nonlinearly, *J. Comput. Phys.*, **55**, 437–460 (1984).

2. Canuto, C., Hussaini, M. Y., Quarteroni, A. and Zang, T. A. *Spectral Methods in Fluid Dynamics* Springer–Verlag, New York, New York, (1987).

3. Chang, H. C. Wave Evolution on a Falling Film, *Annu. Rev. Fluid Mech.* **26**, 103–136 (1994).

4. Chin, R. W., Abernathy, F. H. and Bertschy, J. R. Gravity and Shear Wave Stability of Free Surface Flows. Part 1. Numerical Calculations, *J. Fluid Mech.* **168**, 501–513 (1986).

5. De Bruin, G. J. Stability of a Layer of Liquid Flowing down an Inclined Plane, *J. Eng. Math.* **8**, 259–271 (1974).

6. Floryan, J. M., Davis, S. H. and Kelly, R. E. Instabilities of a Liquid Film Flowing down a Slightly Inclined Plane, *Phys. Fluids* **30**, 983–989 (1987).

7. Ho, L.-W. and Patera, A. T. A Legendre Spectral Element Method for Simulation of Unsteady Incompressible Viscous Free–Surface Flows, *Comput. Methods Appl. Mech. Eng.* **80**, 355–366 (1990).

8. Lin, S. P. Instability of a Liquid Film Flowing down an Inclined Plane, *Phys. Fluids* **10**, 308–313 (1967).

9. Yih, C. S. *Fluid Mechanics* West River Press, Ann Arbor, Michigan, (1977).

# KINETIC INSTABILITIES IN MULTILAYER DOWNFLOWING CREEPING FILMS.

I.L. Kliakhandler[†], G.I. Sivashinsky[†‡]

† *School of Mathematical Sciences, Tel-Aviv University, Tel-Aviv 69978, Israel*

‡ *The Levich Institute, The City College of New York, New York , N. Y. 10031, U.S.A.*

The stability of the creeping flow of multilayer stratified films flowing down an inclined plane is investigated. A new instability type due to surface tension is revealed. A uniform refined description of various long-wavelength instabilities of non-inertial, purely kinetic mode is presented. The notable property of the systems is established: short-wavelength disturbances are suppressed by the viscosity as well as the surface tension. A nonlinear vector evolution equation governing finite-amplitude long waves at the interfaces is derived and examined.

## 1. Introduction

In technological processes such as density currents and liquid extraction, rectilinear flows of several superposed liquid films are frequently encountered. Precision coating of a color film sometimes consists of more than 10 different layers. The impetus for the research of flow stability stems especially from this. Such analyses were carried out with respect to the roles of density and viscosity stratification.

Since the early work of Kao[1-3] it is known that a two-layer heavy-bottom flow may exhibit long-wavelength instability which persists at arbitrary small Reynolds numbers. The corresponding dispersion relation reads (see Sec. 4):

$$Re \ \omega \sim k^2, \quad (k \ll 1) \tag{1}$$

where $Re \ \omega$ is the instability rate of small disturbances ($\sim \exp(\omega t + ikx)$) and $k$ is the wave number.

Despite formal similarity, the above instability differs essentially from that of Yih[4], induced by the inertial forces. Moreover, in contrast to Yih's instability, as is shown in the present study, on slightly inclined planes the typical scale of the waves is controlled mainly by the film viscosity rather than the surface tension.

In three-layer stratified films, as has recently been found by Weinstein and Kurz[5], due to a sort of resonance between the interacting layers, the long-wavelength instability results in a different dispersion relation

$$Re\ \omega \sim k, \quad (k \ll 1). \tag{2}$$

The effect is quite similar to that of Li[6] and Kliakhandler and Sivashinsky[7], obtained in three-layer viscosity stratified Couette and Pouiseulle plane flows, respectively. In the latter work a somewhat unusual instability due to surface tension was also revealed. The same type of instability also appears in multilayer flows, leading to

$$Re\ \omega \sim k^4, \quad (k \ll 1). \tag{3}$$

The number of parameters in theoretical description of these systems is very large (ratios of viscosity, density, etc.). So simple analytical models representing essential qualitative properties of these stratified flows are especially important. The present study is devoted to the capture of the main instability types of purely kinetic, non-inertial mode in its simplest Stokes' statement. There turned out to be several important types of instability in the systems. A nonlinear vector evolution equation describing finite amplitude long waves at the interfaces is derived and examined numerically.

## 2. General formulation of the problem

To capture the essential features of kinetic instabilities consider a simple, Stokes' model (that corresponds to negligibly small inertial forces, or small Reynolds numbers) of a viscous stratified two-dimensional multi-layer creeping flow of incompressible Newtonian liquids with viscosities $\mu_1$, $\mu_2$, etc. and densities $\rho_1$, $\rho_2$, etc. for the the first, second, etc. fluids, respectively, flowing down an inclined plane in a gravitational field. The flow geometry is shown in Fig. 1.

In terms of suitably chosen dimensionless variables (defined below), the flow equations can be written as follows:

$$\frac{m_i}{r_i} \frac{\partial^2 u_i}{\partial y^2} = \frac{2}{r_i} \frac{\partial p_i}{\partial x} - 2 \tag{4}$$

$$\frac{\partial p_i}{\partial y} = -r_i \cot \theta \tag{5}$$

$$\frac{\partial u_i}{\partial x} + \frac{\partial v_i}{\partial y} = 0. \tag{6}$$

Here the index $i = 1, 2, \ldots, n$ denotes the layers $H_2 < y < H_1$, $H_3 < y < H_2$, $\ldots$, $0 < y < H_n$ correspondingly, where $H_i$ is the $i$th interface locus; $x, y$ are the streamwise and spanwise coordinates, respectively, in units of $d$, the unperturbed film thickness; $u, v$ are the corresponding components of the flow velocity, referred to as $U = gd^2 \sin\theta \rho_1 / 2\mu_1$, where $g$ is the gravitational acceleration; $t$ is the time referred to as $d/U$; $p$ is the pressure in units of $\rho_1 g d$; $m_i = \mu_i / \mu_1$ and $r_i = \rho_i / \rho_1$ are the ratio of viscosities and densities, respectively.

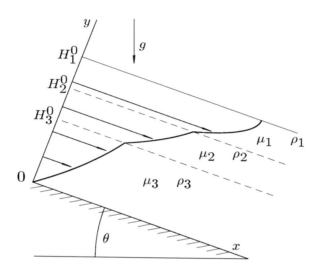

Figure 1: Stratified three-layer film flow down an inclined plane. The diagram shows the undisturbed unidirectional velocity profile. Locations of the inner interfaces are marked with dashed lines.

On the plane $y = 0$ and at the interfaces $y = H_i(x, t)$ the problem (4) - (6) is augmented by the following boundary conditions:
(i) non-slip conditions at the plane

$$u_n = v_n = 0 \quad \text{at} \quad y = 0 \tag{7}$$

(ii) continuity of the velocity at the inner interfaces

$$u_{i-1} = u_i, \quad v_{i-1} = v_i \quad \text{at} \quad y = H_i, \quad i = 2, 3, \ldots, n \tag{8}$$

(iii) continuity of the tangential stress at the inner interfaces

$$m_{i-1} \frac{\partial u_{i-1}}{\partial y} = m_i \frac{\partial u_i}{\partial y} \quad \text{at} \quad y = H_i, \quad i = 2, 3, \ldots, n \tag{9}$$

($iv$) continuity of the normal stress at the inner interfaces

$$p_{i-1} = p_i + \gamma k_i \frac{\partial^2 H_i}{\partial x^2} \quad \text{at} \quad y = H_i, \quad i = 2, ..., n \tag{10}$$

($v$) continuity (vanishing) of the tangential stress at the free surface

$$\frac{\partial u_1}{\partial y} = 0 \quad \text{at} \quad y = H_1 \tag{11}$$

($vi$) continuity of the normal stress at the free surface

$$p_0 = p_1 + \gamma k_1 \frac{\partial^2 H_1}{\partial x^2} \quad \text{at} \quad y = H_1 \tag{12}$$

($vii$) impermeability of the interfaces (kinematic conditions)

$$\frac{\partial H_i}{\partial t} + u_i \frac{\partial H_i}{\partial x} - v_i = 0 \quad \text{at} \quad y = H_i, \quad i = 1, 2, ..., n \ . \tag{13}$$

Here $p_0$ is the atmospheric pressure; $\gamma$ is some typical surface (interfacial) tension in units of $\rho_1 g d^2$ and $k_i$ is the surface (interfacial) tension on the $i$th interface in units of $\gamma$; further, by "surface" tension, we also mean the interfacial one.

Employing equation (6) and boundary conditions (7), (8), the kinematic conditions (13) may be rewritten in the convenient conservative form

$$\frac{\partial H_i}{\partial t} + \frac{\partial}{\partial x} \sum_{k=i}^{n} I_k = 0, \quad I_k = \int_{H_{k+1}}^{H_k} u_k \, dy \ . \tag{14}$$

Here $H_{n+1} = 0$ corresponds to the rigid wall. $n$ equations (4) of second order and $2n$ equations (5)-(6) of first order together with the $4n$ conditions (7)-(12) now pose the moving boundaries problem (14) for $H_i(x,t), \quad i = 1, 2, ..., n$.

# 3. Derivation of the interface evolution equations

For briefness we shall first derive the vector evolution equation and then investigate its stability.

The formulation corresponds to the quasi-one-dimensional and quasi-steady limit which may be extracted from the Stokes' creeping flow model by the following scalings:

$$x = \varepsilon^{-1} \xi, \quad y = \zeta, \quad \tau = \varepsilon^{-1} t,$$
$$u_i = U_i, \quad v_i = \varepsilon V_i, \quad p_i = P_i, \quad H_i = H_i,$$
$$\varepsilon \ll 1 \ . \tag{15}$$

Here $\varepsilon$ corresponds to the shallow water parameter. We emphasize that at this stage order of $\gamma$ remains undefined since various types of long-wavelength instabilities arise for different characteristic values of $\gamma$, as will be shown below. For all these types of instabilities and the damping mechanisms of short waves, the order of parameter $\varepsilon$, given a statement of typical longitudinal length scale of the system, will be explicitly expressed through its physical parameters.

All new variables are assumed to be of order 1. The flow equations (4)-(6) and the boundary conditions (7)-(12), (14) remain unchanged (one has just to replace $x, y, t, u, v, p$ by $\xi, \zeta, \tau, U, V, P$, respectively) except for the terms $\gamma k_i \partial^2 H_i / \partial x^2$, which will now be $\varepsilon^2 \gamma k_i \partial^2 H_i / \xi^2$.

The velocity components and the pressure in the layers are written in terms of asymptotic expansions

$$(U_i, V_i, P_i) = (U_i^{(0)}, V_i^{(0)}, P_i^{(0)}) + \varepsilon (U_i^{(1)}, V_i^{(1)}, P_i^{(1)}) + O(\varepsilon^2) + \dots \ , \quad i = 1, 2, \dots, n \quad (16)$$

Here the upper index denotes the order of approximation. Equations (16) are substituted into equations (4)-(6) and conditions (7)-(12) to obtain a sequence of linear differential equations subject to boundary conditions for successive $U_i^{(i)}, V_i^{(i)}, P_i^{(i)}$. The main stages of these computations and the final results will be described briefly without all the rather simple calculations, which were effectively carried out by the famous MATHEMATICA software.

For the zeroth order problem, Eq. (5) and condition (10) give

$$P_i^{(0)} = p_0 - r_i \cot \theta \zeta + \varrho \cot \theta \sum_{j=1}^{i} f_j H_j - \gamma \varepsilon^2 \sum_{j=1}^{i} k_j H_{j \xi\xi}, \quad H_{i+1} < \zeta < H_i \quad (17)$$

Here $\varrho$ is some typical difference between the densities in the layers in units of $\rho_1 g d$ and $f_i = r_i - r_{i-1}$ in units of $\varrho$ with $r_0 = 0$ being the density of atmosphere. A primary velocity profile is defined by:

$$\frac{\partial^2 U_i^{(0)}}{\partial \zeta^2} = -\frac{2}{m_i} r_i \ , \qquad U_i^{(0)} = -\frac{r_i}{m_i} \zeta^2 + A_i \zeta + B_i \ . \tag{18}$$

Use of boundary conditions (7)-(9), (11) results to explicit expressions of coefficients $A_i$ and $B_i$ through $r_i$ and $m_i$.

For the first order problem only $U_i^{(1)}$ is needed. Eq. (4) gives:

$$\frac{\partial^2 U_i^{(1)}}{\partial \zeta^2} = \frac{2}{m_i} \frac{\partial P_i^{(0)}}{\partial \xi} = \frac{2}{m_i} \left( \varrho \cot \theta \sum_{j=1}^{i} f_j H_{j\xi} - \gamma \varepsilon^2 \sum_{j=1}^{i} k_j H_{j\xi\xi\xi} \right) , \tag{19}$$

$$U_i^{(1)} = \left( \varrho \cot \theta \sum_{j=1}^{i} \frac{f_j}{m_j} H_{j\xi} - \gamma \varepsilon^2 \sum_{j=1}^{i} \frac{k_j}{m_i} H_{j\xi\xi\xi} \right) \zeta^2 + C_i \zeta + D_i \ . \tag{20}$$

Conditions (11), (9) and (8) give the coefficients $C_i$ and $D_i$. They are linear functions with regard to $H_{j\xi}$ and $H_{j\xi\xi\xi}$ whose coefficients are simple rational functions depending on $m_i$, $r_i$, $f_i$ and $H_i$. The procedure is straightforward: condition (11) gives $C_1$; conditions (9) give $C_2, C_3, ..., C_n$; condition (7) gives $D_n = 0$, conditions (8) give $D_{n-1}, D_{i-2}, ..., D_1$.

As a result, the velocity components $U_i^{(1)}$ are represented as linear functions with regard to $H_{i\xi}$ and $H_{j\xi\xi\xi}$ whose coefficients are square functions of $\zeta$. Substituting $U^{(0)}$ and $U_i^{(1)}$ into the kinematic conditions (14) and integrating over the thicknesses of the layers lead these conditions to the following divergent form:

$$\frac{\partial H_i}{\partial \tau} + \frac{\partial}{\partial \xi} Q_i + \varepsilon \frac{\partial}{\partial \xi} \left[ \varrho \cot \theta G_{ij} \frac{\partial H_j}{\partial \xi} + \gamma \varepsilon^2 F_{ij} \frac{\partial^3 H_j}{\partial \xi^3} \right] = 0. \tag{21}$$

In original, nonstretched physical variables Eq. (21) acquires the natural form

$$\frac{\partial H_i}{\partial t} + \frac{\partial}{\partial x} Q_i + \frac{\partial}{\partial x} \left[ \varrho \cot \theta G_{ij} \frac{\partial H_j}{\partial x} + \gamma F_{ij} \frac{\partial^3 H_j}{\partial x^3} \right] = 0. \tag{22}$$

Here we are using the sum convention for repeated indices.

The coupled vector equation (22) yields the sought for system describing the nonlinear evolution of the interacting interfaces on an inclined plane. Functions $Q_i$ and elements of the matrices $G_{ij}$ and $F_{ij}$ are the homogeneous polynomials of order three with regard to all $H_i$ (analogous with the Benney equation[8]). All these functions also depend on $r_i$, $k_i$, $f_i$, $m_i$, $k = 1, 2, ..., n$. If the density stratification is the heavy-bottom one ($r_1 = 1 < r_2 < ... < r_n$, or $f_i > 0$ for all $i$) a numerical inspection shows that the eigenvalues of matrix $G_{ij}$ have a negative real part and, hence, short waves are suppressed due to buoyancy. In the opposite case small short-wavelength disturbances of the plane interfaces grow with time in the absence of surface tension. In addition, an unusual phenomenon of full linear stability of heavy-top flow will be presented, but almost throughout we assume that the density of the system grows with its depth.

Eigenvalues of matrix $F_{ij}$ have a positive real part for all values of $m_i$, $r_i$, $H_i^0$ and $k_i$, as follows from a numerical check. This means that surface tension suppresses short-wavelength disturbances for all plane flows. So in these systems there exist two damping mechanisms of suppression of short-wavelength disturbances: heavy-bottom stratification and surface tension.

To make the system more tractable one may consider its weakly nonlinear approximation assuming perturbations of the planar interfaces are small

$$h_i(x, t) = H_i - H_i^0, \quad i = 1, 2, ..., n . \tag{23}$$

Here $H_i^0$ correspond to the undisturbed interfaces, Fig. 1. Approximations $Q_i$, $G_{ij}$ and $F_{ij}$ by their Taylor expansions near $(H_n^0, H_{n-1}^0, ..., 1)$ lead to

$$\frac{\partial h_i}{\partial t} + \alpha_{ij} \frac{\partial h_j}{\partial x} + \beta_{ijk} \frac{\partial h_j h_k}{\partial x} + \varrho \cot \theta \chi_{ij} \frac{\partial^2 h_j}{\partial x^2} + \gamma \sigma_{ij} \frac{\partial^4 h_j}{\partial x^4} = 0 \tag{24}$$

where

$$\alpha_{ij} = \frac{\partial Q_i}{\partial H_j}(H_n^0, H_{n-1}^0, ..., 1), \qquad \beta_{ijk} = \frac{1}{2}\frac{\partial^2 Q_i}{\partial H_j \partial H_k}(H_n^0, H_{n-1}^0, ..., 1),$$

$$\chi_{ij} = G_{ij}(H_n^0, H_{n-1}^0, ..., 1), \qquad \sigma_{ij} = F_{ij}(H_n^0, H_{n-1}^0, ..., 1). \tag{25}$$

The constant tensors $\alpha_{ij}, \beta_{ijk}, \chi_{ij}$ and $\sigma_{ij}$ are determined by the basic shear flow, i.e. by the initial distribution of viscosities $\mu_i$, densities $r_i$ (and hence relative differences between the densities $f_i$), relative thicknesses of the layers $H_i^0$ and relative surface tensions $k_i$ at the interfaces. We emphasize that the matrix $\alpha_{ij}$ cannot be eliminated from Eq. (24) by translation in the moving coordinate system as in the equation describing a single downflowing film[8]. The matrix is an essential part of the model and in this sense the problem differs from many classic ones.

## 4. Analysis of the vector equation

Note that matrix $\alpha_{ij}$ corresponds to the mean drift flow in the layers on the incline; matrices $\chi_{ij}$ and $\sigma_{ij}$ correspond to the interactions of the layers due to buoyancy and surface tension, respectively, whereas orders of the effects are $\varrho$ and $\gamma$, respectively. Note that $\varrho$ may at most be of order 1 while $\gamma$ may be both large, and small. For various types of instabilities, a typical longitudinal physical length scale will be expressed through $\gamma$ and/or $\varrho$.

To verify the possibility of long-wavelength instability, it is clearly sufficient to examine the linearized version of the system (24)

$$\frac{\partial h_i}{\partial t} + \alpha_{ij}\frac{\partial h_j}{\partial x} + \varrho \cot \theta \chi_{ij}\frac{\partial^2 h_j}{\partial x^2} + \gamma \sigma_{ij}\frac{\partial^4 h_j}{\partial x} = 0 \tag{26}$$

letting

$$h_i = a_i \exp(\omega t + ikx). \tag{27}$$

Hence, Eq. (26) yields

$$\omega a_i = P_{ij}a_j, \qquad P_{ij} = -ik\alpha_{ij} + \varrho \cot \theta k^2 \chi_{ij} - \gamma k^4 \sigma_{ij}. \tag{28}$$

In this way the problem of onset of long-wavelength instability is reduced to exploring of the eigenvalues of operator $\mathcal{P}$. At first it is convenient to study the simplest truncated form $P$ of the operator $\mathcal{P}$: $P = -ik\alpha_{ij}$. Since the matrix $\alpha_{ij}$, in general, is non-Hermitian, one may expect the emergence of the complex eigenvalue(s) (depending on the order of the matrix). Numerical check shows that in the two-layer flow however all the eigenvalues of matrix $\alpha_{ij}$ are purely real; but in the three-layer flow in a certain $r_i, m_i, H_i^0$ parameter range the complex eigenvalue of the matrix $\alpha_{ij}$ has a negative imaginary part. This means exciting of long-wavelength instability of the pertinent three-layer flow. So growth rate $Re\ \omega$ of the waves depends on wavenumber $k$ as

$$Re\ \omega \sim k\ . \tag{29}$$

This fact (in some other terms) was established by Weinstein and Kurz[5] in their study of linear stability of a three-layer flow down an inclined plane. For a three-layer Poiseuille flow an analogous result was established recently by Kliakhandler and Sivashinsky[7]; a corresponding vector evolution equation was derived and examined numerically. The instability formally resembles the well-known alpha-effect in three-dimensional hydrodynamics and magneto-hydrodynamics[7]. So the emergence of the complex eigenvalue(s) with a negative imaginary part of matrix $\alpha_{ij}$ is the first source of onset of the long-wavelength instability of the pertinent flow.

Assume now that the eigenvalues $\lambda_i$, $i = 1, 2, ..., n$ of the matrix $\alpha_{ij}$ are real. For small $k$ it may be easily shown that

$$\omega_i \simeq -ik\lambda_i + \varrho \cot \theta k^2 \Lambda_i + \gamma k^4 \Omega_i, \qquad i = 1, 2, ..., n, \tag{30}$$

$$\Lambda_i = -\frac{\alpha_{12}\sigma_{21} - \alpha_{22}\sigma_{11} + \alpha_{21}\sigma_{12} - \alpha_{11}\sigma_{22} + \lambda_j(\sigma_{11} + \sigma_{22})}{\alpha_{11} + \alpha_{22} - 2\lambda_j} \tag{31}$$

$$\Omega_i = \frac{\alpha_{12}\chi_{21} - \alpha_{22}\chi_{11} + \alpha_{21}\chi_{12} - \alpha_{11}\chi_{22} + \lambda_j(\chi_{11} + \chi_{22})}{\alpha_{11} + \alpha_{22} - 2\lambda_j}. \tag{32}$$

where $\Lambda_i$ and $\Omega_i$ are the corrections of eigenvalues $\lambda_i$ due to matrices $\chi_{ij}$ and $\sigma_{ij}$, respectively. The signs of $\Lambda_i$ and $\Omega_i$ define the stability of long-wavelength disturbances. If in a certain $r_i, m_i, H_i^0, k_i$ parameter domain $\Lambda_i$ or $\Omega_i$ becomes positive for some $i$ it implies long-wavelength instability of the pertinent flow due to heavy-bottom stratification (IHBS) or surface tension (IST), respectively. Numerical check shows that the possibilities are actually realized already for two-layer flows. So for instability due to heavy-bottom stratification

$$Re\ \omega \sim \varrho k^2 \tag{33}$$

and for the instability due to surface tension

$$Re\ \omega \sim \gamma k^4\ . \tag{34}$$

For all its counterintuitive nature, such a result is not entirely unexpected. A similar effect was reported by Majda and Pego[9] for second-order parabolic systems with a viscosity-type dissipation. Also, instability due to surface tension was recently revealed by Kliakhandler and Sivashinsky[7] in a three-layer

Poiseuille flow, and instability of a two-layer film flow, in spite of heavy-bottom stratification, was pointed out by Kao[1-3], though the suppression of short-wavelength disturbances was not discussed by him. We reaffirm all the linear results of Weinstein and Kurz, and Kao, but in a far more lucid manner.

Note that the effects emerge only for multilayer flows due to kinetic interaction of the layers. For single film flow, tensors $\alpha_{ij}, \beta_{ijk}, \chi_{ij}$ and $\sigma_{ij}$ are reduced to numbers and the effects disappear.

Consider now the heavy-top stratification. In this case there exists a range where $\Lambda_i$ is negative for all $i$. So for small $k$ the flow is stable, and for larger values of $k$ significant surface tension may suppress all the modes. Thus, in spite of the heavy-top stratification, the flow is fully linearly stable. This unusual effect is one of the manifestations of inherent peculiar dynamics of such systems.

Note some features of the IHBS, which follow from numerical inspection. In particular, in two-layer flows, the effect takes place for a flow with pure density stratification, at the same viscosities of the liquids in layers. The instability domain fills a significant part of the $(m_2, r_2)$ parameter plane. It is surprising that the effect appears even for a sharp stratification ($r_2 = 10$) and the instability branch occupies a sufficient part of the diagram.

Interesting peculiarities of the IST are also worth noticing. In particular, there exists a range of parameters where the liquids in the layers are miscible (i.e., interfacial tensions vanish) and, nevertheless, the flow is unstable due to surface tension on the free surface. Also, for many flows there exists a set of surface tensions such that the flow is unstable. All these qualities of the instabilities warrant the significance of the kinetic effects.

Now establish typical length scales for different kinds of instabilities. These will be characterized by wavenumber $k_c \sim \varepsilon$ (assumed to be small) corresponding to the maximal growth rate $\omega_c$, which gives a typical temporal scale of development of the disturbances. As can be seen, the assessments are as follows:

- For the alpha-effect controlled by the surface tension which is assumed to be large, $k_c \sim 1/\sqrt[3]{\gamma}$, $\omega_c \sim 1/\sqrt[3]{\gamma}$.

- For the alpha-effect controlled by the stable buoyancy stratification, angle $\theta$ assumed to be small, $k_c \sim \theta$, $\omega_c \sim \theta$.

- For the IST, with short-wavelength disturbances being suppressed by the surface tension which is assumed to be large, $k_c \sim 1/\sqrt[3]{\gamma}$, $\omega_c \sim 1/\sqrt[3]{\gamma}$.

- For the IST, where short-wavelength disturbances are suppressed by the stable buoyancy stratification, all the estimates reduce to the previous case.

- For the IHBS, where short-wavelength disturbances are suppressed by the buoyancy, angle $\theta$ is assumed to be small, and the typical relative difference between densities $\varrho$ is of order $O(1)$, $k_c \sim \theta/\varrho$, $\omega_c \sim \theta^2/\varrho^2$.

- For the IHBS, with short-wavelength disturbances being suppressed by the strong surface tension, $k_c \sim \sqrt{\varrho/\gamma}, \quad \omega_c \sim \varrho/\gamma$.

Numerical simulations of the equation show the abundant variety of the dynamics raised by Eq. (24).

One should keep in mind that the above results are based on the creeping flow model (4) - (6) where the inertial effects are omitted. The latter are known to induce $O(Rk^2)$ ($R$ is the Reynolds number) terms in the dispersion relation[4] which, clearly, may quantitatively change the overall picture of stability but does not alter its qualitative mode. So in the multilayer plane flows heavy-bottom stratification and surface tension while providing dissipation of short-wavelength disturbances may become a destabilizing influence in the long-wavelength region. It seems that similar kinds of instabilities should also arise in plane flows with continuous stratification of density and viscosity.

# References

[1] Kao T.W. Stability of two-layer viscous stratified flow down an inclined plane, *Phys. Fluids* **8,** 812-820 (1965) .

[2] Kao T.W. Role of the interface in the stability of stratified flow down an inclined plane, *Phys. Fluids* **8,** 2190-2194 (1965) .

[3] Kao T.W. Role of viscosity stratification in the stability of two-layer flow down an incline, *J. Fluid Mech.* **33,** 561-572 (1968) .

[4] Yih C. Stability of liquid flow down an inclined plane, *Phys. Fluids* **6,** 321-330 (1963) .

[5] Weinstein, S.J. and Kurz, M.R. Long-wavelength instabilities in three-layer flow down an incline, *Phys. Fluids* **A 3,** 2680-2687 (1991) .

[6] Li C.H. Instability of 3-layer viscous stratified fluid, *Phys. Fluids* **12,** 2473-2481 (1969) .

[7] Kliakhandler, I. and Sivashinsky, G. Kinetic alpha effect in viscosity stratified creeping flows, *Phys. Fluids* **8,** 1866-1871 (1995) .

[8] Benney D.J. Long waves on liquid films, *J. Math. Phys.* **45,** 150-155 (1966) .

[9] Majda, A. and Pego, R.L. Stable viscosity matrices for systems of conservation laws, *J. Diff. Eqs.* **56,** 229-262 (1985) .

# AN INVISCID MODE OF INSTABILITY IN TWO-LAYER FLOWS

Yuan C. Severtson and Cyrus K. Aidun
Institute of Paper Science and Technology, and
The George W. Woodruff School of Mechanical Engineering
Georgia Institute of Technology
Atlanta, Georgia

## ABSTRACT

The general stratified two-layer Poiseuille/Couette flows in inclined channels are analyzed to fully explored their stability characteristics. Applying spectral decomposition, the generalized eigenvalue problem of the general two-layer Orr-Sommerfeld equation is solved to obtain all of the critical modes. In addition to the interfacial and shear modes, this analysis reveals a third mode of instability not reported previously for multi-layer Poiseuille/Couette flows. This mode is analogous to the 'inviscid' mode in single layer parallel flows. A necessary condition for the instability of this mode in general two-layer parallel flows in inclined channels is derived and presented. In contrast to single layer flows, this mode can destabilize at small Reynolds number and in the long-wave range.

## 1 INTRODUCTION

Many physical problems, such as flooding in channels, multilayer coating, or air entrainment in coating system, can be modeled as stratified fluid layers in inclined channels. The critical modes in the stratified flows are crucial to the stability of these systems. In this paper, we first review the unstable modes reported by other studies, and then present a new unstable mode in two-layer inclined channel flows.

It is well known that if a viscosity difference exists between two fluids in plane Poiseuille or plane Couette flows, the flow can become unstable even at small Reynolds numbers. This unstable mode, namely interfacial mode[1-7] or sometime surface mode[8,9], is caused by the viscosity stratification at the interface where energy is transfered from the basic flow to the disturbance via the disturbance tangential stress. In the long wave limit, the growth rate is proportional to the square of the wave number of the disturbance[1]. However, the unstable interfacial mode is stabilized in the short wave due to the surface tension. The mechanism of long wave instability is discussed in detail by Smith[6].

Another familiar unstable mode for multi-layer parallel flows is the shear mode[3-5,7] where viscous effects at the boundary walls dominate the instability and energy is transfered from the basic flow to the disturbance through the Reynolds stresses[5].

Although a lot of research has been done on two-layer parallel flows, only two modes of instability have been reported which are the shear and interfacial modes. Are there other modes that may become unstable in such systems? In the next section, we develop a general model for the two-layer parallel flows in an inclined channel to investigate all the possible modes of instability and show that a third mode can become unstable. We will then outline the structure of this mode and the criteria for its existence.

## 2 GOVERNING EQUATIONS

A schematic of the problem is presented in Fig. 1. Two immiscible liquids are bounded by two solid walls. The coordinate is defined as the $x$ axis in the direction normal to the walls and $z$ axis parallel to the walls. Gravity is introduced with an inclination angle $\theta$ to the $z$ direction. One of the walls is moving at a velocity of $rW_{ll}$, while the other at $qW_{ll}$. We use indices 1 and 2 to denote layers 1 and 2, respectively.

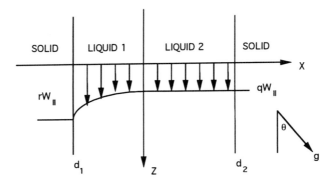

Figure 1. Flow configuration.

The equations are nondimensionalized with respect to the thickness of layer 1, $d_1$, for length, $W_{ll}$ for velocity, $\frac{\mu_1 W_{ll}}{d_1}$ for pressure, and $\frac{\rho_1 d_1^2}{\mu_1}$ for time. This introduces the dimensionless parameters

$$m = \frac{\mu_1}{\mu_2} \quad \text{viscosity ratio} \tag{1}$$

$$e = \frac{\rho_1}{\rho_2} \quad \text{density ratio} \tag{2}$$

$$n = \frac{d_2}{d_1} \quad \text{thickness ratio} \tag{3}$$

$$Re = \frac{\rho_1 d_1 W_{ll}}{\mu_1} \quad \text{Reynolds number} \quad . \tag{4}$$

The base flow is assumed to be steady and fully developed. Thus, it can be expressed as

$$W_1 = A_1 x^2 + a_1 x + b_1 \tag{5}$$
$$W_2 = m A_2 x^2 + a_2 x + b_2 \quad , \tag{6}$$

with

$$A_1 = \frac{\left(dp/dz - \rho_1 g cos(\theta)\right) d_1^2}{2\mu_1 W_{ll}} \tag{7}$$

$$A_2 = \frac{\left(dp/dz - \rho_2 g cos(\theta)\right) d_1^2}{2\mu_1 W_{ll}} \tag{8}$$

$$a_1 = \frac{A_1 - A_2 m n^2 - r + q}{1 + mn} \qquad a_2 = m \, a_1 \tag{9}$$

$$b_1 = \frac{-(A_1 + A_2 n)mn + mnr + q}{1 + mn} \qquad b_2 = b_1 \quad , \tag{10}$$

where $dp/dz$ is the pressure gradient in the $z$ direction, and $g$ is the gravitational acceleration.

Applying the standard linear stability analysis, we impose an infinitesimal velocity field ($\mathbf{u}_i = (u_i, \omega_i)$), and pressure ($p_i$) disturbances on layer $i$ ($i = 1, 2$) as well as a position disturbance ($\eta$) on the interface to perturb the system. The linearized disturbance equations are obtained by neglecting the quadratic and higher order terms,

$$\frac{\partial u_i}{\partial x} + \frac{\partial \omega_i}{\partial z} = 0 \tag{11}$$

$$\frac{\partial u_i}{\partial t} + Re \, W_i \frac{\partial u_i}{\partial z} = -\frac{\partial p_i}{\partial x} \xi_i + \left( \frac{\partial^2 u_i}{\partial x^2} + \frac{\partial^2 u_i}{\partial z^2} \right) \xi_i \varsigma_i \tag{12}$$

$$\frac{\partial \omega_i}{\partial t} + Re \left( W_i \frac{\partial \omega_i}{\partial z} + u_i \frac{dW_i}{dx} \right) = -\frac{\partial p_i}{\partial z} \xi_i + \left( \frac{\partial^2 \omega_i}{\partial x^2} + \frac{\partial^2 \omega_i}{\partial z^2} \right) \xi_i \varsigma_i \quad , \tag{13}$$

where

$$\xi_i = \begin{cases} 1 & i = 1 \\ e & i = 2 \end{cases} , \tag{14}$$

and

$$\varsigma_i = \begin{cases} 1 & i = 1 \\ \frac{1}{m} & i = 2 \end{cases}, \tag{15}$$

subject to the no-slip conditions at the walls, given by

$$u_1(-1) = 0 \qquad w_1(-1) = 0 \tag{16}$$
$$u_2(n) = 0 \qquad w_2(n) = 0 . \tag{17}$$

The boundary conditions at the interface are continuity of velocity in the $x-$ and $z-$ directions, represented respectively by

$$u_1(0) = u_2(0) , \tag{18}$$
$$w_1(0) - w_2(0) = \left[ W_2'(0) - W_1'(0) \right] \eta ; \tag{19}$$

and the continuity in tangential and normal stresses, given by

$$m \left[ \frac{\partial u_1}{\partial z}(0) + \frac{\partial w_1}{\partial x}(0) \right] = \frac{\partial u_2}{\partial z}(0) + \frac{\partial w_2}{\partial x}(0) + \left[ W_2''(0) - m W_1''(0) \right] \eta , \tag{20}$$
$$p_1(0) + \eta F sin(\theta) = p_2(0) + 2 \left[ \frac{\partial u_1}{\partial x}(0) - \frac{1}{m} \frac{\partial u_2}{\partial x} \right] + (Ca)^{-1} \frac{\partial^2 \eta}{\partial z^2} , \tag{21}$$

respectively. The two interfacial parameters, namely the Capillary number, $Ca$, and the Froude number, F, are defined by

$$Ca = \frac{\mu_1 W_u}{\sigma} \qquad F = \frac{\rho_1 g d_1^2 (1 - 1/e)}{\mu_1 W_u} , \tag{22}$$

where $\sigma$ is the surface tension. The kinematic condition is presented by

$$\frac{\partial \eta}{\partial t} = \left[ u_1(0) - W_1(0) \frac{\partial \eta}{\partial z} \right] Re . \tag{23}$$

The equation of continuity (Eq. 11) allows the use of the stream function $\psi_i$ for each layer. The solution to the disturbed system (Eqs. 11-23) is assumed to be periodic in the $z$ direction with wavenumber $k$, and all the disturbances are expressed in normal modes in the form of

$$\begin{pmatrix} \psi_i(x,z,t) \\ p_i(x,z,t) \\ \eta(z,t) \end{pmatrix} = \begin{pmatrix} \hat{\psi}_i(x) \\ \hat{p}_i(x) \\ \hat{\eta} \end{pmatrix} e^{st+ikz} \quad , \tag{24}$$

where $s$ is the growth rate with its positive (negative) real part indicating instability (stability).

## 3 STABILITY ANALYSIS-METHOD

In order to obtain all of the possible modes that could become unstable in the system, we use computational methods along with parametric continuation of the solution to the Orr-Sommerfeld equation for the general two-layer Poiseuille-Couette flow in inclined channels. The numerical scheme applied to this system is the spectral Chebyshev-Tau method[10] which automatically satisfies all boundary conditions. The original coordinate is transfered in order to satisfy the special requirement for the Chebyshev polynomials. The stream function for each layer, $\psi_i(x)$, is expanded in a truncated Chebyshev polynomial series

$$\psi_i(x) = \sum_{j=0}^{N_i} a_{i,j} T_j(x_t) \quad , \tag{25}$$

where $T_j(x)$ are Chebyshev polynomials[10] and $N_i$ are the truncation numbers which are selected based on the consideration of sufficient accuracy of the solution as well as reasonable computational time. Since there are $(N_1 + N_2 + 3)$ unknowns and considering the orthogonal property of the Chebyshev polynomials, a total of $(N_i - 3)$ equations are generated from each layer with $i = 1, 2$, plus 9 equations from the boundary conditions. Therefore, a generalized eigenvalue problem can be defined as

$$\begin{pmatrix} h_{11} & h_{12} & h_{13} \\ h_{21} & h_{22} & h_{23} \\ h_{31} & h_{32} & h_{33} \end{pmatrix} \begin{pmatrix} a_{1,j} \\ a_{2,j} \\ \eta \end{pmatrix} = s \begin{pmatrix} b_{11} & b_{12} & b_{13} \\ b_{21} & b_{22} & b_{23} \\ b_{31} & b_{32} & b_{33} \end{pmatrix} \begin{pmatrix} a_{1,j} \\ a_{2,j} \\ \eta \end{pmatrix} \quad , \tag{26}$$

where the first two rows correspond to layers 1 and 2; and the last raw is an expansion of the interfacial conditions.

In this analysis, we solve the full generalized eigenvalue problem in order to find all the potential modes for instability. If the system are presented as an eigenvalue problem, a shooting method would have been sufficient, which not only requires a good estimate of the solution as the initial guess but also is limited to only one mode. On the other hand, the generalized eigenvalue problem uniquely display the growth rate, $s$, explicitly as an eigenvalue. In this way, all the modes can be traced in order to examine their stability characteristics using pesudoarclength continuation. A detailed description of the transformation of the system into Orr-Sommerfeld equations and then into the form of Chebyshev polynomials are given elsewhere[11] and will not be repeat here.

The neutral stability curve can be obtained through the use of a pseudoarclength continuation method by forcing the real part of the growth rate in the generalized eigenvalue problem to be zero. The strategy is to first rewrite the system in terms of first-order differential equations and then treat them as a dynamical system. The continuation is accomplished by the software AUTO[12].

## 4 STABILITY ANALYSIS-RESULTS AND DISCUSSION

We first obtain the unstable modes of this flow in a horizontal channel. By using the inclination angle as a continuation parameter, we extend the modes obtained for the horizontal layer to the full range of inclination angles from 0 to $\frac{\pi}{2}$ (see Fig. 2). The neutral stability diagram in the long wave region near $\theta = \frac{\pi}{4}$ is amplified in Fig. 3.

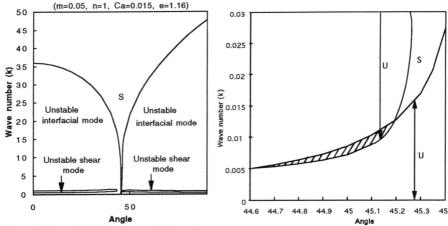

Figure 2. Neutral stability curve with changing inclination angle.

Figure 3. Stability boundaries around 45° inclination angle.

Because of the density stratification, gravity introduces a set of base velocity profiles otherwise not available when $g = 0$ or $cos\theta = 0$ (i.e., $A_1 \neq A_2$ in Eqs. 7 and 8). We explore the stability of the base flow in inclined channels by continuation of the modes previously found at $\theta = 90°$. The stability boundaries for the shear and interfacial modes are shown in Fig. 2. The viscosity, layer thickness, and density ratios for this flow are $m = 0.05$ $n = 1$ and $e = 1.16$. The possible characteristic velocity profile for the base flow, corresponding to the inclination angles ranging from 0 to $\frac{\pi}{2}$, are presented in Fig. 4.

$\theta = 0° \sim 44.6°$       $\theta = 44.6° \sim 45.2°$       $\theta = 45.2° \sim 90°$

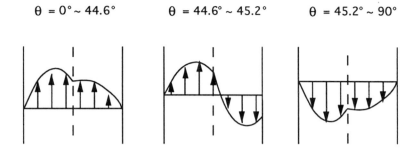

Figure 4. Corresponding base velocity profiles.

The unstable shear mode is limited to the wave number range $0.47 < k < 1.37$ where the interfacial mode $0 \leq k < 47.24$. However, near $\theta = \frac{\pi}{4}$, the shear mode stabilizes and the interfacial mode remains unstable only in the long-wave region, as shown in Fig. 3. A third mode destabilizes in the range of $44.6° < \theta < 45.4°$ and overlaps with the interfacial mode in the shaded area of this figure. This mode behaves similar to the interfacial mode, as shown in Fig. 5, where the growth rate of the new mode, designated as "A", is plotted at $\theta = 45°$. Closer examination of the growth rate near $k = 0$ (see Fig. 6) shows

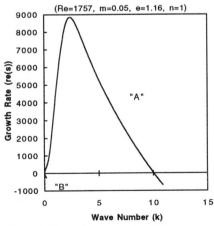

Figure 5. Unstable modes with 45 degree inclined walls.

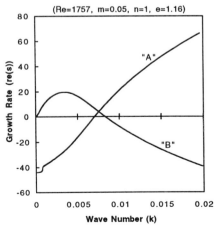

Figure 6. Unstable modes with 45 degree inclined walls at long-wave region .

that the real interfacial mode, "B", which is masked by the scale of the plot in Fig. 5 stabilizes in the long-wave region at a slightly larger wave number where the new mode, "A", destabilizes. The overlapped region correspond to the shaded area in Fig. 3. Mode "A" is not a shear mode; in fact, the shear mode for this flow destabilizes in the range of $.75 < k < 1.3$, as shown in Fig. 2. The structures of the critical disturbance stream functions for mode "A" and the interfacial mode are extracted from the eigenvectors and plotted in Figs. 7. Mode "A" and the interfacial mode show considerable similarity in structure.

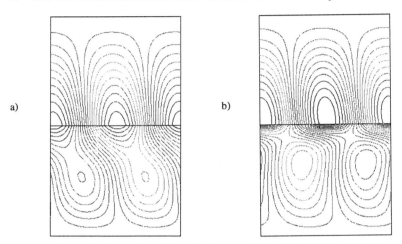

Figure 7. Stream lines of the disturbace of a)mode "A", b)interfacial mode.

Mode "A" can only destabilize for counter-current base state flows where there is a discontinuous inflection point at the interface. That is $W_1''$ and $W_2''$ have different signs. Is this mode analogous to the inviscid mode in parallel boundary layer flows? Although an inviscid mode can not exist in a single layer quadratic flow, by extending Rayleigh's inviscid theory[13,14] to the two-layer parallel flows, we first rescale the growth rate $s$ to $\bar{s}$ by

$$\bar{s} = \frac{i\,s}{k\,Re}\;. \tag{27}$$

By following the same procedure derived by Rayleigh[14], we find the necessary condition for the instability of an inviscid mode in a two-layer Couette/Poiseuille flow to be given by

$$\int_{-1}^{0} \left\{ \frac{W_1''(x)}{|W_1(x) - \bar{s}|^2}|u_1(x)|^2 + \frac{W_2''(x-1)}{|W_2(x-1) - \bar{s}|^2}|u_2(x-1)|^2 \right\} dx = 0\;. \tag{28}$$

Since for the two-layer Poiseuille/Couette flow, the second derivative of the velocity profile, $W_i''$, is constant, Eq. 28 can be satisfied only if $W_1''$ and $W_2''$ have different signs resulting in an inflection point at the interface. From this analysis we conclude that the dominant mode for the inclination angle near $\frac{\pi}{4}$, designated as mode "A", is an 'inviscid' mode which remains unstable as $Re$ becomes infinitely large, as shown by the neutral stability curve of this mode presented in Fig. 8. The critical Reynolds number for this mode depends on the layer thickness ratio and the other parameters. For the case considered here, the critical Raynolds number for instability of this mode is equal to 0.0058.

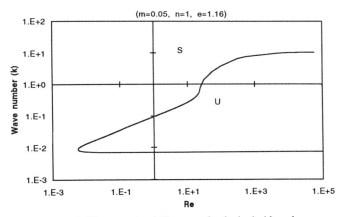

Figure 8. The neutral stability curve for the inviscid mode.

## 5 CONCLUSIONS

The Chebyshev-Tau method is used to obtain the full spectrum of critical modes possible in the two-layer Poiseuille/Couette flows in inclined channels. The interfacial mode is the dominant mode of instability although the shear mode is always present if the Couette component is not too large. The stability characteristic of the two layers at vertical and horizontal orientations are very similar. The interfacial mode is the dominant mode with the shear mode present in a narrow region of the wave length spectrum.

However, at inclination angle about 45°, the shear mode disappears and mode crossing occurs between the interfacial mode and a new mode which is shown to be analogous to the classical inviscid mode of instability in single layer parallel flows. It is interesting to note that the neutral curve for this mode shows instability at very small critical Reynolds number in the long wave region. The mode remains unstable as the Reynolds number becomes infinitely large. A necessary condition for the instability of this mode is derived and presented by

Eq. 28. Only the base states where the second derivative of the velocity in each layer has a different sign can admit this mode. This is equal to existence of an inflection point at the interface.

## ACKNOWLEDGMENTS

This study has been supported by the National Science Foundation through grant CTS-9258667, and by industrial matching contributions. The calculations were conducted, in part, using the National Center for Supercomputing Applications, a resource at Cornell University which is funded by the National Science Foundation, IBM, and New York State.

## REFERENCES

1. Yih, C.S. Instability Due to Viscosity Stratification, *J. Fluid Mech.*, **27**, 337 (1967).
2. Renardy, Y.Y. Instability at the Interface between Two Shearing Fluids in a Channel, *Phys. Fluids*, **28**, 3341 (1985).
3. Joseph, D.D. and Renardy, Y.Y. *Fundamentals of Two-Fluid Dynamics Part I: Mathematical Theory and Applications* Springer-Verlag, (1993).
4. Yiantsios, S.G. and Higgins, B.G. Linear Stability of Plane Poiseuille Flow of Two Superposed Fluids, *Phys. Fluids*, **31**, 3225, (1988).
5. Hooper, A.P. The Stability of Two Superposed Viscous Fluids in a Channel, *Phys. Fluids*, **1**, 1133 (1989).
6. Smith, M.K. The Mechanism for the Long-Wave Instability in Thin Liquid Films, *J. Fluid Mech.*, **217**, 469 (1990).
7. Tilley, B.S., Davis, S.H. and Bankoff, S.G. Linear Stability Theory of Two-Layer Fluid Flow in an Inclined Channel, *Phys. Fluids*, **6**, 3906 (1994).
8. Chin, R.W., Abbernathy, F.H. and Beertschy J.R. Gravity and Shear Wave Stability of Free Surface Flows: Part 1. Numerical Calculations, *J. Fluid Mech.*, **168**, 501 (1986).
9. Floryan, J.M., Davis, S.H. and Kelly, R.E. Stabilities of a Liquid Film Flowing down a Slightly Inclined Plane, *Phys. Fluids*, **30**(4), 983 (1987).
10. Gottlieb, D. and Orszag, S.A. *Numerical Analysis of Spectral Methods* SIAM, 1977.
11. Severtson, Y.C. and Aidun, C.K. Stability of Two-Layer Stratified Flow: Application to Air Entrainment in Coating Systems. *To appear in J. Fluid Mech., 1995.*
12. Doedel, E. and Kerneves, J.P. Auto: Software for Continuation and Bifurcation Problems in Ordinary Differential Equations, (1986).
13. White, F.M. *Viscous Fluid Flow* McGraw-Hill, (1974).
14. Drazin, P.G. and Reid, W.H. *Hydrodynamic Instability* Cambridge University Press, (1985).

# SECTION 3

# Experimental

# Investigations

| **ILFORD** | **Title** | Experimental Methods  1 |
|------------|-----------|-------------------------|

# Experimental Methods for Coating Flows

by

**Peter M. Schweizer**

ILFORD AG

Industriestrasse 15

CH-1701 Fribourg

Switzerland

| **ILFORD** | **Contents** | Experimental Methods  2 |
|------------|--------------|-------------------------|

1. **Value of Experiments**
2. **Tools and Methods for Experimental Investigations**

- Pilot coating equipment
- Statistical experimental design plans
- Quantification of film thickness uniformity
- Sensors and instruments

3. **Applications**

- Film thickness measurement
- Surface topography
- Thermal imaging
- Laser Doppler velocimetry
- Flow visualization
- Picture album of slide coating

| **ILFORD** | **Reference** | Experimental Methods 3 |
|---|---|---|

**Experimental Methods**

by

Peter M. Schweizer

in

*Liquid Film Coating* - **Scientific Principles**

**and Their Technological Implications**

by

Stephan F. Kistler and Peter M. Schweizer, Editors

Chapman & Hall, London and New York

To be published in 1996

| **ILFORD** | **Value of Experimental Methods** | Experimental Methods 4 |
|---|---|---|

**Experiment:** "An operation carried out under controlled conditions in order to discover an unknown effect or law, to test or establish a hypothesis, or to illustrate a known law" (Webster)

Experiments are the basis for the fundamental understanding of physical principles and processes:

- Acquisition of "know-how" and, above all, **"know-why"**
- The value of experiments is undisputed

Experimental methods are most effectively applied in combination with theoretical modeling and mathematical analysis.

| **ILFORD** | **Tools and Methods** | Experimental<br>Methods    5 |
|---|---|---|

**Requirements for Experimentation:**

- Application of experimentation in industry and academia requires
  - ◊ "do the right thing"          ⇒ strategic plan
  - ◊ "do the right thing right"    ⇒ appropriate tools and methods

- Successful experimentation requires
  - ◊ knowledge of experimental tools and methods
  - ◊ knowledge of the field of experimentation
  - ◊ personal qualities of the experimentator:
    - ⇒ innovation, perseverance, attention to details

| **ILFORD** | **Tools and Methods:**<br>**Pilot Coating Equipment** | Experimental<br>Methods    6 |
|---|---|---|

**Purpose of pilot coating equipment:**

Experimentation away from production equipment in order to

- reduce cost of experimentation

- avoid scheduling constraints

- avoid mechanical constraints

- avoid operating constraints

| **ILFORD** | **Tools and Methods:**<br>**Pilot Coating Equipment** | Experimental<br>Methods   7 |
|---|---|---|

**Typical applications for pilot equipment**:

- Effects of solution components on product performance
- Interactions of solution components in multi-layer coatings
- Investigations of unit operations of coating processes
- Improvement of capacity and yield of production equipment
- Determination of coating windows
- Process optimization with regard to product uniformity and manufacturability
- Scale-up from pilot to production stage
- Testing of raw materials

| **ILFORD** | **Tools and Methods:**<br>**Pilot Coating Equipment** | Experimental<br>Methods   8 |
|---|---|---|

**Requirements for Pilot Coating Equipment:**

- Appropriate installations for solution preparation
- Appropriate installations for solution delivery
- Higher web conveyance speed than production machine
- Geometrical flexibility
- Application of different coating methods
- Single-layer and multi-layer capability
- Narrower coating width than production machine
- Light and dark operation (photographic industry)
- Appropriate post-coating processes (setting, drying, curing, etc.)
- Reproduction of product properties as on production machine

| **ILFORD** | **Tools and Methods:** <br> **Pilot Coating Equipment** | Experimental <br> Methods    9 |
|---|---|---|

**Requirements for Pilot Coating Equipment (cont.):**

The simultaneous fulfillment of all requirements will lead to a "narrow but more flexible production machine" with a very high price tag

⇓

It is advantageous to build and operate **several** pilot coating machines, each of which satisfies only some of the requirements

---

| **ILFORD** | **Tools and Methods:** <br> **Pilot Coating Equipment** | Experimental <br> Methods    10 |
|---|---|---|

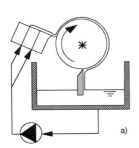 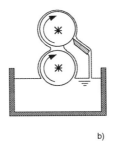 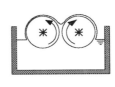

a)          b)          c)

- extrusion coating
- bead coating
- curtain coating

- roll coating
- blade coating
- air knife coating

- dip coating
- film splitting

| **ILFORD** | **Tools and Methods:** **Pilot Coating Equipment** | Experimental Methods    11 |
| --- | --- | --- |

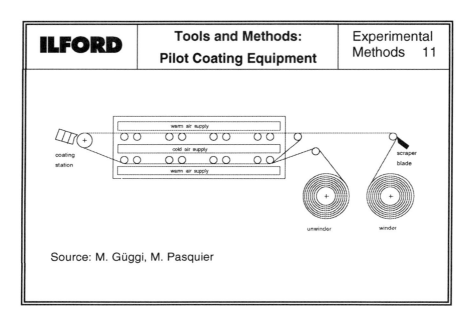

Source: M. Güggi, M. Pasquier

| **ILFORD** | **Pilot Coating Equipment:** **References** | Experimental Methods    12 |
| --- | --- | --- |

Gellrich, P., Schliephake, R., Matter, G., and Sturm, W. 1991. Versuchsbegiessanlagen für Fotofilme and Fotopapier. Coating 3:76-81.

| **ILFORD** | **Tools and Methods:** <br> **Experimental Design Plans** | Experimental <br> Methods　13 |
| --- | --- | --- |

**Purpose of experimental design plans:**

Method for determining the number and placing of experiments in order to obtain the desired information with a minimum of effort

- Results can easily be modeled

- Suitable for laboratory and computer experiments

- Strategy:　　1.　Coarse screening
  　　　　　　　2.　Fine optimization

---

| **ILFORD** | **Tools and Methods:** <br> **Experimental Design Plans** | Experimental <br> Methods　14 |
| --- | --- | --- |

**One-factor Method versus Multi-factor Method**

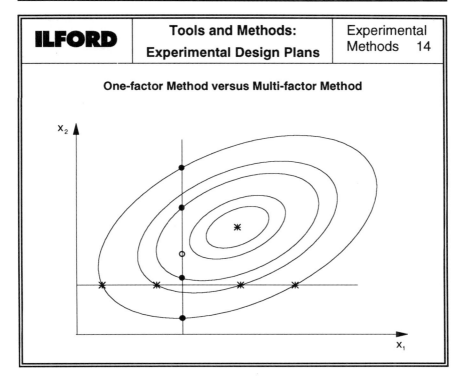

**ILFORD** | **Tools and Methods:** **Experimental Design Plans** | Experimental Methods 15

- Example: *Full Factorial Plan*
  - number of independent variables     = p     (4)
  - number of variable levels     = n     (3)
  - number of experiments     = $n^p$     ( 81)
- Reduction of number of experiments by
  - ◊ use of dimensionless numbers (Π-Theorem)
    - number of basic units     = u     (3)
    - number of dimensionless numbers     = p - u     (1)
  - ◊ use of appropriate experimental plans
- Aid: PC software

---

**ILFORD** | **Tools and Methods:** **Experimental Design Plans** | Experimental Methods 16

**Preventive versus corrective approach (schematic)**

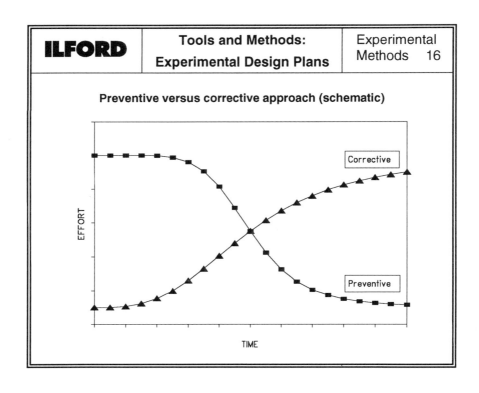

Corrective

Preventive

EFFORT

TIME

| **ILFORD** | **Tools and Methods:** | Experimental |
| | **Experimental Design Plans** | Methods 17 |

### References

- Daniel, C., and Wood, F.S. 1971. Fitting Equations to Data. New York:Wiley.
- Deming, S., and Morgan, S. 1986. Fundamentals of Experimental Design, Short Course of the American Chemical Society, Department of Educational Materials, 1155 16th Street N.W., Washington, D.C. 20036.
- Draper, N.R., and Smith H. 1966. Applied Regression Analysis. New York:Wiley.
- ECHIP Inc., 7460 Lancaster Pike, Suite 6, Hockessin, DE 19707.
- Montgomery, D.C. 1984. Design and Analysis of Experiments. New York:Wiley.
- SAS. 1987. SAS System for Elementary Statistical Analysis. Version 6. SAS Institute Inc., Box 8000, SAS Circle, Cary, NC 27512-8000.
- Statgraphics STSC Inc., 2115 East Jefferson Street, Rockville, MD 20852.
- Torgerson, W.S. 1958. Theory and Methods of Scaling. New York:Wiley.

| **ILFORD** | **Tools and Methods:** | Experimental |
| | **Quantification of Uniformity** | Methods 18 |

**Old approach:** Visual comparison of film thickness uniformity with a set of standard samples

**New approach:** Representation of film thickness uniformity in terms of amplitude, frequency, wave length, or other statistical parameters

Treatment of periodic signals in the frequency space by using Fourier analysis, filter algorithms, and other suitable image processing algorithms

**Examples:** Lowpass filter → cross profile (long wave)
Bandpass filter → diffuse bands (short wave)

| **ILFORD** | **Tools and Methods:** | Experimental |
|---|---|---|
| | **Quantification of Uniformity** | Methods 19 |

**Raw signal of crossweb film thickness profile**   (Source: P.A. Rossier)

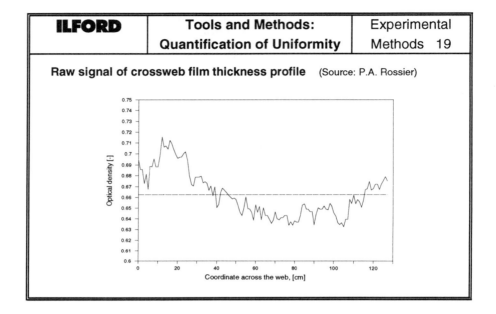

| **ILFORD** | **Tools and Methods:** | Experimental |
|---|---|---|
| | **Quantification of Uniformity** | Methods 20 |

**Lowpass filter: Cross profile**
Source: P.A. Rossier

**Bandpass filter: Diffuse bands**
Source: P.A. Rossier

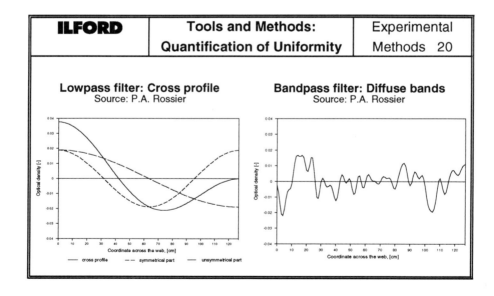

| **ILFORD** | **Tools and Methods:** | Experimental |
|---|---|---|
| | **Quantification of Uniformity** | Methods   21 |

**Example of model for quantification of periodic signal** (reduction of film thickness uniformity into single parameter)

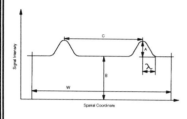

$$QI = \left(\frac{A}{B}\right)\left(\frac{A}{\lambda}\right)\left(\frac{W}{C}\right) = \frac{A^2 W}{BC\lambda}$$

QI  = quality index

A/B  = visibility

A/$\lambda$  = contrast

W/C  = number of periods


Quality increases with decreasing quality index

| **ILFORD** | **Tools and Methods:** | Experimental |
|---|---|---|
| | **Quantification of Uniformity** | Methods   22 |

**References**

- LabWindows. 1991 Version 2.1. National Instruments Corporation, 6504 Bridge Point Parkway, Austin, TX 78730-5039

- Lyne, M. Bruce, and Parush, A. 1983. A survey of offset and letterpress newsprint print quality. Journal of Pulp and Paper Science, Transactions of the Technical Section 9(5):1-7

- Williamson, Samuel, J., and Cummins, Herman, Z. 1983. Light and color in nature and art. New York:Wiley

| **ILFORD** | **Tools and Methods:** **Sensors and Instruments** | Experimental Methods 23 |
|---|---|---|

**Requirements for sensors and instruments:**

- contactless measurements
- point-, one- or two-dimensional measurements
- on-line or off-line measurements
- motion in downweb and crossweb direction (scanner)
- high spatial and temporal resolution, e.g. detection of small coating defects, transients at start-up, onset of instabilities
- high sensitivity, e.g. detection of film thickness variation < 1%

**Examples:** see applications below

| **ILFORD** | **Applications** **Film Thickness Measurement** | Experimental Methods 24 |
|---|---|---|

**Requirement:**  High resolution of film thickness, e.g. < 1% for a dry film thickness of 10 µm in the photographic industry

**Measurement location:**
- on-line or off-line
- wet or dry layer
- point measurement
- one- or two-dimensional
- zigzag (scanner)

**Method of measurement:** Many different techniques are described in the literature. The most important class of contactless methods uses a point source which emits radiation of a specified frequency. The radiation is partially absorbed by the coated layer and received by a detector.

| **ILFORD** | **Applications** | Experimental |
|---|---|---|
| | **Film Thickness Measurement** | Methods    25 |

The Beer-Lambert law
describes the absorption
of radiation across
transparent films:

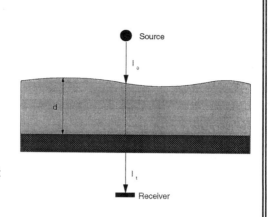

$$I_t = I_0 e^{(-Kd)}$$

$I_0$ = incident radiation
$I_t$ = emitted radiation
$K$ = absorption coefficient
$d$ = layer thickness

---

| **ILFORD** | **Applications** | Experimental |
|---|---|---|
| | **Film Thickness Measurement** | Methods    26 |

**Performance comparison of various methods**

| Method | Range<br>{μm} | Accu-<br>racy<br>{%} | Measur.<br>Interval<br>{s} | Measur.<br>Spot Size<br>{mm} | Scan<br>Speed<br>{m/min} |
|---|---|---|---|---|---|
| **Infrared** | 2 - 150 | > 0.5 | > 0.2 | ⌀ 10 | < 4.2 |
| **Microwave** | 1000 | > 1 | > 1 | ⌀ 5 - 10 | |
| **X-Ray<br>Fluorescence** | de-<br>pends | 1 - 10 | >5 - 100 | ⌀ 30 | < 7.5 |
| **Capacitance<br>Gauge** | < 500 | > 0.1 | > 0.001 | ⌀ 5 | < 3 |

| **ILFORD** | **Applications** | Experimental |
| | **Film Thickness Measurement** | Methods 27 |

### Example: Infrared Absorption

**Phenomenon:**

Selective absorbance of near
infrared energy by many
substances, e.g. PE, water

**Measurement Principle:**

- Use 2 selected wave lengths
  (measurement, reference)

- Difference of reflected energy
  is proportional to substance
  content in sample

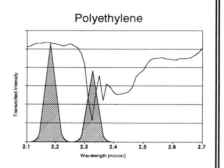

Polyethylene

| **ILFORD** | **Film Thickness Measurement** | Experimental |
| | **Infrared Absorption** | Methods 28 |

### Water Content Cross Profile inside Dryer

Parameter: Number of master roll

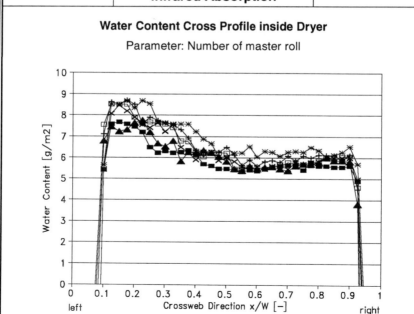

| **ILFORD** | **Applications** | Experimental |
|---|---|---|
| | **Film Thickness Measurement** | Methods 29 |

### References

- ASOMA Instruments, 1992. X-ray fluorescence - Application summary for coating weight measurement, 11675 Jollyville Road, Austin, TX 78759, USA.
- Edgar, R.F., and Stay, B.J. 1985. Techniques for suppressing optical interference errors in infrared film gauging. Infrared Technology and Applications, SPIE Vol. 590:316-320.
- Hindle, P.H. 1984. Thickness measurement - the infrared method. Paper Technology & Industry 25(5).
- Hurley, R.B., Kaufman, I., and Roy, R.P. 1990. Noncontacting microstrip monitor for liquid film thickness. Rev. Sci. Instrum. 61(9): 2462-2465.
- Meyer, W., and Schilz, W. 1980. A microwave method for density independent determination of the moisture content of solids. J. Physics D: Applied Physics 13: 1823-1830.
- Roy, R.P., Ku, J., Kaufman, I., and Shukla, J. 1986. Microwave method for measurement of liquid film thickness in gas-liquid flow. Rev. Sci. Instrum. 57(5): 952-956.

| **ILFORD** | **Applications** | Experimental |
|---|---|---|
| | **Surface Topography** | Methods 30 |

- **Profilometry:** Instruments probe the film surface in a raster-like fashion by way of mechanical or optical sensors. Film thickness must be inferred by calculating the difference between the coated sample and the uncoated reference surface.

- **Interferometry:** Contactless method that produces two-dimensional fringe pattern (equidistant lines). Need for coherent light. Range too small and resolution too high for applications in coating flows.

- **Moiré Technique:** Also called "mechanical interferometry". White light sufficient. Range from 10 μm to 1 m; resolution < 10 μm; size of object from cm to m.

144

| **ILFORD** | **Applications**<br><br>**Moiré Technique** | Experimental<br>Methods  31 |
|---|---|---|

- Superposition of two undisturbed gratings produces straight fringes
- Superposition of disturbed gratings produces distorted fringes that indicate surface topography
- Uneven liquid film surfaces may distort the grating

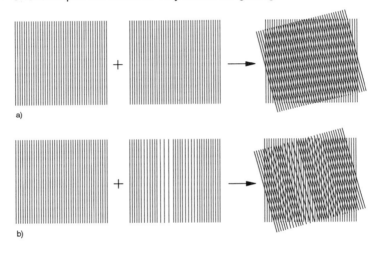

a)

b)

| **ILFORD** | **Applications**<br><br>**Moiré Technique** | Experimental<br>Methods  32 |
|---|---|---|

Various optical arrangements.   Example: **Oblique shadow method**

**Resolution** (distance between adjacent fringes):

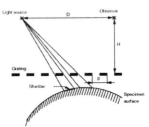

$$\Delta z = \frac{pH}{D}$$

| **ILFORD** | **Applications**<br>**Moiré Technique** | Experimental<br>Methods   33 |
| --- | --- | --- |

**Example:**

Contour moiré fringes of a
human body; taken from
Takasaki, 1979

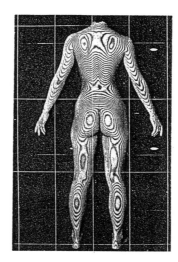

| **ILFORD** | **Applications**<br>**Moiré Technique** | Experimental<br>Methods   34 |
| --- | --- | --- |

**Coating Flow Applications**

- Leveling of a wavy horizontal liquid film

- Growth of surface waves in a film flowing down an inclined plane

- Deformations in a liquid curtain due to ambient disturbances

- Local thinning of a liquid film due to surface tension gradients

- Determination of the critical coating speed at the inception of ribbing
  instabllity

146

| **ILFORD** | **Applications**<br>**Surface Topography** | Experimental<br>Methods  35 |

**References**

- Breuckmann GmbH, Industrielle Bildverarbeitung und Automation, Kunkelgasse 1, D-7758 Meersburg, Germany.
- Breuckmann, B., and Thieme, W. 1985. Computer analysis of holographic interferograms using phase shift method. Applied Optics 24:2145-2149.
- Kheshgi, H.S. and Scriven, L.E. 1983. Measurement of liquid film profiles by moiré topography. Chemical Engineering Science 38(4):525-534.
- Kheshgi, H.S. and Scriven, L.E. 1991. Dewetting: Nucleation and growth of dry regions. Chemical Engineering Science 46(2):519-526.
- Maidhof, A., and Schiller, M. 1992. Topometrie - Hochauflösende optische 3-D Messtechnik. Messtechnik (1):12-15.
- Meadows, D.M., Johnson, W.O. and Allen, J.B. 1970. Generation of surface contours by moiré patterns. Applied Optics 9(4):942-947.
- Takasaki, H. 1970. Moiré topography. Applied Optics 9(6):1467-1472.
- Takasaki, H. 1979. The development and the present status of moiré topography. Optica Acta 26(8):1009-1019.
- Theocaris, P.S. 1967. Moiré topography of curved surfaces. Experimental Mechanics 7(7):289-296.

| **ILFORD** | **Applications**<br>**Thermal Imaging** | Experimental<br>Methods  36 |

**Characteristic features**

- One- or two dimensional scanners providing thermal images and contactless temperature measurements

- Spectral response: 2 - 5 µm or  8 - 14 µm

- Temperature range: - 20°C to + 500°C (up to 2000°C)

- Thermal sensitivity: 0.1°C at 30°C;  accuracy: ± 1°C

- Detector cooling: cryogenic or thermoelectric (Peltier effect)

- Image processing and analysis (real-time imaging, 12-bit recording, color maps, temperature profiles, etc.)

| **ILFORD** | **Applications** **Thermal Imaging** | Experimental Methods 37 |

**Applications of thermal imaging:**

- Control of isothermal operating conditions
- Cross-web temperature uniformity in setting and drying zones
- Effects of air convection on uniformity of film flow
- etc.

| **ILFORD** | **Applications** **Laser Doppler Velocimetry** | Experimental Methods 38 |

**Characteristic features:**

- Contactless electro-optical system for measuring velocities of scattering particles
- High spatial and temporal resolution
- Direction-sensitive
- No calibration
- Wide velocity range ($10^{-3}$ to $10^2$ m/s)
- Simultaneous measurement of three velocity components
- Expensive
- Not-so-easy to use

148

| **ILFORD** | **Applications** <br> **Laser Doppler Velocimetry** | Experimental <br> Methods   39 |
|---|---|---|

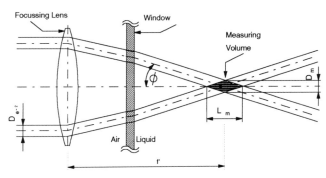

Fringe spacing:

$$\Delta X = \frac{\lambda_0}{2n\sin\phi}$$

Particle velocity:

$$V = \frac{f_D \lambda_0}{2n\sin\phi}$$

$\lambda_0$ = wave length of laser light in vacuum
$f_D$ = Doppler frequency
$\phi$ = half angle of intersecting laser beams
$n$ = refractive index of fluid

| **ILFORD** | **Applications** <br> **Laser Doppler Velocimetry** | Experimental <br> Methods   40 |
|---|---|---|

**Spatial resolution:**     $D_m > 20 \ \mu m$     $L_m > 40 \ \mu m$

**Applications:**
- Flow in liquid curtains
- Flow in meniscus of slide coating
- Flow inside coating dies
- Web speed

| **ILFORD** | Applications<br>**Laser Doppler Velocimetry** | Experimental<br>Methods    41 |
|---|---|---|

**References:**

- Durst, F., Melling, A., and Whitelaw, J.H. *Principles and practice of laser-Doppler Anemometry.* London:Academic Press, 1976
- Durst, F., Koo, J.B., Wagner, H.G., and Walter C. 1993. Experimental study of the boundary layer development at the edge guides in curtain coating, Lehrstuhl für Strömungsmechanik, University of Erlangen, Germany
- Mues, W., Hens, J., and Boiy, L. 1989. Observation of a dynamic wetting process using laser-Doppler velocimeter. AIChE Journal, 35(9):1521-1526
- POLYTEC GmbH, Polytec-Platz 5-7, D-76337 Waldbronn, Germany. Prospectus for *Laser Surface Velocimeter*

| **ILFORD** | Applications<br>**Flow Visualization** | Experimental<br>Methods    42 |
|---|---|---|

**Purpose**

Visualization of small-scale (< 1 mm) two-dimensional flows with free surfaces, interfaces (multi-layer film flows), static and dynamic wetting lines, flow separation (vortices), etc.

### *Method: Optical sectioning*

- Visualization of relevant flow cross section by way of light sheet
- Injection of tracers in form of hydrogen bubbles and/or dye streams
- Optical access to flow field through transparent flow boundary (glass window)
- Magnification with microscope
- Recording of images with still or video camera
- Digital image processing

150

| **ILFORD** | **Applications** <br> **Flow Visualization** | Experimental <br> Methods 43 |
|---|---|---|

**Problems in practice**

- Optical distortions through curved liquid interfaces

- Optical reflections

- Difference in refractive indices in multi-layer film flows

- Need for transparent fluids

- Corrosion of coating equipment


| **ILFORD** | **Applications** <br> **Flow Visualization** | Experimental <br> Methods 44 |
|---|---|---|

**Methods of tracer injection**

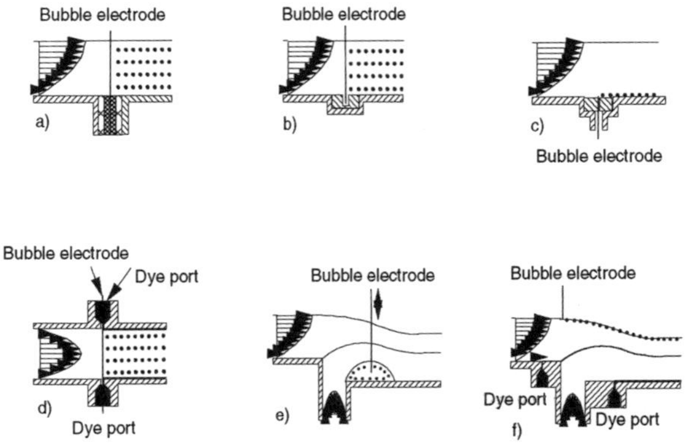

| **ILFORD** | **Applications** **Flow Visualization** | Experimental Methods  45 |
|---|---|---|

**Picture album of slide coating**

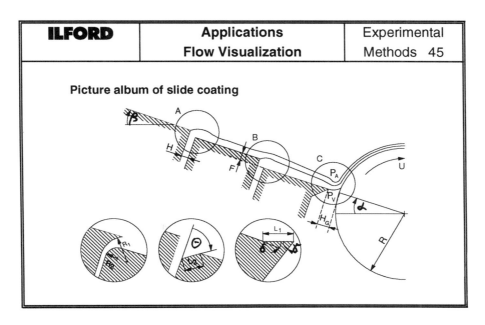

| **ILFORD** | **Flow Visualization** **Picture Album** | Experimental Methods  46 |
|---|---|---|

**Example:  Flow Fields in Slide Coating**

1.  Effect of flow rate on location of static wetting line in film flow on slide

2.  Effect of slot exit design on formation of vortices

3.  Effect of meniscus vacuum on location of static and dynamic wetting lines in bead flow field

4.  Effect of die lip design on bead flow field

5.  Effect of application angle on bead flow field

| **ILFORD** | **Applications** <br> **Flow Visualization** | Experimental <br> Methods   47 |
|---|---|---|

### References

- Bausch & Lomb, MonoZoom-7 Optical System, Bausch & Lomb Inc., Optical Systems Division, P.O. Box 450, Rochester N.Y. 14692-0450.
- Clutter, D.W., and Smith, A.M.O. 1961. Flow visualization by electrolysis of water. Aerospace Engineering 1:24-76.
- Hassan, Y.A., Blanchat, T.K., and Seeley Jr., C.H. 1992. PIV flow visualization using particle tracking techniques. Meas. Sci. Technol. 3:633-642.
- Merzkirch, W. 1974. Flow Visualization. New York: Academic Press.
- Chen, K.S.A., and Scriven, L.E. 1992. Multilayer slide coating: Flow visualization and theoretical modeling.  Paper 41i read at 6th International Coating Process Science and Technology Symposium, AIChE Spring National Meeting, 29 March - 2 April 1992, New Orleans, LA
- Blake, T.D., Clarke, A., and Ruschak, K.J. 1994. Hydrodynamic assist of dynamic wetting. AIChE Journal, 40(2):229-242.
- Schweizer, P.M. 1988. Visualization of coating flows. J. Fluid Mechanics 193:285-302.

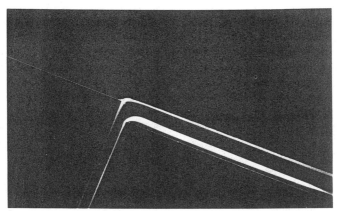

Fig 7·20a

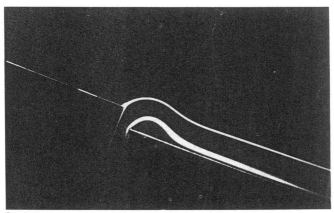

Fig 7·20b

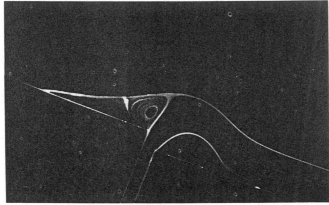

Fig 7·20c

154

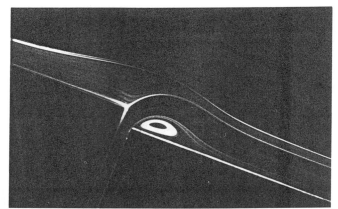

Fig. 7-21a

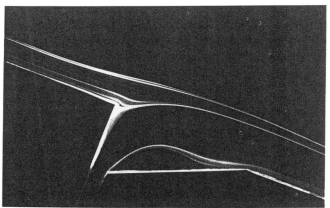

Fig. 7-21b

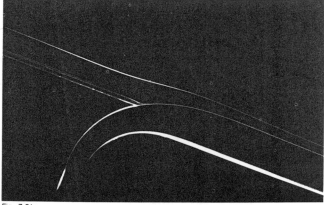

Fig. 7-21c

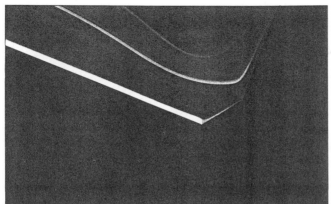

Fig. 7-22a

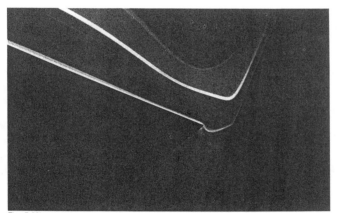

Fig. 7-22b

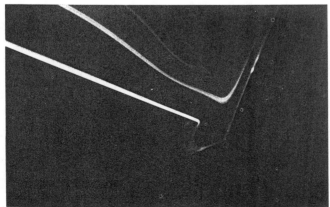

Fig. 7-22c

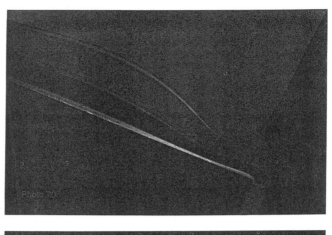

Photo 70

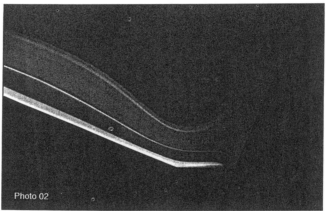

Photo 02

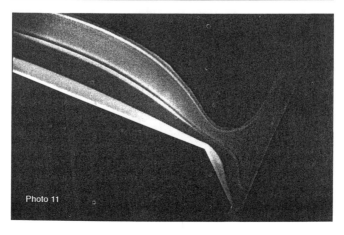

Photo 11

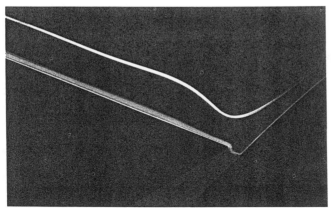

Fig. 7-24 a

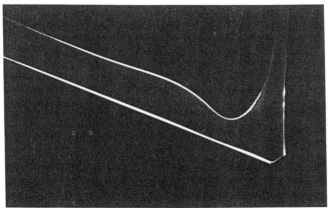

Fig. 7-24b

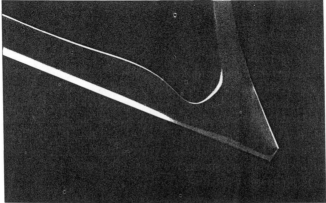

Fig. 7-24c

# MENISCUS CONTROL BY A STRING IN A ROLL COATING EXPERIMENT

J.-M. Buchlin , M. Decré, E. Gailly and P. Planquart
von Karman Institute for Fluid Dynamics
Ch. de Waterloo, 72 Belgium - 1640 Rhode-St-Genèse

## 1. SUMMARY

The ribbing instability is a hindrance to the roll coating process because it yields wavy coating films. The present experimental and numerical study deals with a technique proposed recently to prevent ribbing, whereby a string is placed parallel to the rolls, in contact with the unstable meniscus. In the present study, a free-surface visualisation technique and a film thickness measurement probe are applied. Numerical simulations are also performed with the spectral-element code NEKTON. The ribbing can be eliminated under conditions at least 20 times its natural onset. The string position influences the total and partial flow rates. The string proves to be an efficient concept to get rid of ribbing, and a reliable way of fixing the coating film thickness, without resorting to traditional gap width and roller speed selection. The numerical simulations performed with NEKTON are in good agreement with the experimental observations.

## 2. INTRODUCTION

Roll coating is a process whereby liquid flows into a narrow gap between two horizontal rotating cylinders, the surface of which move either in the same direction (forward) or opposite directions (reverse). In asymmetric forward roll coating, the cylinder radii are different, as are usually their tangential velocities. Some of the liquid is entrained in the narrow gap of width $h_o$, separating the rollers and splits downstream into two thin films, each coating one of the solid surfaces. These films are connected by a meniscus. Figure 1 sketches the physical situation. Such a method is extensively used in painting, photography, tape recording industry for covering a large surface area with one or several uniform layers as well as metallic strip coating in the metallurgic industry. Industrialists wish both to control the amount of liquid applied on the substrate and to obtain a uniform coating layer. Unfortunately, this technique suffers from severe lubrication and free-surface flow problems. At high speed, a flow instability appears, in which the coating becomes ribbed and uneven [1-5]. Further above its onset threshold, the ribbing undergoes secondary instabilities leading to complex spatio-temporal behaviours and eventually chaos [6-7]. Such a behaviour is unbearable in a coating process, and so the study of this instability is of considerable importance. However, under operating conditions compatible with current production requirements, roll speeds are such that ribbing is always present. A research program involving experimental and numerical aspects is currently carried out at the VKI to study the fundamentals of these process limitations and to investigate concepts that prevent ribbing from appearing whatsoever.

158

Recently, the idea to place a string along the cylinders so as to touch the unstable meniscus to eliminate ribbing has been proposed [8]. The present paper deals with the pioneer development of the string concept. It aims at quantitatively understanding the correlation between the position of the string $x_{str}$, $z_{str}$ in the orthonormal coordinate system defined in figure 1, and the basic coating mechanisms such as the ribbing, total flow rate, partial flow rates on individual cylinders and contact line properties. To this end, a free-surface visualisation technique of the meniscus controlled by a string, already developed to measure natural, stable coating meniscus, is presented [9]. A large range of capillary number combinations otherwise displaying ribbing, is covered. Quantitative evidence of the efficiency of the string to eliminate ribbing is demonstrated by measuring film thickness profile along the cylinder with a triangulation laser device. The effect of the position of the string relative to the cylinders on the flow pattern and meniscus characteristics is also numerically investigated. The predictions are compared to the experimental observations.

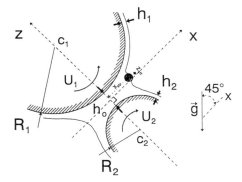

Figure 1: *A cross-sectional representation of the roll coating rig*

## 3. EXPERIMENTAL APPARATUS AND PROCEDURE

The experimental apparatus and the optical setup are described in detail in previous papers dealing with the free meniscus behaviour without string [10-12]. Only the main features relevant to the string experiments are reported. Two stainless steel cylinders, 260 mm long, with radii $R_1 = 67.5$ mm and $R_2 = 27.5$ mm, respectively (see figure 1) are driven by an independent direct current motor, providing tangential speeds $U_1$ and $U_2$ from 0.007 to 0.5 m/s. A pair of transversing mechanisms wearing arms controls accurately the position of the string with respect to the cylinders with a precision of 0.01 mm along $x$ and $z$, yielding a parallelism better than 0.01%. The string of diameter $d= 0.92$ mm, is a nylon P077 wire coated with a LW 432 H (Henkel) ethoxyled product. The string is well stretched but its tension is not controlled. The simulating Newtonian coating fluid is a Rhodorsil 47V500 silicone oil, with a density $\rho=970$ kg/m$^3$, a dynamic viscosity $\mu =0.57$ Pa.s and a surface tension $\sigma = 0.021$ N/m.

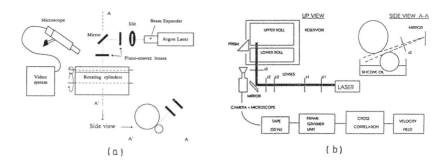

Figure 2: *Schematic of the visualisation technique*

Two similar visualisation techniques are used. They are illustrated by the sketches in figure 2. Both utilise an Argon-laser light sheet of 0.1 mm thick and 25 mm wide. It is created by a series of lenses and directed by a mirror at the locus of interest to visualize a thin slice of liquid equidistant from both cylinder edges. Some methine dye (Yellow Oracete 8GF) is diluted in the oil to provide fluorescence and improve, therefore, the visualisation. In the first configuration shown in figure 2-a, a CCD camera mounted on a large focal length (100 mm) microscope, views obliquely the bright interface. In the second version, the microscope is looking through a 90°-prism that acts as a mirror as indicated in figure 2-b. The face of the prism parallel to the sides of the two rolls is in perfect contact with the surmount of liquid escaping the gap at the edge of the cylinders. That arrangement allows to avoid optical distortion. The laser sheet is located at 40 mm from the edge of the cylinders to be free of end-effects. Such an approach gives an internal optical access to all the flow field in the nip and allows the application of particle tracking and PIDV technique [11-12]. Both techniques enable either on-line image processing on a PC or video recording for later handling and needs careful calibration procedure [12]. Typical absolute accuracy on the experimental meniscus profiles is about 30 μm.

The thickness profile along the top generatrix of the upper cylinder is measured by means of a commercial optical displacement sensor of ODS$^{TM}$ type. It functions on the principle of the optical triangulation of a IR laser beam. The voltage output is proportional to the distance between a reference and the reflecting object. In these experiments, the fluid used is an opaque latex paint to ascertain sufficient reflection of the interrogating beam. Two ODS $^{TM}$ are generally scanning the generatrices, one along the upper cylinder, the other at 45° along the lower cylinder. The ODS$^{TM}$ resolution is 0.5 mm and the accuracy associated to the film thickness measurement is about 6 μm.

## 4. NUMERICAL SIMULATIONS

To simulate the flow pattern in the nip and the meniscus shape and position to the experimental findings, the spectral element code NEKTON Version 2.95 is used. The two-dimensional computational domain is sketched in figure 3. It reproduces only a part of the experimental field. The inlet is placed at the minimum gap $h_o$ and is simulated by a parabolic velocity profile satisfying the experimentally measured normalised flow rate $Q/Uh_o = 1.33$ where $U$ is the mean roll velocity. Two outlets are defined; the upper film is calculated up to an angular position of $12°$ and the lower film simulated until the angular position of $31°$. The free surface conditions along the meniscus and across each film outlet are characterised by null normal traction $T_n = 0$ and a null tangential velocity component $u_t=0$ in the local coordinate system. The presence of a string is simulated by inserting a circular wall near the meniscus and by specifying the contact angles $\gamma_1$ and $\gamma_2$ measured during the experiments. The contact position of the meniscus on the string is free and computed by the code. The local wall velocity is set equal to the roll velocity. The fluid properties duplicate the experiment. Advection terms are neglected since the Reynolds number is low. The code is configured to solve a steady Stokes flow problem, taking gravity into account. The meshing is obtained from spectral elements and Legendre polynomials of $5^{th}$ order are used as trial functions in each macro-element.

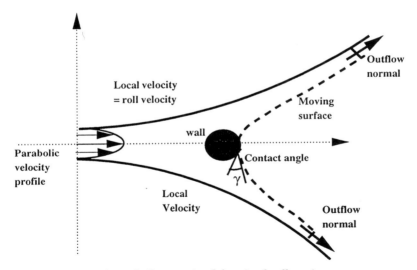

Figure 3: *Computational domain of roll coating*

The initial shape of the free surface is a guess from the experimental observations. A remeshing algorithm computes the modification of the meniscus shape and position up to convergence defined by a normalised change of the mesh coordinates lower than $10^{-4}$. For mean capillary number $Ca = \mu U/\sigma = 0.4$, the viscous forces are of the same order of magnitude then the surface tension effect so that a kinematic algorithm is used for the computation of the free surface position.

## 5. RESULTS

The first step of this study is to investigate how efficient the string is in eliminating ribbing. The scenario of ribbing removal is illustrated in the sequence of photographs given in figure 4. The working parameters are $Ca_1 = 1.0$, $Ca_2 = 3.0$ and a dimensionless gap $G = h_o/R = 0.0163$ where $R$ is the radius of an equivalent symmetric roll arrangement. These operating conditions are well above (at least three times) the ribbing threshold as demonstrated by the top picture, figure 4a, corresponding to a situation without string. On the second view, figure 4-b, the string is just coming into contact with the upper ribbed film. The third view, figure 4-c, corresponds to the situation when the string is dipped in the nip; very uniform coated surfaces are produced. It can be stated that the ribbing elimination is approximately achieved when the string reaches the trough of the ribbing waves. The quantitative demonstration is obtained by the differential thickness distributions measured with the ODS$^{TM}$ probe along a generatrix of the rolls, located sufficiently close to the string to minimise the levelling mechanism.

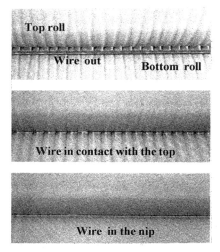

Figure 4: *Visualisation of the string effect on ribbing*

Typical profiles measured for successive positions of the string are plotted in figure 5. It is clearly seen in figure 5-b that, to the accuracy of the probe, ribbing is fully removed. The tests are extended to very extreme conditions corresponding to 20 times above threshold. In all the cases, it is possible to remove ribbing by means of the string and without any sign that it should not be possible at even higher values. The stabilized interface consists of two independent films attached to the string at an apparent contact line. As the string goes deeper into the gap, the contact lines get closer to each other and eventually meet to re-create a free, ribbed meniscus, while the string is fully immersed. The experiments testify there exists a range of string positions within which the string, controlling the meniscus, removes ribbing. It is therefore worth investigating the flow behavior flow within these conditions.

The internal visualisations show that the films leaving the string undergo a rapid relaxation to achieve a nearly constant thickness, as in the free meniscus case. It is possible to determine by DIP, the thickness $h_1$ on the top and $h_2$ on the bottom along each film, and, to calculate the dimensionless partial flow rates $\Omega_i = Q_i / h_o U_i$ from the dimensionless thicknesses $\lambda_i = h_i / h_o$, by taking gravity into account [13]:

$$\Omega_i = \lambda_i - \frac{\lambda_i}{3} \cdot \frac{Bo}{Ca_i} \qquad (1)$$

where $Bo = \rho \, g \, h_o^2 \backslash \sigma$ is the local Bond number. The dimensionless total flow rate is then calculated from its definition :

$$\Omega = \frac{2Q}{h_o(U_1 + U_2)} = 2\frac{(S\Omega_{1+} + \Omega_1)}{S+1} \qquad (2)$$

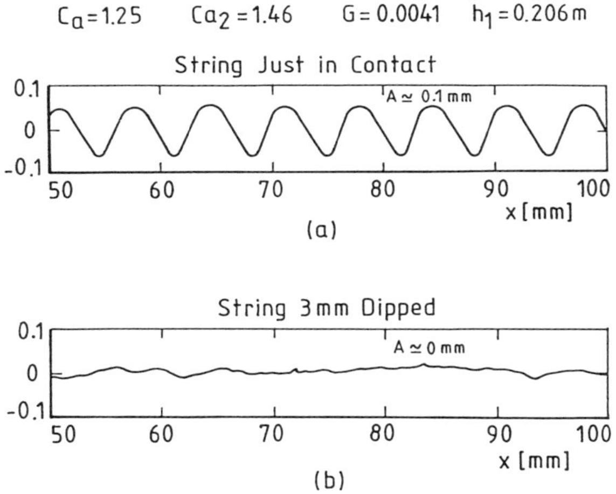

$Ca = 1.25 \qquad Ca_2 = 1.46 \qquad G = 0.0041 \qquad h_1 = 0.206 \, m$

String Just in Contact

(a)

String 3 mm Dipped

(b)

Figure 5: *Thickness profile along the top cylinder: Ribbing elimination*

$\Omega$ is theoretically equal to 1.333 for fully immersed rolls [14]. The predictions with models taking the interface into account never depart much from this [15-16]. Experimental results are less decisive, as $\Omega$ values up to 1.52 could recently be attributed to indirect feeding effects [9]. The variation of $\Omega$ with the streamwise position of the string $x_{str}$ for $Ca_1 = Ca_2 = 2.0$ and $G = 0.0163$ are presented in figure 6; the precision on $\Omega$-estimate is 0.05. The total dimensionless flow rate starts with the value of 1.57 at $x_{str} = 8.8$ mm and decreases when the string goes deeper towards the gap, getting closer to the standard 1.333 value for the free meniscus at $x_{str} = 6.8$ mm , but never reaching it. Although no final demonstration is available to justify this flow rate excess, it is evident that the presence of the string modifies the flow. A major result that concerns the control of the partial flow rates applied on each cylinder is shown in figure 7; the ratio of the partial flow rates $\Omega_1 / \Omega_2$ is plotted as a function of the relative string-to-cylinder distance $d_1/d_2$, where $d_i$ is the distance separating the surface of the string and the cylinder $I$. Since string monitors the films through contact lines, the flow rate is smaller on the cylinder to which the string is closer. Moreover, all the measurements attest that the total flow rate depends only marginally on $z_{str}$ while the partial flow rates are pretty much independent of $x_{str}$.

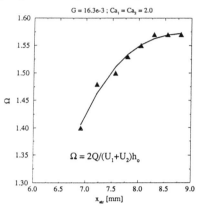

Figure 6: *Effect of the downstream string position on the total flow rate*

A typical NEKTON simulation of the flow field in presence of the string is illustrated by the vector plot presented in figure 8. The operating conditions are $Ca_1 = Ca_1 = 0.4$ and $G=0.0163$. The string position is $x_{str}= 8.89$ mm and $z_{str}$, = -0.27 mm. Stagnation and accelerating flow regions are well identified. A comparison between the predicted meniscus and the experimental observation is proposed in figure 9. A good agreement is found for the meniscus shape and position. In particular, the discrepancy between the film thickness on both cylinders predicted and measured does not exceed 1%.

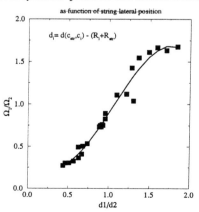

Figure 7: *Effect of the transversal string position on the partial flow rates*

Table 1 : Typical experimental-numerical comparison

| Test # | $\alpha_1$ | | $\alpha_2$ | | $\Omega_1 / \Omega_2$ | |
|---|---|---|---|---|---|---|
| | Exp. | Num. | Exp. | Num. | Exp. | Num. |
| 3 | 39 | 37 | 42 | 43 | 0.75 | 0.75 |
| 12 | 31 | 30 | 62 | 60 | 1.67 | 1.9 |
| 17 | 47 | 48 | 45 | 45 | 0.306 | 0.31 |

Table 1 displays a comparison of typical macroscopic data obtained from experiments and numerical simulations. It can be concluded that NEKTON is able to reproduce experimental findings, in particular the angular positions of the contact point $\alpha$ on the string and the flow rate partitioning with good agreement.

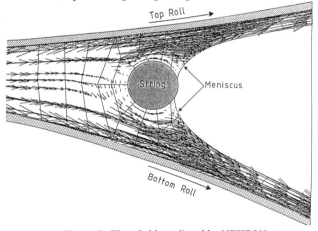

Figure 8: *Flow field predicted by NEKTON*

## 6. CONCLUSIONS

A quantitative study of the effect of a string brought into contact with the meniscus of an asymmetric roll coating arrangement is presented. The determination of the meniscus shape and position are performed using microscopic visualisation techniques. Complementary measurements of the film thickness are done with a laser beam triangulation probe.

The efficiency of the string technique to remove the ribbing instability is demonstrated. Suitable positioning of the string allows to monitor not only the partial flow rates but also the total flow rate. The experimental data validate the NEKTON code that predicts successfully the meniscus shape and the effect of the string on the flow rate partition.

Such results can be of potential interest to industrial applications, as they confirm a simple way to eliminate ribbing and set coating film independently of the speed ratio.

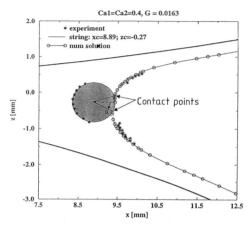

Figure 9: *Experimental-numerical comparison of the meniscus.*

## 7. REFERENCES

[1]     Benkreira, H., Edwards, M.F. and Wilkinson, W.L.:" *Ribbing instability in the roll coating of Newtonian fluids.*", Plast. Rubb. Process. and Appl **2** (2), 137 (1982).

[2]     Savage, M.D.:"*Mathematical model for the onset of ribbing.*", AIChE J. **30** (6), 999 (1984).

[3]     Carter, G.C. \& Savage, M.D.:"*Ribbing in a variable speed two-roll coater.*", Math. Engng. Ind. **1** (1), 83 (1987).

[4]     Coyle, D.J., Macosko, C.W. and Scriven, L.E.:"*Stability of symmetric film-splitting between counter-rotating cylinders.*", J. Fluid. Mech. **216** , 437 (1990).

[5]     Rabaud, M., Michalland, S. and Couder,Y.:"*Dynamical regimes of directional viscous fingering: spatiotemporal chaos and wave propagation.*", Phys. Rev. Lett. **64** (2),  184 (1990).

[6]     Michalland, S., Rabaud, M. and Couder,Y.:" *Transition to chaos by spatio-temporal intermittency in directional viscous fingering.*". Europhys. Lett **22** (1), 17 (1993).

[7]     Decré, M., Gailly, E., Buchlin, J-M and Rabaud, M.:" *Spatio-temporal Intermittency in a Roll Coating Experiment.*", in Spatio-Temporal Patterns, Ed. P.E.Cladis and P.Palffy-Muhoray, SFI studies in the Sciences of Complexity, Addison-Wesley (1995), pp. 561-571.

[8]     Hasegawa, T.and Sorimachi, K.:" *Wavelength and depth of ribbing in roll coating and its elimination.*", AIChE J. **39** (6), 935 (1993).

[9]     Decré, M., Gailly, E. and Buchlin,J-M. :"*Meniscus shape experiments in forward roll coating.*", Phys. Fluids **7** (3), 458 (1995).

[10]    Decré, M. and Buchlin, J-M.:" *An Extra Thin Laser Sheet Technique Used to Investigate Meniscus Shapes by Laser Induced Fluorescence.* ", Exp. Fluids **16** (5), 339 (1994).

[11]    Salimbeni, E.:"*Investigation of the nip of a forward roll coating configuration.* ", VKI-DC Report 1994-09, June 1994

[12]    Decré, M.:"*Etude expérimentale des comportements de l'interface dans l'enduisage par rouleaux.*", thèse de doctorat, U. Paris VI (1994).

[13]    Moffatt, H.K.:" *Behaviour of a viscous film on the outer surface of a rotating cylinder.*" Journal de Mécanique **16** (5), 651 (1977).

[14]    Ballal, B.Y. and Rivlin, R.S.:"*Flow of a Newtonian fluid between eccentric rotating cylinders : inertial effects.*", Arch. Rat. Mech. Anal. **62**, 237 (1976).

[15]    Savage, M.D.:" *Mathematical models for coating processes.*", J. Fluid. Mech. **117,** 443 (1982).

[16]    Coyle, D.J., Macosko, C.W. and Scriven, L.E.:"*Film splitting flows in forward roll coatin.* " J. Fluid. Mech. **171**, 183 (1986).

# THEORETICAL AND EXPERIMENTAL INVESTIGATION OF GAS-JET WIPING

J-M. Buchlin*, M. Manna*, M. Arnalsteen*, M.L. Riethmuller*
and M. Dubois**,

* von Karman Institute for Fluid Dynamics
Ch. de Waterloo, 72  Belgium - 1640 Rhode-St-Genèse
* * Gestion procédés et méthodes
Cockerill Sambre Branche Phenix
Quai du Halage, 10  Belgium - 4400 Flemalle

## 1. SUMMARY

The paper deals with theoretical and experimental modelling of gas-jet wiping. Such a technique allows the control of the final film thickness in continuous coating dip processes. Different theoretical approaches are compared and a simple zero-dimensional model is proposed. The experiments are conducted on a water facility. The experimental observations validate the mathematical modelling and show that the occurrence of splashing hinders the wiping performance. A dimensionless splashing correlation is established. Its applicability is illustrated through a typical exercise in predicting the maximum web speed of a galvanisation line.

## 2. INTRODUCTION

Gas-jet wiping or gas-knife stripping is a precise technique to meter an excess amount of liquid withdrawn by a moving surface as sketched in figure 1. It is the decisive operation in several coating processes such as the hot-dip galvanisation of metal strips and wires as well as the deposition of photographic emulsions. The careful control of the wiping parameters has a strong bearing on the thickness and the uniformity of the emerging liquid film. Because cost reduction remains the driving force of the competition between products, the maximum line speed at which the gas wiping process can be used is a key determinant. That requires a detailed study of the interaction between the liquid film and the impinging gas jet issued from a slot nozzle.

The present paper deals with the modelling of the gas-jet wiping technique and emphasises the physical limit of the process. Its objective is twofold: in a first chapter, an analytical gas-knife model is proposed. Such a model enables a fast prediction of the final film thickness with a sufficient accuracy so that it can be easily implemented on a feedback loop to control a continuous coating line; in a second chapter, the maximum web speed that can be attained, before the wiping mechanism fails is sought for. For this aim, an experimental investigation is carried out to validate the model and generate basic data pertinent to any type of liquid such as water or galvalume.

## 3. THEORETICAL MODELLING

### 3.1. Gas-Jet Characteristics

The principle of jet-wiping is to produce a gas-knife (air or nitrogen ) to impede a coated layer of liquid deposited by a dipping process on a substrate moving upward at a constant velocity $U$ as depicted in Figure 1. The impinging gas-jet induces a runback flow of the layer and part of the liquid returns to the dip. The final

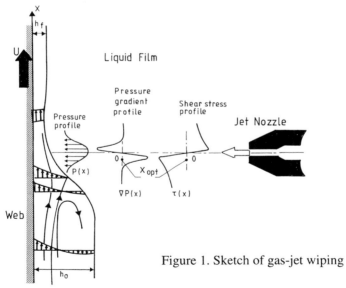

Figure 1. Sketch of gas-jet wiping

result is a thinner film of thickness $h_f$ which coats the moving surface. At the impingement, the gas-knife develops profiles of pressure $P(x)$ and shear stress $\tau(x)$ along the strip direction $Ox$. Generally, the shape of the pressure profile fits a Gaussian distribution for a normal impinging jet [1-2].

$$P(x) = P_{max} e^{-0.693\left[\frac{x}{b}\right]^2}$$

(1)

where $P_{max}$ is the peak pressure at $x=0$ and $b$ the value of $x$ where $P=0.5P_{max}$. The pressure gradient distribution is readily derived from Eq.(1):

$$\nabla P(x) = -\frac{1386}{b} \cdot \left[\frac{x}{b}\right] P(x) \text{ with a maximum value } \nabla P_{max} = 0.714 \frac{P_{max}}{b} \qquad (2)$$

In the impingement region, the surface shear stress $\tau(x)$ increases from zero at the stagnation point to a maximum value $\tau_{max}$. Its expression can be derived using the similarity of static pressure [2]:

$$\tau(x) = \tau_{max}\left[erf(0.833\frac{x}{b}) + 0.144b\nabla P(x)\right] \qquad (3)$$

In the wall-jet region the $\tau$-profile can be approximated by the following relation [3]:

$$\tau_{wj}(x) = \pm 0.26\frac{\Delta P_N}{Re_j} \cdot \frac{1}{\frac{x}{b}+4} \cdot \frac{D}{b} \qquad (4)$$

where $\Delta P_N$ is the nozzle pressure, $Re_j$ the jet Reynolds number and $D$ the gap between the nozzle lips. Typical pressure gradient and shear stress profiles due to the gas-jet impact on a flat plate are illustrated in Figure 1.

## 3.2. Film Equation

The theoretical description of jet-wiping relies on the lubrication approach that assumes negligible inertia . The resulting $Ox$ momentum equation for the liquid film states that the viscous shear stress balances the weight and pressure terms.

$$\mu_\ell\frac{\partial^2 u}{\partial y^2} = \rho_\ell g + \frac{d P_\ell}{d x} \qquad (5)$$

The film is sufficiently thin to consider the liquid pressure as the sum of the jet pressure and surface tension effect due to the surface deformation. Adopting the small slope approximation for the interface curvature, the pressure gradient in the liquid film along the web direction can be expressed as:

$$\frac{d P_\ell}{d x} = \frac{d P_a}{d x} - \sigma_\ell\frac{d^3 h}{d x^3} \qquad (6)$$

The final form of the $Ox$-momentum equation is obtained by inserting Eq.(6) in Eq.(5):

$$\mu_\ell \frac{\partial^2 u}{\partial y^2} = \rho_\ell g + \frac{d P_a}{d x} - \sigma_\ell \frac{d^3 h}{d x^3} \qquad (7)$$

The boundary conditions applied to Eq.(7) express the adherence of the liquid film to the web and the continuity of the surface shear stress at the liquid-gas interface:

$$u = U \quad \text{at} \quad y = 0 \qquad (8a)$$

and

$$\mu_\ell \frac{\partial u}{\partial y} = \tau(x) \quad \text{at} \quad y = h(x) \qquad (8b)$$

The integration of Eq.(7) with the boundary conditions Eqs. (8a-b) leads to the velocity profile $u(x,y)$ from which the volumetric liquid flow rate $q$ is determined through the continuity equation:

$$q = \int_0^{h(x)} u \cdot dy \qquad (9)$$

The evaluation of the flow rate leads to the liquid film equation whose dimensionless form can be:

$$\Gamma \frac{d^3 \hat{h}(X)}{d X^3} = 1 + \nabla \hat{P}(X) + \frac{2Q - 3\hat{h}(X) - 1.5T(X)\hat{h}^2(X)}{\hat{h}^3(X)} \qquad (10)$$

The normalised variables and non-dimensional physical groups are defined as:
- The abscissa normalised by the nozzle slot $X = x / D$

- The dimensionless film thickness $\hat{h} = h / h_o$ with $h_o = \sqrt{\dfrac{\mu_\ell U}{\rho_\ell g}}$ where

  $\rho_\ell$ and $\mu_\ell$ are the density and the dynamic viscosity, respectively, of the liquid and $g$ is the acceleration due to gravity.
- The surface tension group $\Gamma = \dfrac{\sigma h_o}{\rho_\ell g D^3}$ which is the inverse of the Bond number; $\sigma$ is the liquid surface tension.
- The normalised flow rate $Q = \dfrac{q}{q_o}$ where $q_o = \dfrac{2}{3} U h_o$
- The dimensionless jet pressure gradient $\nabla \hat{P} = \dfrac{\nabla P}{\rho_\ell g}$
- The dimesionless jet shear stress $T = \dfrac{\tau}{\tau_o}$ with $\tau_o = \sqrt{\mu_\ell U \rho_\ell g}$

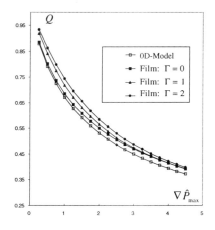

Figure 2. Modelling effect of gas-jet wiping

To solve Eq.(10) different techniques have been developed [4-5]. In the present study, the film profile equation is discretised by a finite difference scheme. The discretisation is organised to lead to a non-linear algebraic system of N equations with N unknowns among which N-1 are the nodal values of the film thickness $h(x)$ and the last one is the liquid flow rate $Q$. The solution is the zero of a system of N non-linear functions in N variables obtained by a modified version of the Powell hybrid method coded in the NAG-Library [6].

Simplified models can be derived from Eq.(10). Anticipating the weak effect of the surface tension, the film equation reduces to:

$$\left[1+\nabla\hat{P}\right]\cdot\hat{h}^3 - 1.5T\hat{h}^2 - 3\hat{h} + 2Q = 0 \qquad (11)$$

The solution of Eq.(11) is obtained by solving a cubic equation for each $x$-value [7]. A zero-dimensional model is now proposed. It considers that the surface tension has no effect, the wiping mechanism is the result of the maximum pressure gradient and maximum shear stress and that both effects act at the same location $X_{op}$ (see figure 1) [3]. In such an approximation, the liquid film equation reduces to the following simple algebraic equation:

$$\left[1+\nabla\hat{P}_{\max}\right]\cdot\hat{h}_{op}^{\ 3} - 1.5T_{\max}\hat{h}_{op}^{\ 2} - 3\hat{h}_{op} + 2Q_{op} = 0 \qquad (12)$$

Since there are two unknowns, $\hat{h}_{op}$ and $Q_{op}$, a second equation is derived by stating that the wiping efficiency corresponds to the optimum of the final net flow rate. Such a condition demands that the optimal film thickness satisfies the following relation:

$$\hat{h}_{op} = \frac{T_{\max} + \sqrt{T_{\max}^2 + 4\left[1+\nabla\hat{P}_{\max}\right]}}{2\left[1+\nabla\hat{P}_{\max}\right]} \qquad (13)$$

Once the effective thickness $\hat{h}_{op}$ is calculated, the flow rate $Q_{op}$ is determined from Eq.(12). The 1D- and 0D-models are compared in figure 2. The dimensionless liquid flow rate after wiping is plotted versus the normalised maximum pressure gradient. These typical results indicate that the surface tension effect becomes less and less significant when the pressure gradient increases. Moreover, figure 2 shows that the 0D-model is a very reliable approximation of more elaborate approaches and deserves to be validated by experimental data.

## 4. EXPERIMENTAL SIMULATION

### 4.1. Experimental Facility

The VKI-experimental facility designed to simulate air-jet wiping in the hot dip galvanising process is sketched in Figure 3. The set-up consists of a moving rubber belt, 5m long and 0.5 m wide, entrained by rollers and passing in a liquid bath. The configuration of the two lower cylinders corresponds to the real industrial arrangement. The simulation liquid is water. Although, the physical properties of liquid zinc and water are different, Table I shows that the controlling non-dimensional geometrical and operating numbers vary in a similar range between the laboratory facility and the industrial plant.

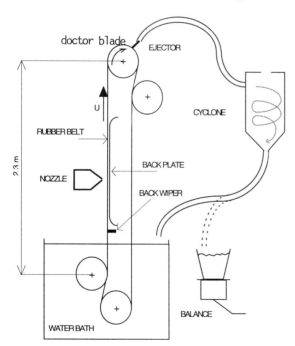

Figure 3. The VKI-water wiping facility

The rubber belt is driven by an electric motor connected to the upper cylinder. The maximum linear velocity at which the strip remains stable is 4 m/s. For sake of simplicity, the wiping is studied only on one side. To maintain the web flat at the jet impingement, an aluminum backplate on which the belt is sliding is placed on the opposite side. The nozzle is designed to allow a uniform even pressure distribution along the width of the strip. The nozzle slot is made of two movable parallel anodised-steel lips with an adjustable gap ranging from 0.7 to 2.1 mm. Air plenum can be varied up to 10000 Pa.

The study of the wiping process requires the measurement of the final coating thickness. On the experimental facility, the resulting mass flux is measured by collecting and weighting the total amount of water that reaches the top of the driving cylinder. A special doctor blade mounted on an aspirating device connected to a cyclone separator is used for that purpose. The average coating thickness is estimated assuming that the liquid film is moving at the bulk velocity $U$. This method is very simple compared to the measurement of the film thickness by a capacitive sensor or by laser triangulation [11,12]. The absence of edge effect in the case of single-jet wiping justifies such an average measurement technique.

Table I : Fluids characterization

|  | WATER | ZINC |
|---|---|---|
| $\hat{Z}$ | 6 - 18 | 7 - 25 |
| $\nabla \hat{P}_{max}$ | 2 - 150 | 5 - 200 |
| $T$ | 1 - 10 | 1 - 5 |
| $\Gamma$ | 1 - 10 | 1 - 20 |

## 4.2. Experimental Results

The set-up is used to validate the jet wiping model in a wide range of operating conditions. Typical data obtained on the laboratory water facility and galvanizing line involving different nozzle designs are compared to the predictions yielded by the 0D air-knife model in figure 4. The dimensionless mass flux $Q$ is plotted versus the normalised maximum pressure gradient. The very good agreement found between experiments and theory demonstrates that a simple engineering model may have the capability to predict satisfactorily air-jet wiping performance. The model provides for any strip velocity and operating parameters such as the slot gap $D$ and the nozzle-to-strip distance $Z$, the nozzle pressure $\Delta P_n$ needed to achieve the prescribed coating weight.

The minimum achievable coating weight at a prescribed line speed and the maximum line speed at which the air wiping remains efficient are important pieces of information for the process engineer. Some experiments carried out on the laboratory facility have brought to light unusual film behaviour. At extreme working conditions, experiments reveal that the wiping model does not incorporate all the physics of the jet-liquid film interaction. Figure 5a shows an example of such a deviation. The final thickness of the water film is plotted versus the nozzle pressure $\Delta P_n$ for different line speed $U$, all the other parameters being kept constant.

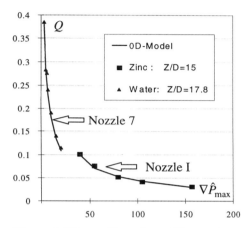

Figure 4. Theory-Experiment Comparison

At moderate line speed, the coating weight decreases when the nozzle pressure increases as predicted by the air-knife model. At higher values of the strip velocity, the measurements indicate a sudden deficiency of the wiping depicted by a sharp augmentation of the coating thickness. The experimental observation shows that the occurrence of this wiping failure is closely related to a severe detachment of the liquid runback flow as visualised for water on the photograph presented in Figure 5b. The origin of this flow behaviour, named *splashing*, is the formation of a spray of tiny droplets dislodging from the liquid surface. The operating conditions of such an event are indicated by arrows in figure 5a . This phenomenon starts generally at the strip edges and spreads rapidly towards the center as the line speed increases, to lead finally to the detachment of the runback flow. At the early stage of the droplet formation, the thickness and quality of the coating layer are not yet affected. On the contrary, full splashing hinders the free development of the gas jet and the wiping performance decreases accordingly. In extreme conditions, the splashing film may even cling to the nozzle and obstruct partially the slot. The final thickness becomes completely uncontrolled and very high nozzle pressures are needed to restore some wiping effect. However, the film quality is severely degraded. The splashing phenomenon is also observed on the industrial galvanising line demonstrating that the phenomenon is a shortcoming of the wiping process and puts an upper limit on the line speed of the process.

To predict the conditions of spray generation, the application of the Hinze's model developed for break-up of liquid film by turbulent gas flow have been attempted [5]. A general correlation expressed in terms of Weber and Reynolds numbers based on the jet velocity $V_j$ can be expected.

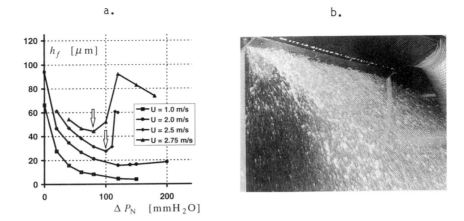

Figure 5. The onset and visualisation of splashing

In the present study, it is postulated that the onset of the splashing happens when the shear stress produced by the downward gas wall-jet $\tau_w$ becomes stronger than the stabilizing surface tension term modelled here by $\sigma/h_0$. Expressing $\tau_w$ in terms of the typical dynamic pressure of the wall jet $0.5\rho U^2_{wj}$ and evaluating the ratio of the dominating forces controlling the splashing mechanism lead to the following effective Weber number $We = \rho_a U^2_{wj} h_0 / \sigma$ . Figure 6 shows that the critical Weber number $We_s$ at which the incipient splashing occurs is correlated to the film Reynolds number $Re$ based on the final coating thickness and the web velocity, $Re = \rho_l U h_f / \mu_\ell$. $We_s$ decreases as $Re$ increases and the correlation fits a power law.

### 4.3. Illustrative Exercise

This phenomenological model can be easily combined with the simple 0D-air-knife model to estimate the splashing threshold and defines the operating window for safe and stable wiping. To exemplify such a capability, an illustrative exercise is conducted. Typical conditions of a continuous galvanising line are selected. One seeks the upper limit of the line speed of an industrial process characterised by a zinc load of 135 g/m$^2$ per side, a stand-off distance Z=10 mm and a nozzle slot D=0. 85 mm. The results are plotted in Figure 7. The wiping model predicts that the nozzle pressure $\Delta P_{N,W}$ needed for the prescribed coating increases almost linearly with the strip velocity $U$.

On the contrary, the splashing correlation points out that the nozzle pressure $\Delta P_{N,S}$ which provokes the incipient wiping failure is inversely proportional to $U$. Figure 7 shows that there may exist an operating condition at which the desired wiping coincides with the occurrence of splashing. In the present case, $U_{max}$ is equal to 200 m/mn and corresponds reasonably well with observation on galvanising lines. Above this line speed, it is impossible to achieve the expected coating since the integrity of the upstream liquid film is no longer ensured .

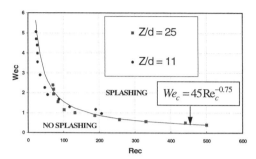

Figure 6. Splashing correlation

## 5. CONCLUSIONS

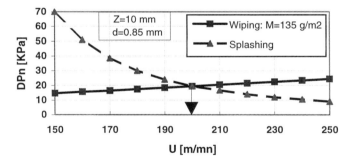

Figure 7. Window of gas-jet wiping

A theoretical and experimental study of gas jet wiping is presented. The mathematical modelling shows that the effect of surface tension can be omitted when the gas knife produces a large pressure gradient at the impingement on the liquid film. The zero-dimensional model is a promising predictive tool which can be the basis of a control loop on a continuous line. The experiments conducted on a water facility validate the OD-model and reveal that the splashing phenomenon constitutes a physical limit of the gas-jet wiping. A dimensionless splashing correlation is proposed and, when combined to the OD-model, allows the prediction of the web speed at which splashing is expected to occur.

## 6. REFERENCES

1. Abramowitch, G.N. *Theory of Turbulent Jet*, MIT Press, (1963).
2. Beltaos, S. Oblique impingement of plane turbulent jets, *J. of the Hydraulics Division*, **102**, HY9, 1177-1192.
3. Buchlin, J-M. *Modélisation de l'essorage par jets d'air d'un film de zinc* VKI-CR 1988-15EA/5,von Karman Institute for Fluid Dynamics, Belgium, (1988).
4. Tuck, E.O. and Vanden Broeck, J.M. Influence of surface tension on jet-stripping continuous coating of sheet material, *AIChE J.*, **30**, 808-811.
5. Yoneda, H. *Analysis of air-knife coating* Degree of Master of Science, University of Minnesota, (1993).
6. Powell, M.J.D. 'A hybrid method for nonlinear algebraic equations', *Numerical methods for nonlinear algebraic equations.*Ed. P. Rabinowitz, P., Gordon and Breach, (1970).
7. Ellen, C.H. and Tu, C.V. 'An analysis of jet stripping of molten metallic coating', *8th Australian Fluid Mechanics Conference*, University of Newcastle., N.W., 2C.4-2C.7, 28 Nov-2 Dec, (1983).
8. Schweizer, P.M. 'Experimental methods for coating flows', *The Mechanics of Thin Film Coatings*, First Biennial European Coating Symposium, Leeds, (1995).
9. Van Geffel, B. and Riethmuller, M.L. 'Distance measurements on wavy surfaces using optical techniques', *Sensor 93 Congress*, Nürnberg, Germany, Oct 11-14, (1993).

# EXPERIMENTAL INVESTIGATIONS ON ROLL
# COATING WITH DEFORMABLE ROLLS

O. Cohu and A. Magnin
Laboratoire de Rhéologie (CNRS, UJF, INPG)
BP 53 - 38041 Grenoble Cedex 9, France

An experimental investigation of deformable roll coating of Newtonian fluids is presented, with the primary aim to predict the coating thickness. The mechanical properties of the elastomeric material that covers the deformable roll are investigated with regard to the time-scale of the process, which allows the incertainities resulting from the viscoelasticity of the rubber to be reduced. With both rolls moving at the same peripherical speed, experimental data are compared with available theoretical predictions, and a correct agreement is obtained. The influence of the finite thickness of the elastomeric cover is also discussed. With both rolls moving with different velocities, preliminary results show that the difference in the roll velocities can affect the boundary conditions at the inlet of the nip, leading to a drastic reduction in the flow rate between the rolls.

## 1. INTRODUCTION

Forward roll coating is a very common way of depositing a thin liquid film onto a moving substrate. Usually, controlling the thickness of the coated film is of prime importance. This is done by having the fluid flow between two counterrotating cylinders, the surfaces of which moving in the same direction at the nip. In many cases (*e.g.* coil-coating industry), one of the rolls is rubber-covered, and both rolls are pressed against each other by an external load (fig. 1). The coating thickness then depends on the operating parameters (roll speeds $V_1$ and $V_2$, loading W), the mechanical properties of the fluid, and the mechanical properties of the elastomeric cover.

Coyle[1] analyzed the coating flow of a Newtonian fluid using an accurate description of the two-dimensional, elastic deformation of the rubber cover. He obtained numerical results for the flow rate between the rolls — which is directly related to the film thickness $e$ if the rolls surfaces move at the same velocity. His results can be fitted by the dimensionless analytic equation

$$\frac{e}{R} \approx 0.4 \left( \frac{\eta V}{ER} \right)^{0.6} \left( \frac{W}{ER} \right)^{-0.3} , \tag{1}$$

where R the roll radius, V is the average roll speed and $\eta$ is the viscosity of the fluid, which is assumed to be Newtonian. $E_1$ and $\nu$ being the Young modulus and the Poisson ratio of the rubber that covers the deformable roll, respectively, the effective elastic modulus of the solid materials involved is $E = E_I/(1 - v^2)$, assuming that the steel roll is much harder than the rubber-covered one. The thickness $b$ of the elastomeric layer does not enter in Eq. (1), which is in agreement with the theoretical results of Hooke[2], who showed that this parameter has a very low influence as long as the contact half-width $\delta/2$ is smaller than the cover thickness, that is, $2b/\delta > 1$.

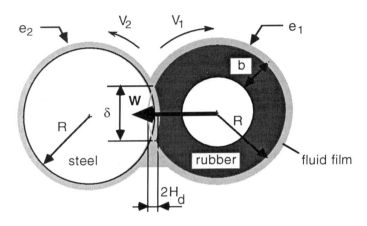

*Fig. 1 : Definition sketch of deformable roll coating.*

Experimental investigations on roll coating involving a rubber-covered roll have been reported by several authors[1,3,4], but significant discrepancies between the various data can be observed. Coyle[5] attributed these discrepancies to the possible viscoelastic behaviour of the rubber. Indeed, the viscoelastic properties of the rubbers used in the experiments were not controlled, nor was examined the actual influence of the rubber-cover thickness.

## 2. EXPERIMENTAL SET-UP

The experiments were conducted in Sollac's Research Centre (Montataire, France). A three-roll pilot coater, supplied by GFG (Milwaulkee, Wisconsin), was used. The apparatus is schematically depicted in fig. 2. The back-up roll simulates the web to be coated. A relatively thick film of fluid emerges from the feed nip between the first pair of rolls. The coated film is then metered by the flow between the pick-up roll (rigid) and the applicator roll (deformable), which are held against each other by movable mechanical stops. Finally, the film carried out by the applicator roll is wiped off totally by the back-up roll which moves in the opposite direction at the nip.

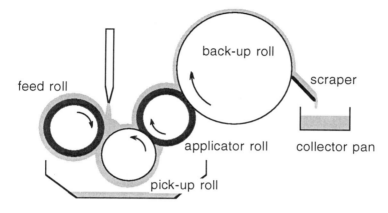

*Fig. 2 : Schematic drawing of the pilot coater used in the experiments.*

The back-up roll was equipped with a doctor blade and a collector pan so that the thickness of the film on this roll could be determined by weighing the collected fluid during a known period of time. Therefore the thickness $e$ of the film carried out by the applicator roll could be obtained, since film transfer between these rolls is complete. Newtonian glycerin/water solutions were used as test fluids. A Rheovisco ELV8 viscosimeter (Rheo, France) was used for viscosity measurements. Loads were measured with force transducers located at the ends of the rolls. A digital tachymeter (Compact instruments, UK) was used for roll speed measurements. The ranges of parameters considered in this work are given in table 1. About 300 data points were obtained.

*Table 1 : Parameter ranges in experimental investigations.*

| Fluid viscosity | $\eta$ | 0.08 | - | 1.2 | ± 10 % | Pa.s |
|---|---|---|---|---|---|---|
| Average roll speed | V | 0.17 | - | 1.67 | ± 3 % | m/s |
| Applied load | W | 1932 | - | 19963 | ± 5 % | N/m |
| Elastic modulus of the rubbers [= f(ω)] | $E_1$ | 0.94 | - | 3.25 | ± 10 % | MPa |
| Roll radius | R | | 118 | | | mm |
| Rubber cover thickness | b | 12 | - | 25 | | mm |

Three polyuretane (PU) materials denoted rubber 1 to 3 were selected as covers. Rubbers 1 and 2 had roughly the same hardness index (55 Shore A), but were supplied by different companies. Rubber 3 was softer, with a Shore A hardness of about 30. Since the duration of the stresses undergone by the cover in a roll nip is very short (typically 10 ms), the mechanical properties of these materials were investigated in a high frequency dynamic regime (1 - 200 rad/s) using a Carrimed Weissenberg Rheogionometer. The Young modulus $E_1$ of the three PU materials is depicted in fig. 3 as a function of the angular frequency $\omega$. It can be seen that $E_1$ depends actually on the frequency. When the frequency tends to zero, the results are close to that given by the suppliers of the rubbers using usual mechanical tests of long time-scales. Conversely, the Young modulus that is to be considered in the roll coating process (under frequencies up to 100 rad/s) can be two times greater, as is the case with rubber 3.

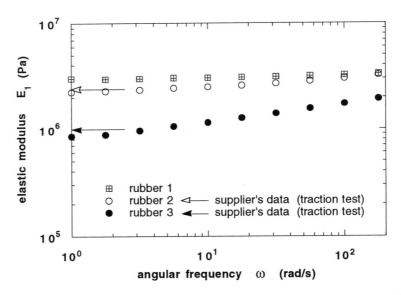

*Fig. 3 : Elastic modulus as a function of the frequency for three types of rubber.*

In order to compare experimental measurements of the coating thickness with previous theoretical results, the elasticity modulus entering into the theoretical equation, Eq. (1), was determined using the results showed in fig. 3, assuming that the duration $t = 1/\omega$ of the stresses undergone by the cover equals $\delta/V$. The contact width $\delta$ was estimated from the deflexion $2H_d$ of the deformable applicator roll against the rigid pick-up roll, which was measured with dial indicators located between the axis of the rolls. In this way, the elasticity modulus does not depend only on the material properties but also on the operating parameters (speed, load — or deflexion — and radii).

# 3. RESULTS AND DISCUSSION

## 3.1 Film thickness in the symmetric case ($V_1 = V_2$)

The experimental results obtained with a relatively thick cover ($b = 25$ mm) will be discussed first. Roll deflexion measurements ($2H_d$) showed that with $b = 25$ mm, the contact width $\delta$ verified

$$0.9 < 2b/\delta < 5 \ . \tag{2}$$

Therefore, according the results of Hooke[2], the rubber cover thickness was expected to have a little influence in this case.

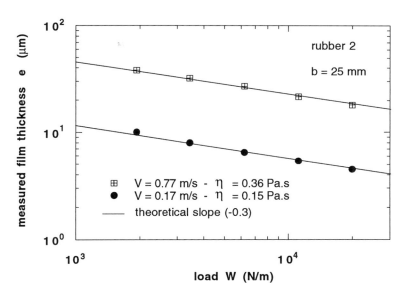

*Fig. 4 : Influence of the loading on the coating thickness ("thick" covers).*

The change in film thickness $e$ as a function of the applied load $W$ is shown in fig. 4. The data follow the slope -0.3 in the log-log plot, as predicted by Coyle[1]. Similarly, it can be seen in fig. 5 that the film thickness is proportional to the product [viscosity $\times$ speed] raised to the power 0.6, which is also in agreement with the theoretical predictions. The frequency-dependent elastic modulus of the rubber materials used in the experiments make the proper influence of E difficult to point out so plainly. Similarly, the influence of the roll radius could not be verified explicitly. Nevertheless, the *quantitative* comparison in terms of film thickness between the experimental data and the theoretical predictions of Coyle[1],

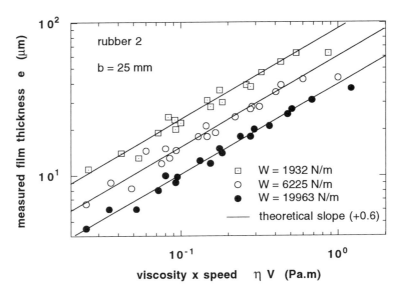

*Fig. 5 : Influence of the viscous forces on the coating thickness.*

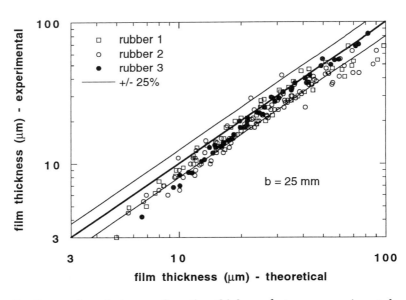

*Fig. 6 : Comparison in terms of coating thickness between experimental and theoretical[1] data ("thick" rubber covers ; $V_1 = V_2$).*

Eq. (1), shows a quite satisfactory agreement since the discrepancy is in most cases less than 25 % (fig. 6). Obviously, at least a part of these discrepancies is related to the uncertainities which affect each of the relevant parameters of the process (see table 1).

Experiments were also performed using a thinner cover than in previous investigations ($b$ = 12 mm). A soft rubber material, of Shore A hardness 30, was used (rubber 3). This allowed us to increase the roll deflexion, leading to larger values of the contact width. Consequently, relatively low values of the ratio $2b/\delta$, down to 0.6, were obtained.

The qualitative influence of the finite thickness of the rubber cover is depicted in fig. 7, which should be compared with fig. 4. The coating thickness $e$ is plotted as a function of the load W. For $2b/\delta$ < 1, the results diverge from the straight line of slope -0.3 which was observed with a thicker cover. The deviation is greater under high loads, since high loads correspond to low $2b/\delta$ ratios. Finally, this shows that decreasing the cover thickness increases the influence of the external load W on the film thickness. This can explain the discrepancies between previous experimental data and the theoretical results of Eq. (1).

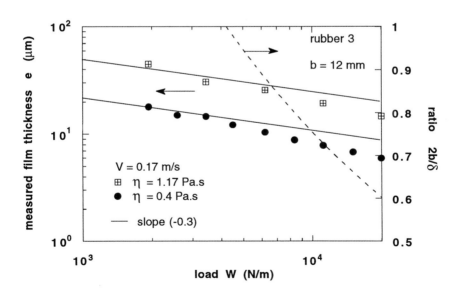

Fig. 7 :  Influence of the loading on the coating thickness for "thin" rubber covers.

## 3.2 Nip starvation in the asymmetric case $(V_1 \neq V_2)$ : experimental observations

The influence of the speed-ratio $\bar{v} \equiv V_1/V_2$ on the flow rate between the rolls was also investigated. Speed-ratio from 1 to 6 were imposed, the applicator deformable roll (denoted with the subscript 1) being faster than the pick-up rigid roll, as is the case in most practical operation. Since we could only measure the thickness $e_1$ of the film carried out by the applicator roll, the total flow rate $e_1V_1 + e_2V_2$ was estimated considering that the speed-ratio dependence of the film split is approximately the same as in rigid roll coating[6], as shown experimentally by Cohu[7].

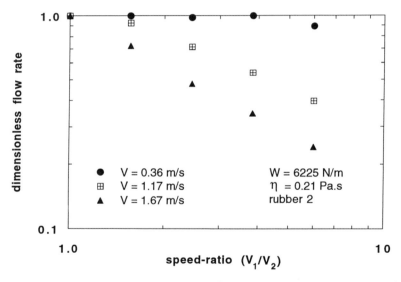

*Fig. 8 : Dimensionless flow rate vs. speed-ratio for various average speeds.*

The results are plotted in fig. 8, where the film thicknesses are normalized with the symmetric case values ($\bar{v} = 1$). For low average roll speeds (up to $V = 0,36$ m/s), the flow rate does not depend on the speed-ratio but only on the average roll speed, which is in agreement with the lubrication theory that governs the flow (see *e.g.* reference[8]). Conversely, for higher average roll speeds, increasing the speed-ratio leads to a severe reduction in the film thickness (all other variables, including the average roll speed, being kept constant).

It is important to note that the decrease of the coating thickness for high values of the speed-ratio and under high average speeds (that is, for large speed difference $V_1$—$V_2$) was always associated with the visual observation of the imprint of the scraper on the rolls at the exit of the nip, as depicted in fig. 9 (the scraper width was indeed smaller than the roll lenght). This visual observation shows that large speed differences affect the boundary conditions at the inlet of the nip, where the arriving film on the pick-up roll comes into contact with the dry surface of the applicator roll. In this case, the coating flow is then partially starved, which can explain the drastic reduction in the film thickness.

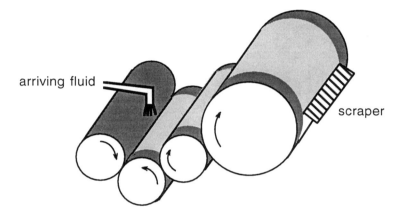

*Fig. 9 : Observation of the imprit of the scraper at the exit of the nip between the pick-up roll and the applicator roll.*

## 4. CONCLUSION

Experimental studies of forward roll coating of Newtonian fluids with a deformable roll have been conducted. First, we have pointed out that the elastic modulus of the rubber materials is often time-dependent, and must therefore be determined according to the time-scale of the process. Having done this, experimental data were compared in terms of coating thickness with previous theoretical results. In the symmetric case (speed-ratio equals unity), a satisfactory agreement was obtained with the theoretical correlation of Coyle[1], Eq. (1), but this agreement is limited to relatively thick rubber covers for the deformable roll. Indeed, it was shown that rubber covers thinner than the contact half-width tend to decrease the coating thickness significantly by increasing the influence of the external load. The finite thickness of the rubber cover and its time-dependent mechanical properties may both explain why previous experimental correlations disagree significantly with available theoretical predictions.

188

Preliminary results were also obtained in the asymmetric case, where both rolls move at different peripherical speeds. It was observed that the speed-ratio has a very low influence on the flow rate as long as the feeding of the nip can be ensured. Unfortunately, it was shown that large speed differences affect the boundary conditions at the inlet of the nip, leading to nip starvation which decreases the flow rate (and then the coating thickness) drasticly. Nevertheless, it is believed that this phenomenon is restricted to the roll configuration that have been studied here, that is when gravity acts against the flooding of the nip[7].

## ACKNOWLEGMENTS

The authors are indebted to Claude BONNEBAT (SOLLAC - C.E.D., Montataire, France) who initiated this project and provided helpful advice and technical support. Many thanks to all the people in SOLLAC-Montataire who made the use of the pilot coater so much easier. The encouragements from Pr. Jean-Michel PIAU were also very much appreciated. This work was supported by SOLLAC and by the ANRT (Association Nationale de la Recherche Technique).

## REFERENCES

1. Coyle, D.J. Experimental studies of flow between deformable rolls, *AIChE Spring Nat. Meet., New Orleans (LA)*, Paper No 3d (1988).

2. Hooke, C.J. The elastohydrodynamic lubrication of a cylinder on an elastomeric layer, *Wear*, **111**, 89 (1986).

3. Smith, J.W. and Maloney, J.D. Flow of fluids between rotating rollers, TAPPI J., **49**, 63 (1966).

4. Kang, Y.T., Lee, K.Y. and Liu, T.J. The effect of polymer additives on the performance of a two-roll coater, *J. Appl. Polym. Sci.*, **14**, 34 (1991).

5. Coyle, D.J. 'Roll Coating', *Modern coating and drying technology*, Ed. E. Cohen and E. Gutoff, VCH publishers New-York, (1992) 63-116.

6 Coyle, D.J., Macosko, C.W. and Scriven, L.E. Film-splitting flow in forward roll coating, *J. Fluid Mech.*, **171**, 183 (1986).

7 Cohu, O. Rhéologie des peintures et procédé de couchage au rouleau, *Thesis*, Institut National Polytechnique de Grenoble (France), (1995)

8 Herrebrugh, K. Solving the incompressible and isothermal problem in elastohydrodynamics through an integral equation, *J. Lub. Tech.*, **90**, 262 (1968).

# SURFACE WAVE INSTABILITY ON AIRCRAFT DE/ANTI-ICING FLUID FILMS

F. Cunha, M. Carbonaro
Aeronautics/Aerospace Department
von Karman Institute, Belgium

Summary

The motion of a thin film of de/anti-icing fluid, deposited on a flat plate and subjected to a shearing airflow of increasing strength is studied. The different instabilities that appear on the fluid surface are of an extreme importance because they cause an increase in the surface roughness as seen from the airflow point of view. When such fluids are applied to the aircraft wings, this increase of surface roughness causes an increase in drag and a decrease in lift. Experiments were conducted on the CWT-1 cryogenic facility at subfreezing temperatures. The variation of the fluid film thickness with space and time is measured. The observed wave patterns (with increasing air velocity) are quite obvious for the de-icing fluids, with Newtonian behaviour, while different patterns are encountered for the initial instability of anti-icing fluids, with non-Newtonian behaviour. It is interesting to verify that the final residual layer thickness is independent of the initial fluid film thickness. A qualitative description of the surface waves encountered is given. With this information it is possible to have a better estimation of the influence of the fluid film on the aerodynamic characteristics.

## 1 INTRODUCTION

Aircraft should operate independently of the atmospheric conditions. During winter, these conditions are mostly adverse with an accumulation of ice or snow on the aircraft surfaces. These accumulations can be of different types: from powder snow, to water freezing on the airfoil surfaces, creating a solid rough layer. These layers have a very negative effect on the aerodynamic properties of the aircraft wing. Even a small thickness of about half of a millimetre can produce a reduction in lift of about 30% [1]. This change can produce catastrophic results, with many accidents to demonstrate it [1]. As Brumby concludes "there is no such thing as a little ice". A lot of information is already available concerning the effect of an ice/snow layer on the aerodynamic properties of an airfoil [2].

Two types of fluids are available for the ice/snow removal: De-icing fluids, with a 60% minimum glycol, have Newtonian properties. They are sprayed hot on the surfaces and should flow off when the aircraft performs its takeoff manoeuvre. Although these fluids have a low freezing point the hold-over period is not very large. This hold-over period is very important because it defines the time that the fluid is capable of protecting the surface against refreezing. Anti-icing fluids contain a minimum of 40% glycol and several manufacturer-specific additives. They are designed to protect the surface, and as such, they have a much higher hold over time than the de-icing fluids. Therefore the anti-icing

189

fluids have non-Newtonian properties, behaving like a very viscous fluid when the aircraft is parked (and therefore with a negligibly small motion under the action of the gravity) and like a fluid film when the aircraft makes the take off ground run (with the increased shear between the air and the fluid).

The study of the effect of de/anti-icing fluids on airfoils' properties was started by Boeing [3] and soon after taken over at VKI. For an overview of the research carried out at the VKI from 85-94 please refer to [4] and [5]. The culmination of these studies was the development of a new, innovative, aerodynamic acceptance test procedure for de/anti-icing fluids [6]. Although a quantification of the aerodynamic effect of de/anti-icing fluids can be obtained by these studies, a more careful analysis of the causes is needed, and therefore the reason for this study.

## 2 THEORETICAL BACKGROUND

Existing literature describes several studies conducted with co-flowing gas-fluid systems. Both gas and fluid are considered as Newtonian fluids. An overview of these past studies is presented in [9] with more recent studies [11] [12] extending the available instability models. The fluid film surface instability is due to the imposition of pressure fluctuations on the fluid surface by the airflow. When the pressure variation in phase with the wave slope exceeds the stabilising effect of the gravity and surface tension, a regular 2D pattern appears. With increasing velocity, irregular solitary waves appear due to the pressure variations in phase with the wave height. Although all these works have covered the theoretical and experimental stages of the initiation of the 2D pattern [7] [8] [12] and the initiation and propagation of solitary waves [10] [11], they used the confined flow inside small diameter tubes. One of the first of such works [7] used a 0.3x0.2 section, but most of the past work was conducted in large aspect ratio section (0.3x0.02-0.1m). In these studies the liquid film has a certain mass rate that is set independently of the air flow permitting this way to established a stability map based on the air Reynolds number and on the liquid Reynolds number. Also the viscosities of the Newtonian fluids used were on the range 10-80 mPa.s. The test duct length was sufficiently large so that a steady state existed in the test section .

This is basically different from the conditions existing on the present study. First the fluid film is at rest and only moves under the influence of the shearing airflow. Second, there is no fluid recirculation, and this causes a decrease of the average fluid film thickness with time. Third the viscosity of the fluids used is one order of magnitude higher than those used in the previous studies. Fourth, the airflow over a de/anti-icing fluid film is not confined, with an airflow boundary layer developing over the fluid film. And last and more important, anti-icing fluids have non-Newtonian properties that influence their fluid behaviour.

## 3 EXPERIMENTAL SET-UP:

This study was performed in the closed loop, thermally insulated facility VKI CWT1 (Cold Wind Tunnel one fig 1). It's cooled by the injection of liquid nitrogen, or by the use of a refrigerator unit. Tests' temperatures of -35°C can be obtained with the first method and

-15°C with the second. The tunnel motor is computer controlled with a feed back system. In the aerodynamic acceptance tests [6] the fluid is deposited as a thin layer on the test section bottom wall and is subjected to an accelerating airflow, that simulates a commercial aircraft take-off.

The test section bottom and top wall are made out of double glass to allow optical access into the test section. Three pressure taps connected to two pressure transducers give information about the airflow velocity and boundary layer displacement thickness (see [6] for more details). Four thermocouples measure the temperature of the air, fluid and bottom wall.

The fluid is pre-cooled to the test temperature and deposited on the test section bottom glass wall. With a specially made scraper, a fluid film of constant thickness is obtained on the whole width and length of the test section. The tunnel runs at idle speed (below 5 m/s) before each test, for a sufficient time to establish a uniform temperature both in the air and in the fluid.

The fluid film thickness measurements were conducted using a non-intrusive measurement technique based on the light absorption. Using the well-known relation:

$$I = I_0 e^{-\lambda h}$$

where h is the film thickness $I_0$ the incident light, I the transmitted light and ● the light absorption coefficient we can obtain the following expression:

$$\log\left(\frac{I_0}{I}\right) = \lambda h$$

This gives the thickness of the film based on measurements of the incident and transmitted light, if the light absorption coefficient for the particular fluid is known. This is obtained using a wedge, made out of two glass plates with a certain angle between them as a calibration device. The fluid thickness between the plates is the same as the distance between them and can be measured by mechanical means. From these thickness measurements and knowing the incident and transmitted light the value of the light absorption coefficient can be calculated. The incident light is obtained reflecting and diffusing light from below the test section. The images are acquired with a CCD camera positioned above the test section. The images were stored directly on a PC (direct digitalisation from the camera) or on a video tape. The first case gives better noise-free images while the second enables a more complete view of the test, without having to take into account the image acquisition frequency of the first method. Test conditions were the same for all fluids. Due to the heating from the illuminating lamps, the lowest possible test temperature was of -20⊕ Celsius. Tests temperatures of -10⊕ and 0⊕ Celsius were also used. Tests were conducted with the whole test section width inside the field of view of the CCD camera. Some of these tests gave the overall fluid film flow during a normal transport aircraft take-off (constant acceleration for the first 25 seconds, and then constant speed of 65 m/s for 35 seconds). Further tests were conducted with the camera position closer to the test section for better spatial resolution. These tests were also conducted with different accelerations and speeds.

## 4 TEST RESULTS

Rheological measurements (surface tension, viscosity and density) were conducted on each fluid at each test temperature used. Non-Newtonian anti-icing fluids have a power law behaviour (fig 2) with a variation of viscosity of one order of magnitude, with the shear stress normally encountered in the tests. The behaviour of a Newtonian fluid with a constant viscosity with varying shear stress is also plotted.

Several de-icing and anti-icing fluids were studied. De-icing fluids have similar characteristics. They show an increase of viscosity with a decrease of temperature, and a surface instability pattern that was already described in several papers [8], [12]. The first instability to appear on a de-icing fluid film is a 2D sinusoidal wave pattern (figs 3,4). It covers the whole width of the test section and has a clear spatial frequency as it can be seen on a spatial frequency spectrum (fig. 5a). This leads to a roughening of the fluid film surface as can be seen in (fig 3b). The amplitude of these waves is quite small compared with the fluid film mean thickness.
Of extreme importance is the spatial frequency of the film surface waves. Fig. 5b shows the variation of this parameter with the film thickness. This behaviour is also quite clear from figures 3a and 4a.
This surface instability only appears for a very short air velocity interval and with a small airflow velocity increase this 2D instability breaks down to a 3D pattern (fig 6). This breakdown was identified in [15] as a herringbone pattern and is due to a sub harmonic instability.
With a further velocity increase solitary waves appear, propagating very rapidly and transporting a considerable amount of fluid. Contrarily to the sinusoidal wave these solitary waves have a much higher amplitude and propagate on top of the fluid film (fig 7). With increasing velocity and decreasing thickness these solitary waves form ripple like structures that depending on the fluid film thickness can propagate with very low velocities (fig 8). These waves were already observed in [14]. If the thickness of the residual film is very small the film breaks up into several streaks that ondulate in the streamwise direction.
Another important observation is that the residual thickness does not depend on the initial fluid film thickness. As it can be seen in fig. 9 after a certain time (normally even before the rotation manoeuvre is started) the fluid film has the same average thickness that is independent of the initial thickness.

Anti-icing fluids have a much different behaviour. First the viscosity variation depends on the fluid itself, with some fluids having their lowest viscosity at negative temperatures.
A different pattern is encountered for the non-Newtonian fluid. The initial 2D pattern is not present at all. In some anti-icing fluids a pattern resembling this 2D instability is present but not on the whole extension of the test section, as encountered for the Newtonian case. This pattern is encountered in small individual areas of the fluid film and with a small increase of velocity solitary waves appear, as for the Newtonian case. Some differences are obvious between de-icing and anti-icing fluid films. The solitary wave (fig 10) that appears in anti-icing fluids starts at a point and propagates laterally as well as

forward. As it propagates it takes some fluid from the film, at the edges of the wave (fig. 11) . It leaves therefore a triangular wake. The angle of this triangular wake does not vary with time (fig 10a-10d). These waves seem to generate at small irregularities in the initial fluid film surface.

The overall fluid film movement is similar for both kinds of fluids. As the airflow velocity steadily increases the leading edge of the fluid film is pushed downstream creating a higher elevation front. This front is steadily pushed downstream and the fluid film ahead of this front seems to retain the same initial film thickness minus the amount of fluid carried away by the solitary waves. As this "front" leaves the test section it carries with it large amounts of fluid and therefore a sharp decrease of the average fluid film thickness is measured.

## 5 CONCLUSIONS

Experiments were conducted in real scale simulation of an aircraft take-off on several de/anti-icing fluids. These experiments showed the different behaviour of wind-induced surface instabilities for a de-icing (Newtonian) fluids and an anti-icing (non-Newtonian) fluid. Important aspects of the surface instability pattern were studied. This could lead to future determination of the effect of the fluid film on an aircraft airfoil performance using the surface waves height and spatial frequency. The different wave patterns were described and explained, including the case of a solitary wave with triangular shaped wake. This seems to be a newly found wave pattern, although the wave front shows some similarity with the one encountered in the Newtonian case.

## 6 ACKNOWLEDGEMENTS

The first author would like to thank Junta Nacional de Investigação Ciêntifica e Tecnologica of Portugal for the support with fellowships Ciência/BD/1663/91-RM and Praxis XXI/BD/2649/94

## 7 REFERENCES

[1] Brumby R. E., 'The effect of ice wing contamination on essential flight characteristics', AGARD CP 496 France (1991)

[2] AGARD Conference Proceedings CP-496, "Effects of Adverse Weather Conditions on Aerodynamics", (1991)

[3] Nark T., 'Wind Tunnel Investigation of the Aerodynamic Effects of Type II Anti-Icing fluids when applied to Airfoils', Boeing Commercial Aircraft Co., Document D6-37730, (1983)

[4] Carbonaro M.,Cunha F. 'Aerodynamic Effects of De/Anti-icing fluids and Criteria for their Aerodynamic Acceptance', Second SAE ground Deicing Conference, SAE USA, (1993)

[5] Carbonaro M., Cunha F. 'Aerodynamic Effects of De/Anti-icing fluids and Description of a Facility and Test Technique for their Assessment', Aircraft Flight Safety, AGARD France, (1993)

[6] Carbonaro M. Proposed specification for aerodynamic acceptance tests of aircraft de/anti-icing fluids, *AIA TC Project 218-4 / AECMA Meeting*, Amsterdam, (1990).

[7] Craik A. D. D., Wind Generated waves in thin films, *J. Fluid Mech.*, **26**, 369-392, (1965)

[8] Andritsos N., Hanratty T. J., Interfacial instabilities for horizontal gas-liquid flows in pipelines, *Int. J. Multiphase Flow*, **13**, 583-603 (1987)

[9] Hanratty T. J., 'Interfacial instabilities caused by air flow over a thin liquid layer', *Waves on fluid interfaces*, Academic Press, New York, (1983) pp 221-259.

[10] Peng, C.-A., Jurman, L. A., McCready M. J., Formation of Solitary Waves on Gas-Sheared Liquid Layers, *Int. J. Multiphase Flow*, **17**, 767-782 (1991)

[11] Jurman, L. A., McCready M. J., Study of Waves in Thin Liquid Films Sheared by Turbulent Gas Flows, *Phys. Fluids A*, **1**, 522-536 (1988)

[12] Jurman, L. A., Bruno, K., McCready M. J., Periodic and Solitary Waves on Thin Horizontal, Gas-Sheared Liquid Films, *Int. J. Multiphase Flow*, **15**, 371-384 (1989)

[13] Spedding, P. L. and Spence, D. R., Flow Regimes in Two-Phase Gas-Liquid Flow, *Int. J. Multiphase Flow*, **13**, 245-280 (1993)

[14] Asali, J. C. and Hanratty, T. J., Ripples Generated n a liquid Film at High gas Velocities, *Int. J. Multiphase Flow*, **13**, 229-243 (1993)

[15] Jun Liu, Schneider J. B. and Gollub, J. P., Three dimensional instabilities of film flows, *Phys. Fluids*, **7**, 55-67 (1995)

[16] Jun Liu and Gollub, J. P., Solitary Wave Dynamics of Film Flows, *Phys. Fluids*, **6**, 1702-1712 (1994)

[17] Jun Liu, Jonathan D. P. and Gollub, J. P., Measurements of the primary instabilities of film flows, *J. Fluid Mech.*, **250**, 69-101 (1993)

[18] Carbonaro M., Experimental study of the flow of a film of aircraft anti-icing fluid during a simulated take-off at subfreezing temperatures, *VKI Contract Report 1986-22* (1986)

[19] Barnes H.A., Hutton J.F. and Walters K., *An introduction to rheology*, Rheology Series 3,

[20] Harri, J., *Rheology and non-Newtonian flow*, London-New York, Longman (1977)

[21] McSpadden, C. Visualisation and depth measurement of a fluid flow over a. flat surface, *VKI Stagiaire Report 1984-17* (1984)

[22] Carbonaro M., Locatelli A., Mantegazza C., et. all.: Experimental study of the flow of a film of aircraft de-icing fluid during simulated take off at subfreezing temperature, *VKI Contract Report 1985-02* (1985)

[23] Benschop T. Visualisation and measurement of the thickness of a de-icing fluid on a flat surface, *VKI Stagiaire Report 1986-01* (1986)

[24] Ditchburn R. W., *Light*, Volume II, Academic press, (1976)

[25] Yih, C.-S., Wave formation on a liquid layer for de-icing airplane wings, *J. Fluid Mech*, **212**, 41-53 (1990)

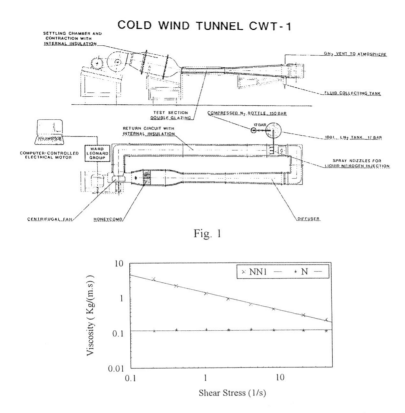

# COLD WIND TUNNEL CWT-1

Fig. 1

Fig. 2 Typical Viscosity vs. Shear Stress variation of a De-Icing fluid and of an Anti-icing fluid.

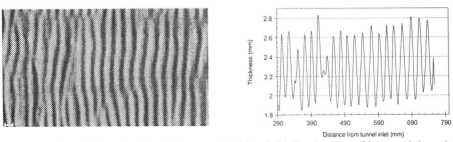

Fig 3 2D sinusoidal surface instability on a De-Icing fluid film (average thickness 2.3 mm). The graph on the left corresponds to a horizontal cut (streamwise direction) on the picture. The small white bar in the image corresponds to 20mm. Flow from left to right. Image grey scale dynamically enhanced.

196

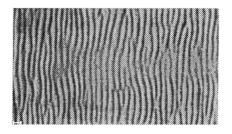 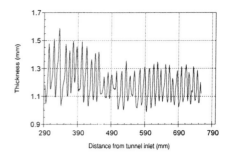

Fig 4 2D sinusoidal surface instability on a De-Icing fluid film (average thickness 1.2 mm) The graph on the left corresponds to a horizontal cut (streamwise direction) on the picture. The small white bar in the image corresponds to 20mm. Flow from left to right. Image grey scale dynamically enhanced.

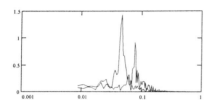 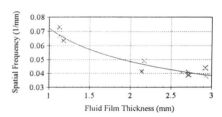

Fig 5 (a) Spatial frequency domain from fig 3 & 4 (arbitrary unit vs. 1/mm) . (b) Variation of the spatial frequency with fluid film thickness for a de-icing fluid.

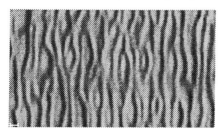 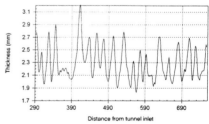

Fig. 6 3D breakdown surface instability on a De-Icing fluid film (average thickness 2.3 mm). The graph on the left corresponds to a horizontal cut (streamwise direction). The small white bar in the image corresponds to 20mm.Flow from left to right. Image grey scale dynamically enhanced.

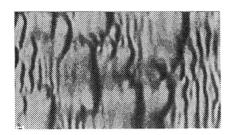 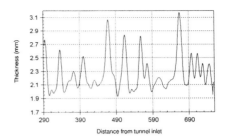

Fig 7 Solitary waves surface instability on a De-Icing fluid film (average thickness 2.3 mm). The graph on the left corresponds to a horizontal cut (streamwise direction). The small white bar in the image corresponds to 20mm. Flow from left to right. Image grey scale dynamically enhanced.

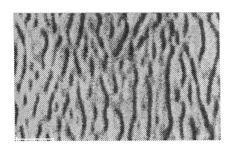 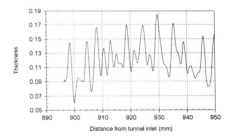

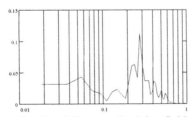

Fig 8 Final residual layer surface instability on a De-Icing fluid film. The graph on the left corresponds to a horizontal cut (streamwise direction). The small white bar in the image corresponds to 20mm. Flow from left to right. Image grey scale dynamically enhanced. Lower graph corresponds to the frequency domain of the curve showed (arbitrary unit vs. 1/mm)

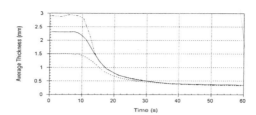

Fig 11 Average fluid film thickness variation with time for a de-icing fluid. Anti-icing film show a similar behaviour

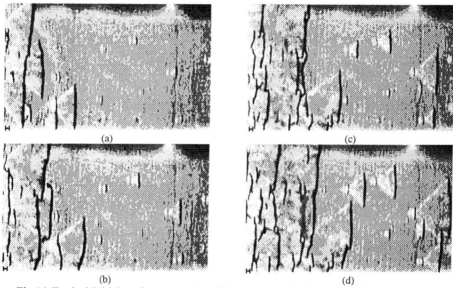

(a)

(c)

(b)

(d)

Fig 10 Typical Initial surface wave instability on an Anti-icing fluid. Flow from left to right. Image grey scale dynamically enhanced. The small back bar corresponds to 20mm.

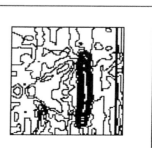

Fig 11 Zoom on a anti-icing solitary wave and it's wake. Iso-Thickness plot.

# FEED CONDITION EFFECTS IN FORWARD AND REVERSE ROLL COATING

G. E. Innes, J. L. Summers and H. M. Thompson

*Department of Mechanical Engineering,*
*University of Leeds, Leeds, UK LS2 9JT.*

## SUMMARY

The effects of inlet flow rate on flow structures in forward and reverse roll coating are investigated both experimentally, using a two-roll apparatus, and numerically via the finite element method. The two techniques of dye injection and particle imaging are used to visualise the evolution of fluid transfer jets and eddies as the flow rate is increased. A range of roll coating regimes from meniscus (ultra-starved) to classical (flooded) is covered. The sequence of flow transitions observed experimentally is found to be consistent with those predicted by the finite element method.

## 1. INTRODUCTION

The discovery of complex flow structures in apparently simple coating flows has prompted a number of investigations into the conditions necessary for their existence. Notable experimental studies of flow structures in coating are, for example, those of Schweizer[1] and Malone[2] for slide and roll coating respectively while numerous numerical investigations, based on the finite element method, have been reported covering a wide range of coating processes – see Kistler and Scriven[3], Christodoulou and Scriven[4] and Gaskell, Savage, Summers and Thompson[5]. This paper is concerned with flow transformations in a two-roll coater, operating in either forward (both rolls moving in the same direction through the nip) or reverse (opposite) mode. This extends the previous work of Pitts and Greiller[6] and Schneider[7] for the forward case and that of Malone for both cases. A key feature of the results presented here is the existence of eddies and transfer jets in roll coating, particularly under "ultra-starved" conditions.

The flow structures in forward and reverse roll coating undergo a series of transitions according to changes in several non-dimensional parameters[5]. The key parameters are found to be the inlet flow rate $\lambda$ and the roll speed ratio $S$, which are varied to produce maps showing where the different flow structures exist. The main purpose of the present experimental work is to verify the sequence of changes predicted by the finite element method as the inlet flow rate is increased.

### 1.1. Experimental Apparatus

The apparatus shown in Figure 1 consists of a pair of stainless steel rolls 50 mm in diameter, arranged one above the other. The rolls are cantilever mounted to a frame and driven by electric motors via 30:1 gearboxes. The maximum surface speed used in this study is 50 mm/s, equivalent to a capillary number of $Ca = 0.015$. The upper roll is mounted on a precision slide with micrometer to allow the gap between the rolls to be changed. The lower roll dips into a reservoir of light mineral oil and supplies inlet fluid to the coating bead held in the nip between the rolls. This fluid splits, some of it continuing with the lower roll as it leaves the bead and the rest transferring to the upper roll. The upper roll is wiped by a polyethylene wiper to keep it nearly dry, thus imitating the industrial process of coating a dry substrate.

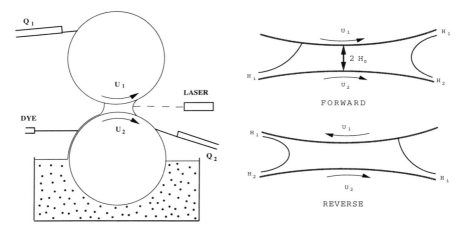

Figure 1: Two-roll apparatus showing wipers, dye injection and laser light source.

Figure 2: Forward and reverse modes of roll coating.

A video camera and monocular microscope are used to observe and record the flow patterns within the coating bead. A zoom lens allows a seven fold range of magnification factor and the microscope may be translated to view different sections of the bead. A grid superimposed on the recorded image is calibrated to indicate either 1 mm or 2 mm spacings. A perspex plate is mounted against the roll faces to prevent side flow and to allow an undistorted view of the flow inside the coating bead. Dyed fluid is injected upstream of the nip to highlight the presence of fluid transfer jets. A diode laser positioned downstream provides optical sectioning of the flow field and the addition of fine, neutrally buoyant particles enables the eddies to be seen.

Post-processing of the video recordings is done on a Silicon Graphics Indigo 2 workstation capable of frame grabbing and image editing. The Figures shown below were converted to black and white "negative" form and the contrast improved to show the dye traces more clearly. The neutrally buoyant particles appear dark in this form, as do the dye traces. Stray reflections from the upstream and downstream menisci have also been removed for the sake of clarity.

## 1.2. Nondimensional Variables

Figure 2 shows the two modes of roll coating and the definitions of the film thicknesses $H_i$, $H_1$, $H_2$ and roll speeds $U_1$, $U_2$. The forward mode differs from the reverse mode in that the two roll surfaces move in the same direction. The upper film $H_1$ is drawn *upstream* in the reverse mode. The physical parameters measured in these experiments may be grouped into the following nondimensional forms:

$$Ca = \frac{U\mu}{\sigma} \tag{1}$$

$$Ca^* = Ca \left(\frac{R}{H_o}\right)^{1/2} \tag{2}$$

$$S = \frac{U_1}{U_2} \tag{3}$$

$$\lambda = \frac{H_i}{2H_o} \tag{4}$$

where $Ca$ is the capillary number, $Ca^*$ modified capillary number, $R/H_o$ geometry ratio, $S$ speed ratio, $\lambda$ flow rate, $U_1$ and $U_2$ upper and lower roll speeds, $H_1$ and $H_2$ upper and lower film thicknesses, $H_i$ inlet film thickness, $H_o$ minimum semi-gap, $\mu$ dynamic viscosity and $\sigma$ surface tension.

The modified capillary number (2) represents the ratio of hydrodynamic to surface tension forces. For the meniscus regime, it is usually less than about 0.15[5] and the feed condition is ultra-starved, i.e. $\lambda \ll 1$. The dominant force is that of surface tension and the flow structures contain one or more eddies. For typical geometry ratios of at least 100, $Ca$ is an order of magnitude smaller than the modified capillary number. For the classical regime, the hydrodynamic forces become important, the feed condition becomes flooded and a very different set of flow structures exists within the coating bead.

It is noted that the nondimensional flow rate $\lambda$ may be changed by either changing $H_i$ (and hence the lower roll speed and $Ca$) or by adjusting the gap $2H_o$. The latter method also changes the geometry ratio, but has the advantage that the capillary number can be held constant. Both methods are used to produce changes in flow structure and hence to explore $\lambda - S$ parameter space.

## 2. FORWARD ROLL COATING

The forward mode features counter-rotating rolls, which produce a double eddy flow structure within the coating bead. Both roll surfaces move left to right in Figures 3 to 11 below, so that the upper eddy rotates clockwise and the lower eddy counter-clockwise. A fluid transfer jet (primary jet) follows an S-shaped path around the eddies before leaving the bead with the upper roll. The sequence of changes to this structure is first demonstrated by the numerical predictions and then verified by images obtained experimentally.

### 2.1. Numerical Predictions

The flow in Figure 3 is "ultra-starved" and in the meniscus coating regime[5] with closed upper and lower eddies, each containing two sub-eddies. The upstream sub-eddies are smaller than those downstream. Fluid is brought in via the lower left corner and travels the full length of the bead. At the lower right corner, some of the fluid leaves the bead with the lower roll while the rest turns to flow up the face of the downstream meniscus. This fluid transfer jet then turns and travels in the *upstream* direction between the counter-rotating eddies. At the upstream meniscus, the jet turns again and travels the length of the bead a third and final time, leaving with the upper roll (upper right corner).

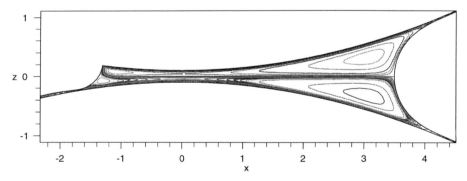

Figure 3: Streamlines in forward meniscus roll coating, $\lambda = 0.17$, $Ca_2 = 1.9 \times 10^{-3}$, $R/H_o = 100$, $S = 1$. The axes are scaled in units of the minimum semi-gap $H_o$.

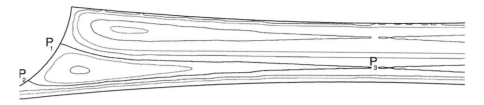

Figure 4: Enlarged view of the upstream flow pattern at the first critical flow rate, $\lambda = 0.1111$, $Ca_2 = 1.9 \times 10^{-3}$, $R/H_o = 100$, $S = 1$. The point $P_3$ is at the nip and joins the separating streamlines of both the upstream and downstream lower subbeddies.

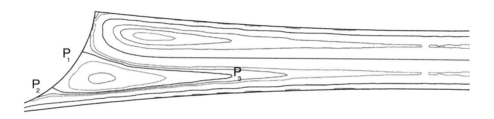

Figure 5: Enlarged view of the upstream flow pattern after the first critical flow rate, $\lambda = 0.13$, $Ca_2 = 1.9 \times 10^{-3}$, $R/H_o = 100$, $S = 1$. The point $P_3$ is no longer at the nip and the secondary jet now passes between the two lower subbeddies. The primary jet continues to exist, passing between the two saddle points (stagnation points) located at the nip.

As the inlet flow rate is increased the primary jet becomes stronger, causing the eddy structures to be squeezed. At a critical flow rate the streamline pattern changes, Figures 4 and 5, whereby the two lower subbeddies become disconnected from each other. This allows fluid to pass directly between them, thus forming a secondary transfer jet to the upper roll.

As the flow rate is increased further, the strength of the primary jet decreases and the flow through the secondary jet increases. At a second critical flow rate the primary jet switches off, Figure 6, and all of the fluid transferred to the upper roll does so via the secondary jet. This event corresponds to a change in the

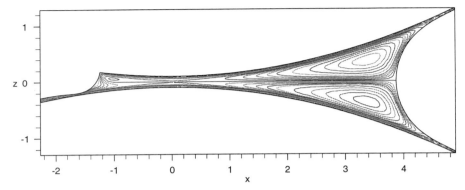

Figure 6: The streamlines at the second critical flow rate, $\lambda = 0.333$, $Ca_2 = 3.2 \times 10^{-3}$, $R/H_o = 100$, $S = 1$. The two saddle points at the nip have been forced together, switching off the primary transfer jet.

streamline pattern whereby the two upper sub-eddies are decoupled. As the flow rate increases beyond the second critical value a region of quasi-unidirectional flow develops in the nip region, Figure 7, and the assumptions of classical roll coating theory hold. Note that although the two critical events always occur in the same order, regardless of speed ratio, the specific flow rate values are affected by $S$, see Figure 8.

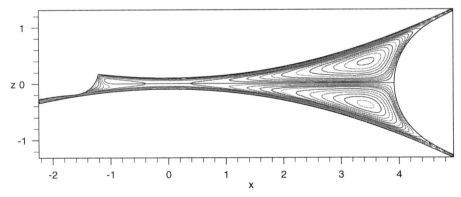

Figure 7: The streamlines after the second critical flow rate, $\lambda = 0.35$, $Ca_2 = 3.6 \times 10^{-3}$, $R/H_o = 100$, $S = 1$. The upstream and downstream eddies are separated by a region of quasi-unidirectional flow at the nip.

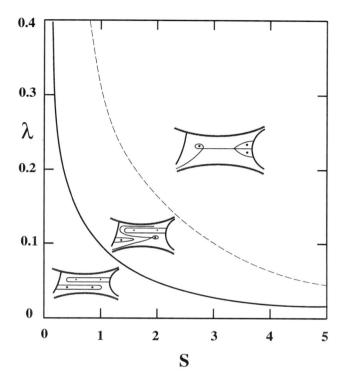

Figure 8: $\lambda - S$ diagram summarising the flow features found in forward roll coating. The solid line represents the first critical event[5] and the dashed line approximately represents the second.

## 2.2. Experimental Observations

Figure 9 shows the coating bead for ultra-starved conditions. The primary jet is highlighted by a dye trace, showing an S-shaped pattern as it travels around the two main eddies. The grid spacing is 1 mm, indicating the typical length scale of these flows. The left side of the image is always fainter because the laser light is blocked by the rolls and is absorbed or reflected as it passes through the bead.

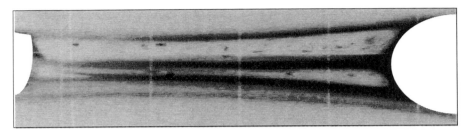

Figure 9: Forward meniscus roll coating showing the primary transfer jet following an S-shaped path around the two main eddies, $S = 1$.

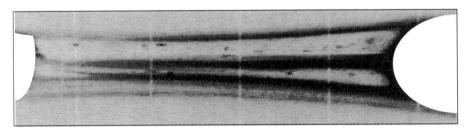

Figure 10: Forward coating showing both the primary and secondary transfer jets, $S = 1$.

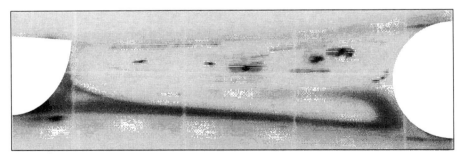

Figure 11: Forward coating showing a stronger secondary jet and internal dye recirculation in the downstream lower eddy, $S = 1.2$.

As the roll speeds are increased, Figure 10, the flow rate increases and the eddies become squeezed. The secondary jet appears at the lower left subeddy, but is much weaker than the primary jet. As the flow rate is increased further, the secondary jet increases in strength and the primary jet disappears, Figure 11. A dye trace still appears within the lower right subeddy due to diffusion

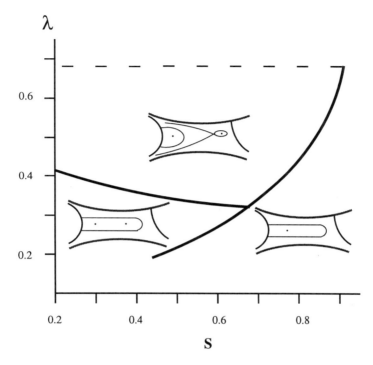

Figure 12: $\lambda - S$ diagram showing the flow features of reverse roll coating.

across the lower boundary and prevents accurate determination of the second critical flow rate.

At higher roll speeds, the bead is subject to an instability known as "bead break", hence no further observations are possible for this fixed gap value. However, the same pattern emerges for smaller gaps although with poorer resolution due to diffusion of the dye.

## 3. REVERSE ROLL COATING

In reverse coating the rolls move in opposite directions through the nip which leads to the existence of only one main eddy within the coating bead. In this section experimental observations are compared with the numerical results reported by Richardson et al.[8] and in which there are two predicted sequences of flow transition depending on the speed ratio, and three categories of flow structure, as summarised in Figure 12.

For ultra-starved conditions there is only a single eddy with no subeddies for

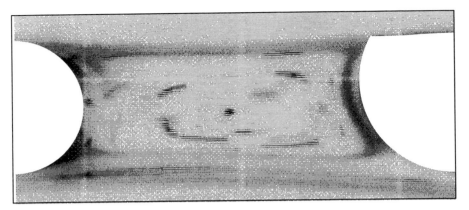

Figure 13: Reverse meniscus coating with a single eddy, $S = -0.25$.

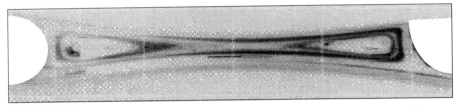

Figure 14: Reverse meniscus coating with a double eddy, $S = -0.25$.

all speed ratios. Fluid is transferred to the upper roll by the primary transfer jet, which travels directly from the lower roll via the face of the downstream meniscus. For small $-S$ values, as flow rate increases beyond a critical value the transfer jet squeezes the eddy, producing two sub-eddies. These remain completely enclosed within the main eddy as the flow rate increases to a second critical value. The streamline pattern then changes and the sub-eddies are disconnected, allowing a secondary transfer jet to pass between them.

For moderate $-S$ values, the only observable critical event is the appearance of a downstream sub-eddy as the flow rate is increased. The secondary jet automatically appears as part of this event.

## 3.1. Experimental Observations

An ultra-starved coating bead at low speed ratio is shown in Figure 13. The upper roller now moves from right to left, and the grid spacing is 2 mm. The presence of the primary fluid transfer jet is indicated by the dye trace. A single eddy is indicated by the particle tracks within the bead, which orbit around a single centre.

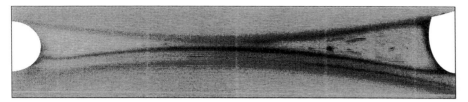

Figure 15: Reverse meniscus coating: secondary jet switched on, $S = -0.25$.

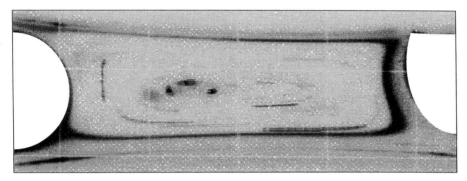

Figure 16: Single-eddy structure in reverse meniscus coating, $S = -0.75$, 1 mm grid.

As the gap between the rolls is reduced, Figure 14, a flow bifurcation occurs and the single centre gives rise to two sub-eddies having centres upstream and downstream of the nip. Dye traces captured within each sub-eddy orbit their respective centres. The primary jet still passes around the perimeter of the main eddy. For a slightly smaller gap, Figure 15, the opposing dye traces at the nip begin to touch, and a secondary jet appears.

For a higher speed ratio the ultra starved bead has a single eddy circumnavigated by the primary jet, Figure 16. As the gap height is decreased, the downstream sub-eddy appears spontaneously along with a secondary jet upstream of the nip, Figure 17. The dye trace indicates that the secondary jet is stronger than the primary jet for the conditions shown.

## 4. CONCLUSIONS

Experimental measurements of the flow transformations that occur in both forward and reverse roll coating when flux is varied are in qualitative agreement with theoretical predictions obtained using the finite element method. Both

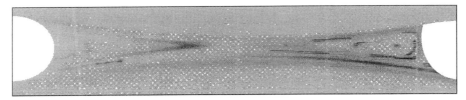

Figure 17: Double eddy structure, $S = -0.75$.

transformations affect the presence of eddies and transfer jets within the bead.

## 5. REFERENCES

1. Schweizer, P. M., Visualization of coating flows, *J. Fluid Mech.*, **193**, 285–302 (1988).
2. Malone, B., An experimental investigation into roll coating phenomena, *Ph.D. thesis*, Dept. Mech. Engg, Univ. of Leeds, Leeds (1992).
3. Kistler, S.F. and Scriven, L.E. 'Coating Flows', *Computational Analysis of Polymer Processing*, Ed. J.R.A. Pearson and S.M. Richardson, Appl. Sci. Publishers London and New York, 243–299 (1983).
4. Christodoulou, K.N., Scriven, L.E., The fluid mechanics of slide coating, *J. Fluid Mech.*, **208**, 321–354 (1989).
5. Gaskell, P.H., Savage, M.D., Summers, J.L. and Thompson, H.M., Modelling and analysis of meniscus roll coating, *J. Fluid Mech.*, **298**, 113–137 (1995).
6. Pitts, E. and Greiller, J., The flow of thin liquid films between rollers, *J. Fluid Mech.*, **11**, 33–50 (1961).
7. Schneider, G. B., Analysis of forces causing flow in roll coaters, *Trans. Soc. Rheology*, **6**, 209–221 (1962).
8. Richardson, C., Gaskell, P.H. and Savage, M.D., Reverse roll coating, *Proc. First Biennial European Coating Symposium*, Leeds (1995).

# SECTION 4

# Roll and

# Gravure Coating

# FILM THICKNESS AND INSTABILITIES WITH FORWARD GRAVURE COATING

H.Benkreira, R.Patel, M. Naheem
Department of Chemical Engineering
University of Bradford, UK BD7 1DP
and
J.M.Leclercq
Faculte Polytechnique de Mons, Belgium

## 1. INTRODUCTION.

Although gravure coating is used widely in thin film applications, particularly when thin coatings (< 10 microns) and high speeds (> 2m/s) are required, research into its fluid mechanics is extremely limited as reported by Benkreira and Patel [1,2]. Gravure coating involves the use of a knurled steel or ceramic roller with a pattern that is either chemically or mechanically engraved on it. The coating liquid is fed underneath a doctor blade which ideally should wipe clean the surplus film formed over the filled cells. The liquid trapped in the cells is then transferred on the substrate in a variety of ways as shown in Figure 1: directly with the

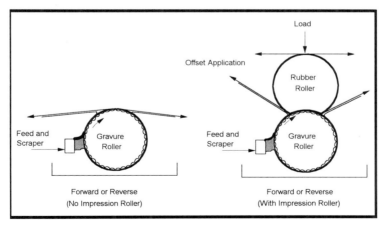

**Figure 1  Gravure coating flows**

substrate moving in the forward or reverse direction to the gravure roller with or without an impression (rubber) roller or indirectly ( offset application ) via a transfer impression roller. Figure 1 also forms a basis for identifying the key gravure coating flows : forward or reverse direct gravure with or without an impression roller. The offset mode is not a key coating flow per se; it is composed of a gravure coating flow and a transfer coating flow and measurement of the film on the substrate does not necessarily provide the data required as some of the fluid could be entrained by the impression roller back to the nip. Flow situations where a an excessive (in comparison with cell depth )surplus film is formed over the cells are strictly not gravure coating flows which are designed specifically with the aim of achieving the thinnest possible films as explained by Benkreira et al[3]. The blade loading over the gravure roller forms thus an important part of the study and this aspect was investigated by Benkreira and Patel [1,2]. The practical difficulty of complete doctoring of the surplus, without excessively loading the blade hence possibly damaging the cells particularly with standard chromed (rather than ceramic) rollers, can explain the use of the impression rubber roller which removes the surplus liquid at entry to the nip. It must be accepted however that there will always be a tiny surplus film over the cells. The implication of these observations is that gravure coating is best suited to low viscosity coatings as these are easily doctored.

Previous studies by Benkreira and Patel [1,2] concentrated on the simple direct reverse mode of operation and led to a remarkably simple rule for the film transfer from the cells to the substrate. It was observed on the basis of a comprehensive experimental programme that about a third of the volume held in the cells was transferred to the substrate regardless of cell geometry , speed, viscosity or surface tension variations. There are no other film transfer data on this coating operation reported in the literature. Pulkrabek and Munter[4] observed ( no data given ) that with an impression roller the transfer amounted consistently to about 59% of the volume of the roll grooves, for a knurl roll. This is nearly twice as much as the 1/3 rule measured by Benkreira and Patel [1,2] for operations without an impression roller.

The objective of the present investigation is to extend knowledge of this coating method by looking at applications when the gravure roller moves in the forward mode with the substrate with or without an impression rubber roller as shown in Figure 1. Film thickness variations with operating conditions and instability limits are the required parameters of this experimental investigation.

## 2. EXPERIMENTAL PROCEDURE

In this study three gravure rollers, 220 quadrangular, 85 trihelical and 60 pyramidal with cell volume per unit area of roller surface, $V_c$, values of 13.25, 28.50 and 43.37 x $10^{-6}$ $m^3/m^2$ respectively were used and are shown in Figure 2. These rollers were 0.1m in diameter and 0.182 m

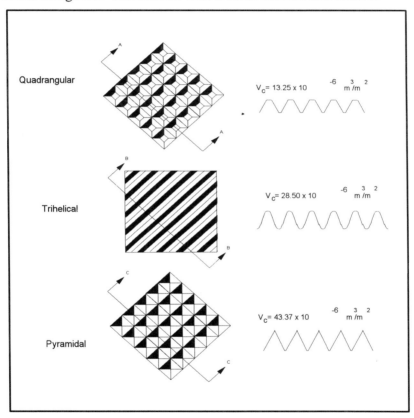

**Figure 2 Gravure cell geometries**

wide. An impression rubber roller of the same size was used; it had a 15 mm thick rubber sleeve of 65° shore EPD and was loaded by tightening the bearing blocks holding it. The coating liquid was fed via a perspex feed slot and underneath a flexible plastic blade which scraped the surplus liquid over the cells. The thickness of film formed on the substrate was measured using an infrared thickness gauge accurate to +/-0.5 micron.

More details on the experimental technique can be found in Patel's[5] work. In the present study seven water based Newtonian solutions with viscosities in the range 1 to 14 mPa and surface tensions in the range 0.030 to 0.067 N/m were used. The viscous properties and the surface tension data of the liquids were obtained using the Brabender Rheotron viscometer and the drop-weight method respectively.

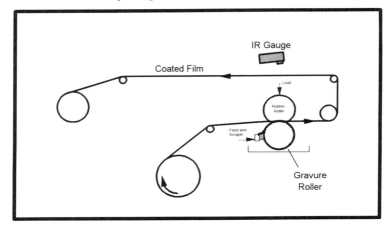

**Figure 3  Pilot coating rig**

## 3. EXPERIMENTAL RESULTS AND DISCUSSION

### 3.1 Direct forward gravure without impression roller

The first set of experiments was designed to establish the stability window of the unloaded (no impression roller) forward gravure coating at different wrap angles. Experiments were thus carried out at constant speed ratio, $u_g/u_s$ from 0.5 to 1.5 in steps of 0.25 for a substrate speed varying from 10 to 40 m/min in steps of 10m/min. With five liquids and three gravure geometries , these produced 300 data points revealing the types of instability produced. Essentially at speed ratios less than 1, lines of fine bubbles developed in the machine direction; at speed ratios of unity stripes in the transverse direction appeared and stable conditions were only possible when the speed ratio exceeded 1, i.e. when the substrate was slower than the gravure roller. These experiments demonstrated the crucial role played by the speed ratio which had to be greater than 1 for any chance of stable flow. Further experiments were carried out to draw precisely the envelope of the stability window. Substrate speeds of 5, 7.5, 10 and 20m/min were used and the gravure roller speeds was increased from 10 to 100m/min in steps of 10m/min. Again with five liquids and three gravure geometries, these experiments produced 420 data points to draw the limit of stable operation for a given wrap angle. The results are remarkably simple as shown in Figure 4 which is typical of all the

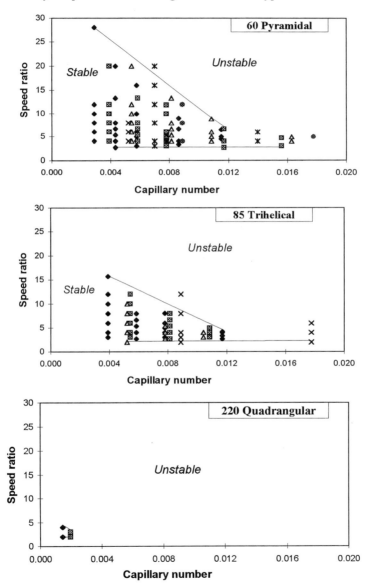

**Figure 4 Stability window with unloaded forward gravure**

data. They confirm the observations made with smooth forward roll coating by Benkreira et al[6] that increasing the capillary number (here defined with respect to substrate speed) or decreasing gap destabilise flow. Here the equivalent cell depth, the cell volume per roller area, has a similar effect to that of a gap.

An additional variable as mentioned earlier was the wrap angle of the substrate around the gravure roller. The above experiments were repeated for 4 wrap angles, 0, 9.0, 20.0 and 29.0 degrees. In these experiments the substrate was wrapped symmetrically about the vertical axis of the roller. The data showed that as the wrap angle was increased the stability window widened. This observation suggests similarity with smooth forward roll coating ( and indeed all other coating flows ) in that the curvature of the separation meniscus has an effect on the onset of instabilities. At large wrap angle a tight separation meniscus was formed and this is less prone to instabilities.

We now turn our attention to the variation of film thickness with operating conditions in the stable regime. This eliminates the consideration of the quadrangular cell structure whose $V_c$ was low ( $= 13 \times 10^{-6}$ m$^3$/m$^2$) and restricted the substrate speed to about 15m/min. At that speed , we observed a film transfer of about 20% of the equivalent cell depth, at most, which remained constant for speed ratios in the range 3 to 6. Compared to the reverse situation studied by Benkreira and Patel[1,2] where a transfer of 33% was observed, forward gravure coating produces thinner films in the stable regime and requires that the substrate travels relatively slowly than the gravure roller.

In conclusion it can be said that the unloaded case offers little practical use with film thickness though very thin (20% at most of equivalent cell depth) stable only at low substrate speeds (< 20m/min). Wrapping the substrate round the roller incresases this speed ( but only marginally) at which instability was observed .

## 3.2 Direct forward gravure with impression roller

The second set of experiments considered an additional variable, that of loading the substrate onto the gravure roller using an impression rubber sleeved roller. The aim is to extend the range of substrate speed which in the situation above was too low. Experiments similar to those described above were conducted with the additional variable of loading in the range up to 1000 N tried in five increments. For simplicity and practicality of operation, the experiments were carried out at speed ratio of unity. The merits of loading are clear from the data summarised in Figure 5. Higher loads enable stable operation at higher substrate speeds and capillary

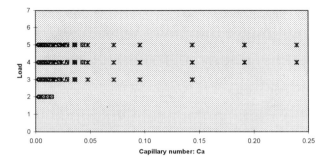

**Figure 5  Widening of stability window with increasing load**

number.  With all geometries, an optimal load appears and extends the stable speed of operation up to 70m/min with the lowest viscosity fluid. This optimal load is essentially constant for the three geometries studied and stands at about 900N.  Above this value, no further widening of the stability limits is observed.  At high viscosities, the gain in pushing the stability limits is only marginal with stable operating speed of 10m/min at most.  Also, the destabilising effect of low equivalent cell depths persists as shown in the data with the 200 quadrangular geometries which under load can operate stable only up to 20m/min with the lowest viscosity fluid. This is however a substantial improvement on the always unstable mode of the unloaded case. Moreover, it demonstrates conclusively the benefit of loading on creating and widening stable operation.  A corollary to these observations is that loading creates favourable stable conditions by virtue of creating an optimal stable flow geometry (a coating bead) without affecting cell depth.

Now as for the film thickness transferred to the moving substrate, the data, typified by Figure 6, reveal remarkable simple results: for a given cell

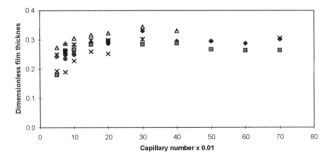

**Figure 6  Typical film transfer data showing the 1/3 rule**

geometry, the film thickness develops quickly into becoming constant at speed above 20m/min independently of capillary number values and slightly less than the 1/3 of the equivalent cell depth is transferred to form a continuous film.    Increasing load up to the optimal (stability) value decreases this transfer only marginally.

## 4. CONCLUSION

This investigation  demonstrates that the stable window of operation can be widened by loading the gravure roller with an impression roller and speeds greater than 70m/min with low viscosity fluids were achieved. Also, the rule of thumb for film formation observed    by Benkreira and Patel[1,2]  for reverse gravure without an impression roller holds to a first approximation for loaded forward gravure:   the film thickness is constant, independent of speeds and physical properties and is about  1/3 of the equivalent cell depth.

## REFERENCES.

1.   Benkreira,  H.  and  Patel,  R     Direct  Gravure  Roll  Coating, *Chem.Engng..Sci., 48, 2329-2335 (1993).*
2. Patel, R. and Benkreira, H.   Gravure roll coating of Newtonian fluids, *Chem.Engng.Sci., 46, 751-756 (1991).*
3.   Benkreira,   H.,   Patel,   R.,   Edwards,   M.F.   and   Wilkinson,W.L. Classification and analyses of coating flows, *Journal of Non-Newtonian Fluid Mechanics, 54, 437-447 (1994).*
4.   Pulkrabek,   W.W.   and   Munter,   J.D.     Knurl   roll   design   for   stable rotogravure coating , *Chem.Eng.Sci.,38, 1309-1314 (1983).*
5  Patel, R. Fluid Mechanics of Direct Gravure Roll Coating, *Phd Thesis, University of Bradford (UK), February 1989.*
6. Benkreira, H.et al.  Ribbing instabilties in the roll coating of Newtonian fluids, *Plastics & Rubber Process & Appl., 2, 137-144 (1982).*

# Deformable Roll Coating:
# Modeling of Steady Flow in Gaps and Nips

M. S. Carvalho and L. E. Scriven

Coating Process Fundamentals Program
Center for Interfacial Engineering and
Department of Chemical Engineering & Material Science
University of Minnesota, Minneapolis, MN 55455

*Abstract*

Of the two rolls that make a forward-roll coating gap or nip, one is often covered by a layer of more-or-less deformable elastomer. Liquid carried into the converging side of the gap (or nip) can develop high enough pressure to deform the resilient cover, which changes the gap geometry and thus alters the velocity and pressure fields and the position of the film split meniscus. This elastohydrodynamic coupled action is not yet well understood. By *gap* here is meant the closest approach, or clearance, between two rolls that would not be in contact or interference were liquid not present; by *nip* is meant the closest approach between two rolls that would be in interference, i.e. a *negative gap*.

In earlier work (Carvalho & Scriven 1994) we modeled this flow with the lubrication approximation, a viscocapillary model based on the Landau-Levich equation, and a set of one-dimensional springs to describe the roll-cover deformation. However, this approach does not give the details of the flow near the meniscus nor describe how the film thickness on each roll varies with the speed ratio and roll-cover properties.

In this work, deformable-roll coating flows were examined by solving the complete Navier-Stokes system with a representation of the resilient roll boundary by elastic elements that deform linearly with the traction force imposed by the liquid flow.

The equation system was solved by the Galerkin/finite element method; the resulting set of non-linear algebraic equations was solved by Newton's method with initialization by pseudo-arc-length continuation as parameters were varied.

Results show how roll deformation affects total flow rate, film split, and forces on the rolls. Some of the results can be compared with measurements and observations of deformable roll coating flows made in our laboratory.

## 1. Introduction

The flow in a deformable-roll coating gap was analyzed by Carvalho & Scriven (1994) by means of the lubrication approximation, a viscocapillary model of the film split, and radially oriented spring elements to approximate the roll deformation. The lubrication approximation describes well the flow between rotating rolls except near the upstream and downstream menisci. The viscocapillary model that uses the Landau-Levich equation of developing film flow as the downstream boundary condition on the lubrication flow region extends to lower capillary numbers the range of parameters in which the combined model yields predictions of both steady flow and flow stability close to those of the Navier-Stokes system. Nevertheless, only with the more accurate Navier-Stokes system can the downstream free surfaces, the presence of recirculations, and the effect of roll deformation on the film split region be analyzed.

Some coating operations employ a succession of film-splits, for example, the coating of ultra-thin silicone release material on paper and polymer films. To predict the thickness of the layer finally coated on the substrate, it is important to know how much liquid is carried away by each roll at each film split (cf. Benjamin et al. 1995). The ratio of the layer, or film, thickness on the two rolls is often found to fit a power-law relation with the ratio of roll speeds, which is a convenient correlation:

$$\frac{h_A}{h_B} = \alpha \left(\frac{V_A}{V_B}\right)^\beta$$

The film-split fitting parameters , the coefficient $\alpha$ and exponent $\beta$, depend on operating conditions, primarily average roll speed, liquid viscosity, rheology and surface tension, distance between the rolls, roll radii, and roll hardness.

Film-splitting flows of Newtonian liquids between rigid rolls have been extensively studied in the past by both theory (Savage 1982, Coyle et al. 1986, Savage et al. 1992, Benjamin et al. 1994) and experiments (Benkreira et al. 1982, Savage 1982, Coyle et al. 1986). The first theoretical analyses were based on the lubrication approximation. With this approach, the relationship between film thickness ratio and speed ratio depends strongly on the boundary condition used at the film split. Savage (1982) and later Coyle et al.(1986) postulated that the liquid splits at the first point (line in three dimensions) of the flow where $u = \partial u / \partial y = 0$. They found a square root relationship between the film thickness ratio and speed ratio, i.e. $\beta = 0.5$. Savage et al. (1992) improved this analysis by assuming that the flow separates at a stagnation point, i.e. where $u = v = 0$. With this assumption, they obtained an algebraic relationship between film thickness ratio and speed ratio:

$$\frac{h_A}{h_B} = \frac{S(S+3)}{3S+1} \quad , \quad \text{where} \quad S \equiv \frac{V_A}{V_B}$$

To study forward-roll film-splitting flow, Coyle et al. (1986) solved the Navier-Stokes system by the Galerkin / finite element method. Their theoretical predictions of film-split ratio were well fitted by the power-law relation with $\alpha = 1$ and $\beta = 0.65$ for flows without gravity; and $0.8 < \alpha < 1$ and $\beta = 0.65$ for flows with the rolls arranged vertically. They did not account for inertia in their calculations.

Benjamin et al. (1994) extended Coyle's (1986) calculations and showed how the meniscus position and the fitting parameters $\alpha$ and $\beta$ vary with Reynolds number, capillary number, Stokes number and clearance between the rolls. They showed that as Reynolds number rises, $\beta$ falls; at $Re = 10$, $\beta \approx 0.45$.

Benkreira et al. (1981) reported that their film thickness ratio measurements were well correlated by a power-law with $\alpha = 0.87$ and $\beta = 0.65$. These results are in close agreement with Coyle's prediction. However, there is no general agreement about the values of $\alpha$ and $\beta$: different researchers report distinct values. Savage's (1982) data from vertically arranged rolls can be fitted to the power-law relationship with $\alpha = 0.96$ and $\beta = 0.41$. As illustrated by Benjamin et al. (1994), some of these differences can be related to the effects of inertia and gap, which usually are not explored in the experiments.

All the prior work addressed the flow between rigid rolls. However, in general, one roll of each pair is covered with a resilient layer that deforms during operation. Experiments with rolls arranged horizontally reported by Carvalho et al. (1994) show that at the conditions explored, when the rolls share the same angular speed, the thickness of the layer attached to the rigid roll is thicker than the one carried away by the rigid roll. The deformation of the roll breaks the symmetry of the flow and this leads to asymmetric film splitting even at a speed ratio of $V_A/V_B = 1$.

In this paper, the effects of the roll deformation on film-splitting flows are explored. The flow rate through the gap, which can be translated to coated layer thickness or coat weight; the position of the film-split meniscus; and the fitting parameters $\alpha$ and $\beta$ for film thickness ratio are predicted as function of center-to-center distance and roll-cover properties at different flow conditions. The Navier-Stokes system for steady, two-dimensional flow is solved by the Galerkin / finite element method. The roll deformation is accounted for by independent radial springs that deform in response to the normal component of the traction force applied at their extremity. This model is a simple approximation of the roll-cover deformation. The main advantage is that the elastic response enters as a boundary condition on the mesh generation equations of the liquid domain and thus the number of algebraic equations is not raised by the presence of the compliant wall. The drawback of using the complete description of the situation, i.e. the non-linear elastic, plane strain formulation of the deformable wall, is that the deformation and velocity fields of the entire solid region have to be computed and therefore the size of the algebraic system of equations increases considerably. The advantages and limitations of both formulations are investigated in the case of flow between two fully submerged rolls by Carvalho & Scriven (1996).

## 2. Governing Equations

The governing equations and boundary conditions used to study forward-roll film-split flows in a deformable gap are summarized in Fig. 1.

The velocity and pressure fields are governed by the Navier-Stokes system, i.e. the momentum and continuity equations, which in dimensionless form are

$$Re \; \mathbf{v} \cdot \nabla \mathbf{v} - \nabla \cdot \boldsymbol{\sigma} - St\mathbf{g} = 0 \quad \text{and} \quad \nabla \cdot \mathbf{v} = 0 \tag{1}$$

$Re \equiv \rho \overline{V} H_0/\mu$ is the Reynolds number, and $St \equiv \rho g H_0^2/\mu \overline{V}$ is the Stokes number. $\boldsymbol{\sigma}$ is the total stress tensor, the sum of pressure and viscous stress. Velocities are measured in units of average roll speed $\overline{V} \equiv (V_A + V_B)/2$. Lengths are written in units of half of the undeformed clearance (or interference) between the rolls $H_0$, as illustrated in Fig. 2.

At the rigid roll surfaces (1), the no-slip and no-penetration conditions apply, viz.

$$\mathbf{v} = \mathbf{V_{Roll}} = \Omega \, R \, \mathbf{t} \tag{2}$$

$\mathbf{t}$ is the unit tangent vector to the roll surface, in the direction of rotation, $R$ is the roll radius, and $\Omega$ is the angular speed of the roll.

Along the inflow boundary (2), the pressure is constant:

$$p = P_{IN} \tag{3}$$

In the present analysis, the inlet pressure was set to zero, although it could be instructive and useful to investigate the effects of pressure difference across the coating bead.

At the deformed roll surface (3), the traction on the liquid balances the elastic force on the springs, which is radial:

$$\mathbf{N_0} \cdot (\mathbf{n} \cdot \boldsymbol{\sigma}) = -K \, \mathbf{N_0} \cdot (\mathbf{x} - \mathbf{X_0}) \tag{4}$$

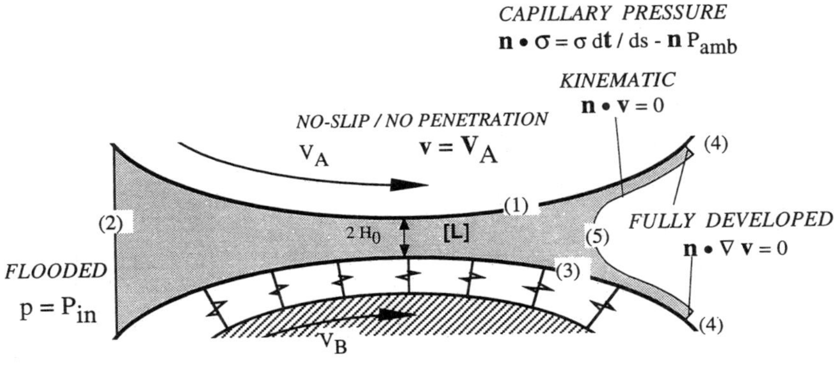

**Figure 1.** *Sketch of flow domain and boundary conditions used for film-splitting flows in a deformable gap*

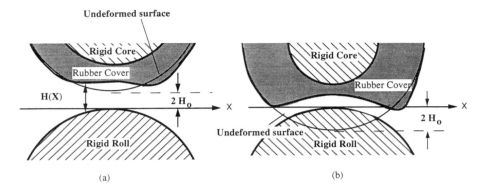

**Figure 2.** *Gap profile with deformable roll. (a) Positive gap, i.e. clearance between unde-formed rolls. (b) Negative gap, i.e. interference between undeformed rolls.*

$N_0$ is the unit normal vector to the undeformed surface, which in this case is the radial direction $(N_0 = e_R)$, $N_0 \cdot (n \cdot \sigma)$ is the radial component of the liquid traction, and $N_0 \cdot (x - X_0)$ is the radial displacement, as indicated in Fig. 1.

The liquid neither slips nor penetrates the cover, so that the velocity components in the horizontal direction $v_x$ and vertical direction $v_y$ are

$$v_x = u = \frac{\Omega R}{n \, N_0} t_x \quad \text{and} \quad v_y = v = \frac{\Omega \times R}{n \cdot N_0} t_y \tag{5}$$

Again $n$ and $N_0$ are the normal unit vectors at each point of the roll surface at the current and undeformed configuration, respectively. For the particular case of roll covers, $N_0$ is the unit vector in the radial direction.

At the outflow boundaries (4), the flow is assumed to be already fully developed. The directional derivative of the velocity in the direction perpendicular to the outlet plane is set to zero:

$$\frac{\partial v}{\partial n} \equiv n \cdot \nabla v = 0 \tag{6}$$

The synthetic outflow plane was located at a horizontal coordinate $X/R = 0.6$. If the outflow was moved further downstream, the flow rate was affected only in the fourth significant figure.

At free surfaces (5), the traction in the liquid balances the capillary pressure and there is no mass flow across the interface:

$$n \cdot \sigma = \sigma \frac{dt}{ds} - n P_{amb}$$

$$\tag{7}$$

$$n \cdot v = 0$$

$\sigma$ is the liquid surface tension.

The dimensionless parameters that govern this flow are

$$\textbf{Reynolds Number :} \quad Re \equiv \frac{\rho \overline{V} H_0}{\mu}$$

$$\text{Stokes Number}: \quad St \equiv \frac{\rho g H_0^2}{\mu \overline{V}}$$

$$\text{Roll Speed Ratio}: \quad S \equiv \frac{\Omega_A}{\Omega_B}$$

$$\text{Dimensionless undeformed clearance (or interference)}: \quad \frac{H_0}{R}$$

$$\text{Modified Elasticity Number}: \quad Ne^* \equiv \frac{\mu \overline{V}}{K R^2}$$

$$\text{Capillary Number}: \quad Ca \equiv \frac{\mu \overline{V}}{\sigma}$$

Gravitational effects are neglected in the cases analyzed here, i.e. $St = 0$.

Elliptic mesh generation was used to map the unknown physical domain into a fixed reference domain and the Galerkin / finite element method was used to solve the governing equation system — Eqs. (1) to (7).

The resulting set of mostly non-linear algebraic equations for the coefficients of the finite element basis functions was solved by Newton's method and pseudo-arc-length continuation. The mesh generation equations and the Navier-Stokes equations were handled together in each Newton iteration. The domain was divided into 352 elements, which corresponds to 7156 algebraic equations. Increasing the number of elements to 780 elements and therefore the number of algebraic equations to 15488 changed the flow rate and film split meniscus position only in the fourth significant figure. All computations were performed with a Cray X-MP computer and each Newton iteration took approximately 12 seconds.

## 3. Results and Discussion

### 3.1. Predictions of flow rate and meniscus position

The flow rate through a forward-roll deformable coating gap can be controlled by the clearance or interference of the rolls, that is, by the center-to-center distance of the rolls. The dependence varies according to the rigidity of the deformable roll used. Figure 3 graphs the flow rate versus gap at capillary number $Ca = 0.1$ and different modified elasticity numbers. From the results presented by Carvalho & Scriven (1996), the spring constant can be roughly estimated to be $K \approx 4 \times E/L$, where $E$ is the Young's modulus of the roll cover, and $L$, its thickness. If the distance between the rolls is large, the traction (pressure plus viscous terms) exerted by the liquid is not strong enough to deform the roll and the deformable gap behaves as if it were rigid. As the rolls are pushed together, the resilient cover deforms, widening the gap. At a given gap, the softer the roll, the larger the flow rate, as expected.

When the flow rate to be metered by a forward-roll rigid gap is small, the distance between the roll surfaces has to be very small. When the rolls are rigid, at small values of $H_0/R$, the flow rate varies greatly with gap: any small variation in the actual distance between the rolls, such as that caused by roll runout, changes the flow rate and therefore the metered coating thickness substantially. The alternative is to use a soft roll and smaller or even negative gaps, i.e. interference between the undeformed roll surfaces. In deformable gaps, the flow rate dependence on gap weakens as the distance between the roll centers diminishes.

However, because flow rate is less sensitive to center-to-center distance, only a small range of flow rates can be obtained at a given modified elasticity number, i.e. with a given roll cover. For example, a dimensionless flow rate of $q^* \equiv Q/2\overline{V}R < 10^{-3}$ cannot be obtained at $Ne^* = 10^{-6}$ (see Fig. 3). In this case, the amount of liquid that flows through the gap can be controlled by the roll hardness. If the desired metered flow rate is smaller than $10^{-3}$, a harder roll has to be used. Thus the roll cover has to be chosen according to the thickness of the coated layer that is desired.

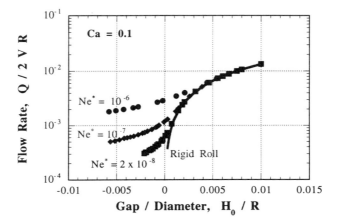

**Figure 3.** Predicted dimensionless flow rate versus gap at capillary number $Ca = 0.1$. The parameter $Ne^*$ is the modified elasticity number, which is smaller the harder the roll cover. The continuous line represents the limiting case of rigid rolls.

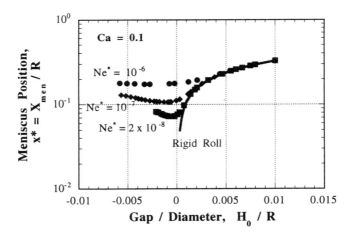

**Figure 4.** Predicted meniscus position versus gap at capillary number $Ca = 0.1$. The parameter $Ne^*$ is the modified elasticity number, which is smaller the harder the roll cover. The continuous line represents the limiting case of rigid rolls. The position is measured from the plane that connects both roll's centers.

In a rigid roll-coating gap, as the rolls are brought closer, the meniscus shifts toward the minimum clearance between them, which is at the line (plane in three dimensions) through the roll centers, as depicted in Fig. 4. The position $x^*$ shown in Fig. 4 is measured from the minimum clearance plane. The relocation of the meniscus heightens the pressure gradient in the liquid beneath it. It is the pressure gradient at the free surface that destabilizes the flow with respect to three-dimensional disturbances and gives rise to ribbing if the leveling action of surface tension is not strong enough in comparison. Therefore, at a given capillary number, there is a critical gap $H_0/R$, below which the flow is unstable and ribbing occurs.

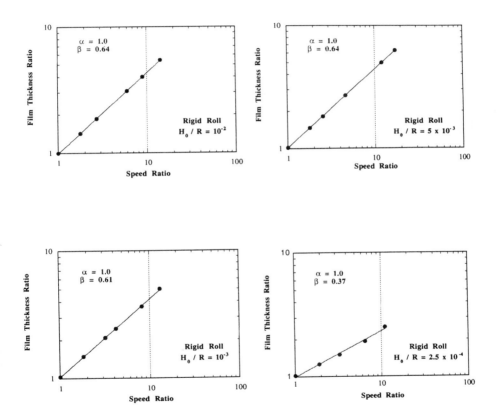

**Figure 5.** *Film-split ratio in flow between two rigid rolls versus speed ratio at different gaps. The line represents the power-law fit, obtained by least squares, of the theoretical predictions.*

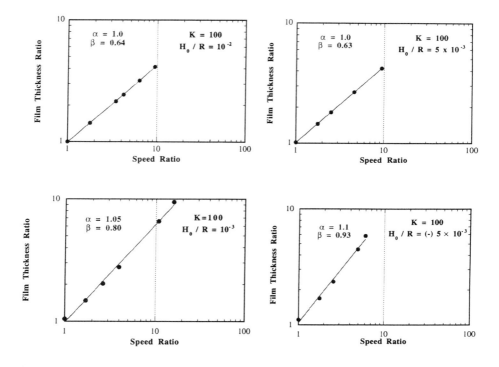

**Figure 6.** *Film-split ratio in flow between a rigid roll and a deformable roll versus speed ratio at $Ne^* = 10^{-6}$ and different gaps. The line represents the power-law fit, obtained by least squares, of the theoretical predictions.*

In deformable roll-coating gaps, the same trends can be observed if there is a clearance between the undeformed roll surfaces. However, when there is not, i.e. at negative gaps, the meniscus starts to shift *away* from the gap after the center-to-center distance has been narrowed past a certain value, as illustrated in Fig. 4. As the rolls are pressed against each other, the liquid filled channel between them grows shorter and then longer. The channel length (in practice often called the *footprint*) increases as the center-to-center distance decreases and the meniscus is pushed away from the plane through the roll centers. Because the meniscus is relocated further and further away from the gap, the pressure gradient at the free surface falls and therefore the flow may become stable with respect to three-dimensional disturbances up to a higher capillary number.

## 3.2. Film-split ratio

To evaluate the film-split coefficient $\alpha$ and exponent $\beta$ in the power-law correlation,

$$\frac{h_A}{h_B} = \alpha \left( \frac{V_A}{V_B} \right)^{\beta}$$

families of solutions were obtained by varying the speed ratio and keeping all other parameters fixed.

Figure 5 shows the variation of film thickness ratio with speed ratio for rigid rolls at capillary number $Ca = 0.1$, and different gaps. The continous line represents the power-law curve-fitting obtained by the least squares method. At all gaps and at speed ratios in the range between 1 and 10, the power-law relation fits the theoretical predictions well (the regression coefficient was always larger than 0.99). However, at small gaps, the fit is not as good as at larger gaps. The regression coefficient at $H_0/R = 10^{-2}$ was 0.999, and at $H_0/R = 2.5 \times 10^{-4}$, 0.995. As expected, if both rolls are rigid, the film split is symmetric, i.e. the film thicknesses on both rolls are the same when they share the same speed. The power $\beta$ falls slightly as the rolls are pushed against each other: The film thickness ratio becomes less sensitive to speed ratio. A similar trend was observed in experiments performed in a bench-top apparatus (see Carvalho et al. 1994).

In Fig. 6, the relation of film thickness ratio is plotted versus speed ratio at a modified elasticity number $\mu \overline{V}/KR^2 = 10^{-6}$, capillary number of $Ca = 0.1$ and different gaps. At a gap of $H_0/R = 10^{-2}$, the pressure is not strong enough to deform the rubber cover: the values of $\alpha$ and $\beta$ are the same as in the case of rigid rolls. As the rolls are pressed against each other, the film split power $\beta$ first falls, following the rigid gap trend, but then rises again. This behavior was also observed experimentally by Carvalho et al. (1994). The coefficient $\alpha$ is close to one at large gaps and increases as the rolls are brought together. At a negative gap (interference) of $H_0/R = (-)5 \times 10^{-3}$, it is $\alpha = 1.1$. The film carried by the rigid roll is 10% thicker than the one dragged away by the deformable roll even if the rolls share the same speed.

## 4. Final Remarks

Flows between a rigid and a deformable roll have now been extensively analyzed by theory. The Navier-Stokes system of equations coupled with the equation system of elastic deformation of the resilient roll cover can be solved by Galerkin / finite element method. A suitable model to describe the roll deformation is a simple elastic spring model. Its merits and disadvantages are discussed elsewhere (Carvalho & Scriven 1996).

The theoretical predictions reveal how the flow rate, meniscus position and the ratio between the thicknesses of the liquid layer, or film, carried away by each roll vary with the roll cover hardness. The results illustrate how deformable rolls can be used to obtain thin coated layers with much less sensitivity to roll runout than those obtained with rigid rolls.

Because the roll deformation affects the flow between the rolls, it also alters the stability of that flow with respect to three-dimensional infinitesimal disturbances, which are ever-present in practice. As capillary number rises, stability is lost and this seems to mark the onset of the coating defect that consists of waviness in the transverse direction and is known as *ribbing, corduroy,* or *pin-striping.* Carvalho & Scriven (1995) have used linear stability analysis to determine the flow conditions at which the flow becomes unstable. The results reveal that the roll deformation delays the onset of ribbing.

## 5. References

[1] BENKREIRA H., EDWARDS M. F. & WILKINSON W. L. 1981 Roll coating of purely viscous liquids *Chemical Engineering Science.* **36**, 429.

[2] BENKREIRA H., EDWARDS M. F. & WILKINSON W. L. 1982 Ribbing instability in the roll coating of Newtonian fluids. *Plastics and Rubber Processing and Applications.* **2**,137.

[3] BENJAMIN D.F., CARVALHO M. S., ANDERSON T.D. & SCRIVEN L.E. 1994 Forward Roll Film-Splitting Theory and Experiment. *1994 TAPPI Coating Conference, San Diego, CA.*

[4] BENJAMIN D.F. 1994 *Roll Coating Flows and Multiple Roll Systems.* Ph.D. Thesis, University of Minnesota, Minneapolis.

[5] BOLSTAD J. H. & KELLER H. B. 1986 A multigrid continuation method for elliptic problems with folds. *SIAM J. Sci. Stat. Comput.* **7(4)**, 1081.

[6] CARVALHO M. S. & SCRIVEN L. E. 1994 Elastohydrodynamics of Deformable Roll Coating. *AIChE Spring National Meeting, Atlanta.* **Paper 7c.**

[7] CARVALHO M. S., ANDERSON T. J. & SCRIVEN L. E. 1994 Experimental Analysis of Film Split in a Deformable Roll Coating Gap. *AIChE Spring National Meeting, Atlanta.* **Paper 3b.**

[8] CARVALHO M. S. & SCRIVEN L. E. 1995 Capillary and viscoelastic effects on elastohydrodynamic lubrication flow and film-splitting in roller nips. *Accepted for publication in Journal of Tribology.*

[9] CARVALHO M. S. & SCRIVEN L. E. 1995 Deformable Roll Coating: Analysis of Ribbing Instability and its Delay. *$1^{st}$ European Coating Sysmposium, Leeds.*

[10] CARVALHO M. S. & SCRIVEN L. E. 1996 Forward Roll Coating with Deformable Roll: Analysis of Elastomer-Hydrodynamic Interaction. *AIChE Spring National Meeting, New Orleans.*

[11] COYLE D. J. 1984 *The Fluid Mechanics of Roll Coating: Steady Flows, Stability and Rheology.* Ph.D. Thesis, University of Minnesota, Minneapolis.

[12] COYLE D.J., SCRIVEN L.E. AND MACOSKO C.W. 1986 Film splitting flows in forward roll coating. *Journal of Fluid Mechanics.* **171**, 183.

[13] COYLE D. J. 1988 Forward roll coating with deformable rolls: A simple one-dimensional elastohydrodynamic model. *Chemical Engineering Science.* **43**, 2673.

[14] COYLE D. J., MACOSKO C. W. & SCRIVEN L. E. 1990 Stability of symmetric film-splitting between counter-rotating cylinders. *Journal of Fluid Mechanics.* **216**, 437.

[15] DERYAGIN B. M. & LEVI S. M. 1959 *Film Coating Theory.* Focal Press Ltd. London and New York.

[16] LANDAU L. & LEVICH B. 1942 Dragging of a liquid by a moving plate. *Acta Physicochim. USSR.* **17**, 42.

[17] SAVAGE M. D. 1982 Mathematical models for coating processes. *Journal of Fluid Mechanics.* **117**, 443.

[18] SAVAGE M. D., GASKELL P. H., MALONE B. & THOMPSON H. M. 1992 The Relevance of Lubrication Theory to Roll Coating. *AIChE Spring National Meeting, New Orleans.* **No.39e.**

## 6. Acknowledgments

M. S. Carvalho was supported by a fellowship from CAPES (Brazilian Federal Government). Further support came from cooperating corporations through the Center for Interfacial Engineering and was supplemented by the National Science Foundation.

# THIN FILM FORMING IN COATING-NIP FLOWS

M. Kodama, K. Adachi
Department of Applied Chemistry
Kyushu Institute of Technology, Kitakyushu, 804 Japan

A simple model of coating flow and thin-film forming at a meniscus region has been analysed and the results are compared with those of Coyne and Elrod[1]. The fundamental feature of film forming flow and its coating application will be discussed.

## 1. INTRODUCTION

The film forming flow in a coating nip consists of two regions of flows. One is the region of one dimensional, thin film flow which can be expressed by a lubrication theory, and the other is the region of two dimensional, film-splitting flow which is usually analysed by numerical simulation. The aim of the present paper is to elucidate the fundamental feature of this film forming flow and to establish simple but useful relations to express the film forming flow.

The relation of film forming (or splitting) flow has been studied by many Investigators. Bretherton[2] investigated both experimentally and theoretically the thickness of the film left behind during the very slow passage of a large air bubble in a liquid-filled capillary tube, and he found the folowing relation:

$$h_\infty/H = 1.34 Ca^{2/3} \qquad (1)$$

Here, $h_\infty$ is the film tickness, H is the half of the tube diameter, Ca is the capillary number defined by $\mu U/\sigma$. Deryaguin and Levi[3] showed that the above relation of Bretherton is equivalent to the result of Landau and Levich[4] for dip coating flow at a low capillary number.

Coyne and Elrod[1] derived a relation between dimensionless film thickness, $h_\infty/H$, and flow condition, Ca, using a film-splitting flow model, and showed a good agreement between their theoretical predictions and experimental results of Bretherton[2] and Taylor[5] over a wide range of Ca number. Rushack[6] showed that the above Coyne and Elrod relation holds for the film-splitting flow between two rolls rotating at the same speed in forward roll coating. Benkreira et al.[7] applied the Coyne and Elrod relation to the film-splitting flow between two rolls rotating at different speeds in forward roll coating.

Rushack[8,9] suggested that the film-splitting ratio between two rolls in forward roll coating can be predicted at a low capillary number from Eq.(1), the Bretherton relation of film-splitting flow, and that the 2/3 power rule of film thickness ratio,

$$h_{\infty 1}/h_{\infty 2} = (U_1/U_2)^{2/3} \tag{2}$$

which was found experimentally by Benkreira et al.[10] over a wide range of Ca number, results from the 2/3 power of Ca number in Eq. (1). However, Equation (1) only holds at a low capillary number, so that it would be reasonable to derived Eq.(2) from the Coyne and Elrod relation or a real universal relation of film-splitting flow.

It will be considered in the present paper if the Coyne and Elrod relation is a universal one or not, and also if Eq(2) can be derived from the Coyne and Elrod relation or not. The present work has been directly motivated by the discussion given by Rushack[9] and Thompson and Davies[11] at ICR vol.1.

## 2. NUMERICAL SIMULATION OF A MODEL OF FILM FORMING FLOW

We investigated the fundamental feature of film-forming flow with the model shown in Fig.1. The flow was assumed to be steady, incompressible and Newtonian, and the Navier-Stokes equation and the quation of continuity were solved simultaneously using the computer software NEKTON which is based on the spectral-element method.

The parallel plane walls of a duct, which are apart by 2H, are moving at two independent speeds, $U_1$ and $U_2$, in the same direction. The liquid in the duct is splitted by a floating, but stationary long air bubble, and entrained by each wall to form a

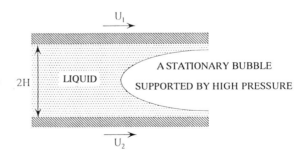

Fig.1 A model of film-forming flow

film of the thickness, $h_{\infty 1}$ or $h_{\infty 2}$.

A parallel flow and a constant pressure are assumed at the upstream boundary, and two parallel flows and the constant atmospheric pressure are assumed at the downstream boundary. It is also assumed that the normal stress difference across the free surface is caused by the surface tension, and that the free surface consists of streamlines. The constant pressure at the upstream boundary has to be adjusted and determined by trial-and-error so as to fix the floating bubble at a given position.Then, the film thickness at the downstream boundary, $h_{\infty i}$ (i=1 or 2), the pressure jump at the tip of the bubble, $\Delta p$, and the pressure gradient, dp/dx, are the important physical quantities to be determined for a given set of flow conditions. These quantities are combined with each other by the relations which can be derived from the lubrication theory, and which hold almost exactly in the region, of one dimensional flow, far upstream from the free surface of a floating bubble. When the film thicknesses, $h_{\infty i}$, and the prssure jamp, $\Delta p$, are given, then all physical quantities mentioned before can be decided.

As shown in Fig.2, the line of the minimum velocity (free shear) at the upstream boundary, which is designated by $u_0$, is at $y=H-H'_1=-H+H'_2$. The dividing line, which does not always locate at the tip of the air bubble, is at $y=H-H_1=-H+H_2$. For $U_1{\neq}U_2$, these lines are not at y=0. Then, the lubrication theory gives the following relations:

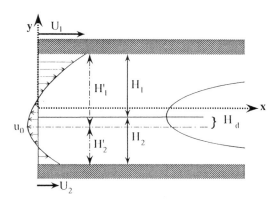

Fig.2 A model of lubrication flow

$$(dp/dx)/2\mu$$
$$=(U_1-u_0)/(H'_1)^2=3U_1(1-h_{\infty 1}/H_1)/\{2(H'_1)^2-H'_1H_d-H_d^2\} \qquad (3)$$
$$=(U_2-u_0)/(H'_2)^2=3U_2(1-h_{\infty 2}/H'_2)/\{2(H'_2)^2+H'_2H_d-H_d^2\}$$

Here, $H_d=H'_1-H_1=H_2-H'_2$.

For $U_1=U_2=U$, $H_d=0$, $H'_1=H_1=H'_2=H_2=H$, and $h_{\infty 1}=h_{\infty 2}=h_\infty$. Then, Eqs.(3) reduce to

$$(H^2/\mu U)dp/dx=3(1-h_\infty/H) , u_0/U=(3/2)(h_\infty/H-1/3) \qquad (4)$$

## 3. RESULTS AND DISCUSSION

The relation between the dimensionless film thicness, $h_\infty/H$, and the flow condition, Ca, for the case of $U_1=U_2$ has been shown in the Fig.3 in comparison with the experimental result of Taylor[5] and the calculated result of Coyne and Elrod[1]. These three results are in a good agreement to each other.

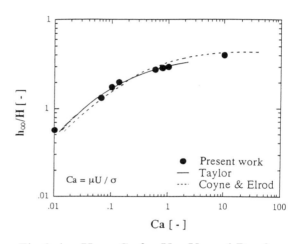

Fig.3 $h_\infty/H$ vs. Ca for $U_1=U_2$ and Re=0.

The pressure jump across the free surface at the tip of the bubble for the case of $U_1=U_2$ has been shown in Fig.4. The dimensionless pressure jump approaches to a constant value since the surface tension becomes dominant as $\mu U$ decreases, and the pressure jump increases linearly since the viscous shear force becomes dominant as $\mu U$ increases.

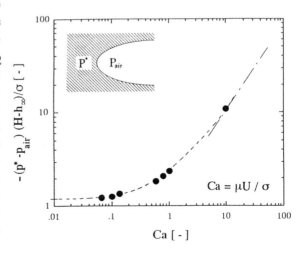

Fig.4 Pressure jump for $U_1=U_2$ and Re=0.

The relation between the dimensionless film thicness, $h_{\infty i}/H_i$, and the flow condition, $Ca_i$ $(=\mu U_i/\sigma)$, for the case of $U_1 \neq U_2$ has been shown in Fig.5 in comparison with the results for the case of $U_1 = U_2$.

The 6 data points of open triangle in the figure has been obtained from three different set of computed data. The two flows separated by a dividing line seem to be independent from each other when the dividing line locates nearby the free shear line ($H_d \approx 0$).

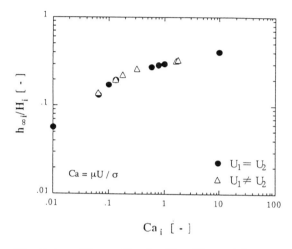

Fig.5 $h_{\infty i}/H_i$ vs. $Ca_i$ for $U_1 \neq U_2$ and Re=0.

We estimated $h_{\infty i}/H_i$ by $h_\infty/H$ at $Ca = Ca_i$ using the correlation chart shown in Fig.3, and we decided $h_{\infty i}$, $H_i$, $u_0$ and $dp/dx$ from a given set of flow conditions using Eqs. (3). The data points of Coyne and Elrod and present work on the film thickness ratio in the film-splitting flow for $U_1 \neq U_2$, which have been shown in Fig.6, were obtained from the data for $U_1 = U_2$. The four solid circles are the results of present direct simulation for $U_1 \neq U_2$. All

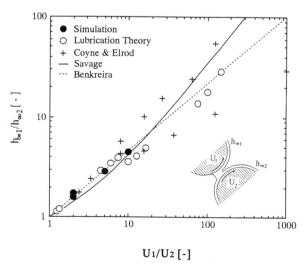

Fig.6 Liquid partition rule between two rolks

these results are in a good agreement to the empirical line of Benkreira et al.[11] expressed by Eq.(2) in a range of speed ratio less than 10. The theoretical prediction of Savage, which is expressed by Eq.(5), is not bad also in this range.

$$h_{\infty 1}/h_{\infty 2} = S(S+3)/(1+3S), \quad S=U_1/U_2 \qquad (5)$$

## 4. CONCLUSIONS

The relation of Coyne and Elrod between $h_\infty/H$ and Ca is not exactly universal, but can be used to estimate the film-splitting ratio for the case of $U_1 \neq U_2$ approximately. The universal correlation for the film splitting flow is expected to have the form of $h_\infty/R=$func.(Ca), or $h_\infty \Delta p/\sigma=$ func.(Ca), where $\Delta p$ is the pressure jump at the tip of the bubble.

## 5.REFFERENCES

1. Coyne, J.C. and Elrod, H.G. Conditions for the Rupture of a Lubricating Film. Part I: Theoretical Model. *J. of Lubrication Technology*, **92**[3], 451-456(1970).
2. Bretherton, F.P. The Motion of Long Bubbles in Tubes. *J. Fluid Mch.*, **10**, 166 -187(1961).
3. Deryaguin, B.V. and Levi, S.M. *Film Coating Theory*. Focal Press, New York(1964).
4. Landau, L. and Levich, B. Dragging of a Liquid by a Moving Plate. *Acta Physcochim. URSS,* **17**, 42-54(1942).
5. Taylor, G.I. Deposition of a Viscous Fluid on the Wall of a Tube. *J. Fluid Mech.*, **10**, 161-165(1961).
6. Rushack, K.J. Boundary Conditions at a Liquid/Air Interface in Lubrication Flows. *J. Fluid Mech.*, **119**, 107-120(1982).
7. Benkreira, H., Edwards, M.F. and Wilkinson, W.L. Ribbing Instability in the Roll Coating of Newtonian Liquid. *Plast. and Rub.Proc. and Appl.*, **2**, 137-144(1982).
8. Rushack, K.J. Coating Flows. *Ann. Rev. Fluid Mech.*, **17**, 65-89(19 85).
9. Rushack, K.J. Technical Note on Forward Roll Coating. *Ind. Coat. Res.*, **1**, 59-62(1991).

10. Benkreira, H., Edwards, M.F. and Wilkinson, W.L. Roll Coating of Purely Viscous Liquids. *Chem. Eng. Sci.*, **36**, 429-434.
11. Thompson, N.E. and Davies, M. J. Review of papers by D.J.Coyle. *Ind. Coat. Res.,* **1**, 63-70(1991).
12. Savage, M.D. Mathematical Models for Coating Processes. *J. Fluid Mech.*, **117**, 443-455(1982).

# A Combined Slot and Roll Coating Technique

S. Krauss, F. Durst, H. Raszillier

*Lehrstuhl für Strömungsmechanik*
*Universität Erlangen-Nürnberg*
*Cauerstr. 4, D-91058 Erlangen (Germany)*

## SUMMARY

A coating technique is presented that employs a slot coater as a feed system for a pair of counterrotating rolls. The premetering of the amount of fluid fed into the roll system allows a simplified analysis of the metering properties of the rolls. These metering properties are analyzed combining simple lubrication theory for arbitrary shear dependent fluids with accurate numerical results.

## 1. INTRODUCTION

The great variety of industrially used thin film coating processes can be divided up into premetered and non premetered processes. Widely used examples of non premetered processes are forward and reverse roll coating. Typically a pan feed system supplies the rolls with a comparatively thick film, which is then reduced to the desired film thickness using multiple film splits between different rolls. The final film thickness is a complicated function of the different roll speeds[1]. A typical example of a premetered coating system is the slot coater. The wet film thickness can be controlled precisely with this system, but it is restricted by a lower bound[2,3]. Surface roughness of the substrate and inaccuracies of the backing roll tend to raise the minimum wet film thickness, because they put constraints on the minimum gap width.

The advantages of the slot coater and the roll coater can possibly be combined, if the slot coater is used as a feed system for the roll coater as sketched in Fig. 1. Such a combination would make it possible to coat very thin films without the necessity of multiple film splits, because the slot coater already supplies a thin film to the roll system. In addition effects of surface roughness should be reduced, because the task of applying the liquid to the substrate is shifted from the slot coater to the roll system. The minimum gap width between the slot coater and the first roll is therefore only restricted by manufacturing inaccuracies of the roll.

From the theoretical point of view the most important feature is that the preme-

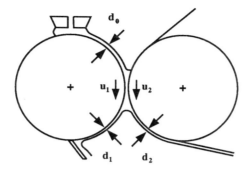

Figure 1: *Combined Slot- and Roll coating*

tering is partly retained in the combined system in the following sense: The fact that the rolls are supplied with a precisely known flow rate, significantly simplifies the prediction of the final film thickness, which will rid the user of the necessity of performing extensive numerical calculations.

The analysis presented here will focus mainly on the metering properties of the rolls. A theory of film splitting will be presented that applies to a wide class of liquids.

## 2. METERING BETWEEN THE ROLLS

### 2.1. Lubrication Theory of Film Splitting for Arbitrary Shear Dependent Liquids

Given a fluid with rheological properties described by a shear dependent viscosity

$$\eta = \eta(\dot\gamma) \tag{1}$$

the flow between the two rolls will be approximated using the well known lubrication approximation:

$$\frac{\partial p}{\partial x} = \frac{\partial \tau}{\partial y},$$
$$\frac{\partial p}{\partial y} = 0. \tag{2}$$

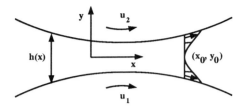

Figure 2: *Coordinate system in the gap*

The flow is subject to the boundary conditions

$$\begin{aligned} u(x,0) &= u_1 \\ u(x,h(x)) &= u_2, \end{aligned} \tag{3}$$

where $h(x)$ is the local width of the gap and $u_1$ and $u_2$ are the speeds of the first and second roll, respectively. As the film that is fed into the roll system is premetered, the flow must integrate to a prescribed flow rate $Q$.

$$Q = \int_0^{h(x)} u(x,y)dy. \tag{4}$$

The whole analysis will be based on the assumption, that the point $(x_0, y_0)$ where the film separates can be identified with the stagnation point of the lubrication flow. With that assumption the equations determining the splitting point can be written as

$$u(x_0, y_0) = \tau(x_0, y_0) = 0 \tag{5}$$

Using the stagnation point condition and momentum balance the velocity field in the plane $x = x_0$ can be expressed as:

$$\begin{aligned} u(x_0, y) &= \int_{y_0}^{y} \frac{\partial u}{\partial y}dy \\ &= \frac{1}{\frac{\partial p}{\partial x}} \int_0^{\tau(y)} \dot{\gamma}d\tau \end{aligned} \tag{6}$$

Substituting the integration variable $y$ in the expression for $Q$ by

$$\xi = \int_0^{\tau(y)} \dot{\gamma}(\tau)d\tau \tag{7}$$

and using the boundary conditions

$$
\begin{aligned}
\xi(0) &= u_1 \frac{\partial p}{\partial x} \\
\xi(h(x)) &= u_2 \frac{\partial p}{\partial x}
\end{aligned}
\tag{8}
$$

the expression for $Q$ can be restated as

$$
Q \sqrt{\frac{\partial p}{\partial x}} = u_1^{\frac{3}{2}} \mathcal{R} \left( u_1 \frac{\partial p}{\partial x} \right) + u_2^{\frac{3}{2}} \mathcal{R} \left( u_2 \frac{\partial p}{\partial x} \right) ,
\tag{9}
$$

where $\mathcal{R}(\xi)$ is defined by

$$
\mathcal{R}(\xi) = \xi^{-\frac{3}{2}} \int_0^\xi \frac{\xi}{\dot{\gamma}(\xi)} d\xi
\tag{10}
$$

and $\dot{\gamma}(\xi)$ is the inverse function corresponding to the definition of $\xi$.

Clearly $\mathcal{R}(\xi)$ is a function that does not depend on any process parameters like the width of the gap or roll speed, but is determined uniquely by the rheological properties of the fluid. Once the pressure gradient at the meniscus has been calculated, the splitting ratio $\rho$ and related properties can easily be expressed in terms of $\mathcal{R}$. The splitting ratio is given by

$$
\rho = \sqrt{\frac{u_2}{u_1}} \frac{\mathcal{R} \left( u_2 \frac{\partial p}{\partial x} \right)}{\mathcal{R} \left( u_1 \frac{\partial p}{\partial x} \right)} .
\tag{11}
$$

Fig. 3 shows the flow curve for a specific latex suspension and the corresponding rheological function $\mathcal{R}$. It is clearly seen, that there is a close relationship between the functions $\eta$ and $\mathcal{R}$. In fact for $\eta$ varying slowly with $\dot{\gamma}$ the function $\mathcal{R}$ becomes roughly proportional to $\sqrt{\eta(\dot{\gamma}(\xi))}$.

The simplest example of a shear dependent liquid is the power-law-liquid, for which

$$
\eta \propto \dot{\gamma}^{-n} .
\tag{12}
$$

In this case $\mathcal{R}$ is calculated easily to give

$$
\mathcal{R}(\xi) \propto \xi^{-\frac{n}{4-2n}} .
\tag{13}
$$

Inserting this relation into the expression for the splitting ratio immediately yields Coyle's[4] classical result:

$$
\rho = \left( \frac{u_2}{u_1} \right)^{\frac{1-n}{2-n}}
\tag{14}
$$

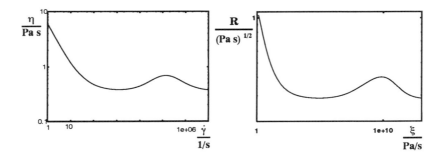

Figure 3: *The rheological properties of a specific latex suspension*

In this case Eq. (9) does not even have to be solved, because the unknown pressure gradient drops out of the expression for $\rho$.

## 2.2. Numerical Calculations for Power-Law-Fluids

It is well known that lubrication theory does not yield quantitatively accurate results for the splitting ratio. The desired accuracy of the predictions can be achieved performing a Finite Element Analysis of the flow. Such calculations are on the one hand relatively easy to perform in principle, on the other hand they may involve some tedious work and are therefore not suitable for industrial purposes. The main problem about the fully numerical calculations is therefore not to perform them, but rather to find an appropriate way to parameterize the results.

Parameterization becomes easy though for power law fluids ($\eta \propto \dot{\gamma}^{-n}$) where the splitting ratio is determined by the speed ratio of the rolls and the power law index.

$$\rho = \rho(\epsilon, n)$$
$$\epsilon = \frac{u_2}{u_1} \tag{15}$$

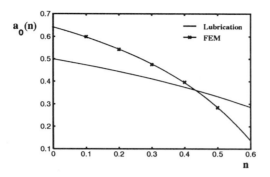

Figure 4: *The exponent corresponding to the power law for the splitting ratio*

If gravity is neglected, the following symmetry property has to hold:

$$\rho\left(\frac{1}{\epsilon}, n\right) = \frac{1}{\rho(\epsilon, n)} \tag{16}$$

This reduces $\rho$ to functions of the form

$$\rho(\epsilon, n) = \epsilon^{a_0(n)} \exp \sum_{k=1}^{\infty} a_k(n)(\ln \epsilon)^{2k+1} \tag{17}$$

Usually only the power law part of this expression is used to parameterize experimental or computational results. Higher terms of the expansion may become important though when inertia terms are no longer negligible. In this work the power law will be used. Fig. 4 shows the exponent $a_0(n)$ obtained from Finite Element Calculations compared to the exponent that lubrication theory yields.

It is interesting to see that lubrication theory underestimates the splitting ratio for newtonian and weakly shear thinning liquids, whereas it overestimates the splitting ratio for strongly shear thinning liquids. This phenomenon can be understood qualitatively noting that on the one hand lubrication theory neglects the pressure gradient in $y$-direction that leads to a significant additional flow towards the faster roll in the meniscus region and on the other hand neglects contributions to the shear rate (more exactly the second invariant of the deformation tensor) that lead to additional shear thinning effects.

## 2.3. The Local Power Law Index

So far a general but quantitatively insufficient lubrication theory of film splitting and accurate numerical calculations in a very limited parameter space have been presented. The idea of the following section will be to combine these two approaches to obtain quantitatively accurate expressions for the splitting ratio for general liquid properties.

This "hybrid" approach is based on the assumption that there is a one to one mapping between the metering properties one obtains from lubrication theory and the true metering. Of course this property will not hold for situations where the validity of the lubrication approximation is greatly restricted like in the meniscus coating regime. Yet if the assumption is valid and a local power law index $n$ is chosen in such a way that

$$\rho(\epsilon, \dot{Q}, \{\eta(\dot{\gamma})\}) = \rho(\epsilon, n) \tag{18}$$

in the framework of lubrication theory, this relation should also hold for the true splitting ratio. We will therefore choose $n$ to satisfy

$$\epsilon^{\frac{1-n}{2-n}} = \sqrt{\epsilon} \, \frac{\mathcal{R}\left(u_2 \frac{\partial p}{\partial x}\right)}{\mathcal{R}\left(u_1 \frac{\partial p}{\partial x}\right)} \tag{19}$$

and then approximate the true splitting ratio by

$$\rho = \epsilon^{a_0(n)}, \tag{20}$$

where $a_0(n)$ is taken from numerical calculations.

As a simple example the splitting ratio for a power law fluid with a nonzero infinite shear viscosity

$$\eta(\dot{\gamma}) = \eta_0 \left(\left(\frac{\dot{\gamma}}{\dot{\gamma}_0}\right)^{-n} + \beta\right) \tag{21}$$

and a speed ratio $\epsilon = 3$ is presented in Fig. 5.

For $\beta \to 0$ the liquid becomes a pure power law liquid with index $n = 0.5$, for $\beta \to \infty$ it effectively becomes newtonian. The comparison of the splitting ratios

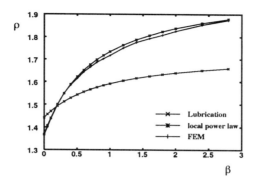

Figure 5: *Comparison of the numerical results with the lubrication and local power law approach*

that we get from lubrication theory with those obtained using the local power law index shows a significant improvement of the achieved accuracy.

## 3. THE FINAL FILM THICKNESS

Finally the film thickness $d_2$ applied to the substrate if the rolls are supplied with an initial film thickness $d_0$ has to be determined.

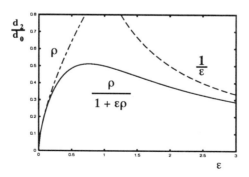

Figure 6: *The final film thickness for a newtonian liquid*

It is easily calculated to give

$$d_2 = \frac{\rho}{1 + \epsilon \rho} d_0, \qquad (22)$$

where $\rho$ is calculated in the way described above. For small $\epsilon$ only a small fraction of the liquid is applied to the second roll and the final film thickness is dominantly determined by the splitting ratio. For large $\epsilon$ most of the liquid is applied to the second roll. Thinning of the film in this case is due to "stretching" of the film (Fig. 6). In the transition region between these two limiting cases the final film thickness reaches a maximum near $\epsilon = 1$. The exact location of this maximum depends on the rheological properties of the liquid.

## 4. CONCLUSION

It has been shown that exact premetering of the amount of fluid supplied to a system of counterrotating rolls allows a precise prediction of the metering properties of the rolls using a semi-analytical approach. In this sense a combination of a slot coater and a roll coating system can be viewed as a "quasi-premetered" coating technique. The fact that the liquid is applied to the substrate by the rolls instead of the slot coater reduces effects of surface roughness of the substrate.

## 5. REFERENCES

1. Dean. F. Benjamin, Todd J. Anderson, and L.E.Scriven, Multiple Roll Systems: Steady-State Operation, *AICHE J.*, Vol. 41, No.5, 1045 (1995)

2. Kwong-Yang Lee, Li-Dah Liu, and Ta-Jo Liu, Minimum Wet Thickness in Extrusion Slot Coating, *Chem. Engng Sci.*, Vol. 47, No. 7, 1703 (1992)

3. Wen-Jen Yu, Ta-Jo Liu, Tsai-An Yu, Reduction of the Minimum Wet Thickness in Extrusion Slot Coating, *Chem. Engng Sci.*, Vol. 50, No. 6, 917 (1995)

4. D. J. Coyle, The Fluid Mechanics of Roll coating: Steady Flows, Stability, and Rheology, Dissertation, Univ. of Minnesota 1984

# SECTION 5

# Spreading, Levelling,

# Surface Tension and

# Gravity Driven Flows

# DYNAMICS OF LIQUID SPREADING ON SOLID SURFACES.

S. Kalliadasis*

*School of Mathematics, University of Bristol, Bristol, UK BS8 1TW.*

## SUMMARY

Using simple scaling arguments and a precursor film model, we show that the appropriate macroscopic contact angle $\theta$ during the spreading of a completely or partially wetting liquid should be described by $tan\theta = \left[tan^3\theta_e - 9\log\eta Ca\right]^{1/3}$ where $\theta_e$ is the static contact angle, $Ca$ is the capillary number and $\eta$ is a scaled Hamaker constant. Using this simple condition, we are able to quantitatively model, without any empirical parameter, the spreading dynamics of several classical spreading phenomena-capilarry rise, sessile and pendant drop spreading.

## 1. INTRODUCTION

The spreading of a liquid over a solid is an intriguing phenomenon that remains difficult to model and understand. In addition to the usual mathematical difficulty involved in the study of free-surface problems, there is the seminal question of stress singularity at the three-phase contact line[1,2]. Over the years, several models have been proposed to alleviate these difficulties. The most popular among these is an apparent dynamic contact angle condition that allows a finite slip velocity at the contact line to remove the singularity[3]. However, the form of the dynamic contact angle boundary condition and the slip coefficient of this approach are often unkown and must be determined empirically. Application of this model to decipher spreading dynamics is also quite complex.

Recently, a more physical approach has been proposed by de Gennes[4] and Teletzke et al.[5]. Long-range intermolecular forces are shown to create a negative pressure (relative to the ambient air pressure). This negative 'disjoining pressure', together with the negative capillary pressure created by a concave cusp-shaped interface, produce a negative pressure gradient that sucks liquid from the bulk into a front-running precursor film that has been experimentally detected recently[6]. Since this film is thin compared to the bulk length scales, it can be shown (as we shall later in the manuscript) that it behaves 'quasi-steadily' relative to the bulk motion - the motion at the edge where the bulk and precursor film meet is at constant velocity as far as the flow dynamics in the film is concerned. The interfacial shape of the precursor film also remains constant and simply translates at a constant speed. However, over a longer time scale, as the bulk liquid spreads significantly, the 'adiabatically slaved' precursor film dynamics also changes accordingly. A quasi-steady precursor film implies that the liquid sucked into the film by the negative disjoining and capillary

---
*Current Address: Department of Chemical Engineering, University of Leeds, Leeds, UK LS2 9JT

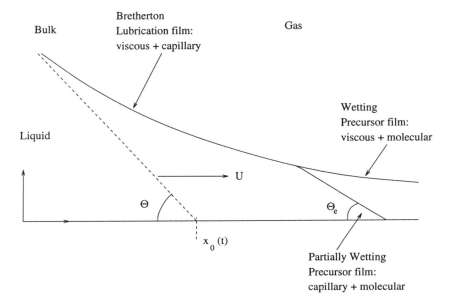

Figure 1: The precursor film ahead of the bulk liquid. The Bertherton matching region has a relative scaling of $(\hat{H}/\hat{L}) \sim Ca^{1/3}$ with the normal and tangential scales for the film being $\hat{H}_w$ and $\hat{L}_w$ for complete wetting and $\hat{H}_p$ and $\hat{L}_p$ for partial wetting, $\hat{H}_p \ll \hat{H}_w$ and $\hat{L}_p \ll \hat{L}_w$. The partially wetting film approaches the solid at the static contact ange $\theta_e$.

pressures overcomes viscous resistance in such a manner that the film moves in a quasi-steady manner with roughly constant shape and speed. The precursor film then necessarily has a cusp of wedge shape because is the only shape that provides the proper curvature to generate the necessary negative capillary pressure that sucks the liquid in and the proper shape near the cusp to allow the disjoining pressure to overcome the increasing viscous resistance (see Fig. 1).

Away from the cusp and towards the bulk, the disjoining pressure is unimportant and the steadily travelling interfacial shape is determined by capillary and viscous forces only. As shown by Bretherton[7], a film driven by these two forces to translate at a constant speed must have a slope that scales as $Ca^{1/3}$ where the capillary number $Ca = (\mu U/\sigma)$ is defined with respect to the quasi-steady edge speed U. This suggests that the outer boundary of the cusp may blow up linearly from the cusp to provide an apparent (macroscopic) contact angle $\theta$, different from the cusp angle, that scales as $Ca^{1/3}$. This would then provide a natural macroscopic dynamic contact angle condition for the bulk region and one can conveniently excise the precursor film from subsequent macroscopic modelling. It also seems promising because measurements by Tanner[8] and Hoffman[9], among others, have suggested that the macroscopic dynamic contact angle $\theta$ does scale as $Ca^{1/3}$. Analysis of the asymptotic behavior of the pertinent equation for the precursor film reveals the existence of a linear segment that can be matched with the bulk region and yield the necessary macroscopic contact angle condition. This matching idea was first forwarded by Joanny[10] for the precursor film model but it is consistent with similar 'excision' ideas based on the slip model (Hocking[11]; Ngan and Dussan[12]).

The appropriate asymptotic behavior of the precursor film away from the solid boundary is $(-s)[\log(-s) - c]^{1/3}$ where $s$ is a coordinate measured from the cusp tip. To obtain this asymptotic behavior and to determine the constant $c$, one needs to integrate the precursor film from the cusp tip. This is where the precursor film approach again encounters the stress singularity asociated with moving contact lines. Hocking[13] has recently inserted a finite slip velocity for the contact line of the precursor film whose value must again be determined empirically. It is also questionable whether the continuum model used by both approaches remain valid at the contact line. However, Joanny[10] has suggested that, under conditions that he did not specify, viscous forces become uimportant at the tip and the cusp approaches a shape determined only by capillary and intermolecular forces at the tip before one reaches the molecular scales. The exact physical mechanism that moves the molecules at the contact line then becomes unimportant. This would also remove the cusp tip from the problem and avoid the need for an unknown slip velocity to remove the stress singularity. In fact, since the Hamaker constant for the intermolecular forces is now measurable for many liquid-solid pairs, this would imply that all the parameters for the apparent dynamic contact angle of a wetting fluid are a priori known. The continuum model remains valid up to the static limit. Joanny uused the de Gennes pancake static shape at the tip. However, as Kalliadasis & Chang[14] have shown, the static shape at the tip determines the precursor film shape and the constant c in its asymptotic bahavior but c is unimportant as far as the macroscopic (apparent) contact angle $\theta$ is concerned. This suggests that the leading order approximation for $\theta$ can be obtained from simple scaling arguments without recourse to detailed analysis.

In the present work, we employ such simple scaling argument to justify some of Joanny's assumptions and to derive an explicit apparent dynamic contact angle condition that can be used in any spreading problem satisfying some weak conditions. Our simple derivation allows us to extend the theory from completely wetting fluids to partially wetting fluids. For the later case of fluids, a static contact angle $\theta_e$ is approached at large times such that the spreading will eventually stop as a static shape is approached. Since $\theta_e$ is also easily measurable, the contact-angle conditions we derive can be applied immediately to several classical spreading problems: pendant and sissile drop spreading and capillary rise. We are able to quantitatively model the dynamics of these spreading phenomena and extend the particular effect of wettability, as measured by $\theta_e$, on the dynamics.

## 2. MACROSCOPIC BULK DYNAMICS

Although the capillary and disjoining pressures can suck fluid from the bulk into the precursor film, it is not always the dominant physical mechanism that drives spreading. Gravity, which is unimportant in the thin precursor film, often drives the spreading in the bulk region and the precursor film drainage only allows the film to quasi-steadily follow the gravity-driven dynamics in the bulk. We hence need to delineate the distinction between gravity spreading and capillary/molecular spreading by analyzing the bulk region. We shall distinguish these two spreading mechanisms via the simplest spreading phenomenon-the gravity current when a thin film coats an inclined surface. This is a simple problem since it is one-dimensional and the lubrication approximation can be applied. The conclusion, however, can be extended to all other spreading problems.

Ignoring, for now, the intermolecular forces in the precursor film, one can describe the bulk-region during the spreading of a cylinder of liquid into a thin film on an inclined plane by the lubrication equation

$$3\mu \frac{\partial H}{\partial T} = -\sigma \frac{\partial}{\partial X}\left(H^3 \frac{\partial^3 H}{\partial X^3}\right) - \rho g' \frac{\partial}{\partial X}\left(H^3\right). \qquad (1)$$

where we have omitted transverse variation and assumed the transverse width remains constant during the spreading. The parameters $\mu$, $\sigma$ and $g' = g \sin \beta$ are the viscosity, surface tension and gravitational accelleration along the incline with an angle $\beta$ with respect to the horizontal. If the initial area of the cylinder is small, we expect the second gravitational term on the right of (1) to be unimportant and the cross-section shape of the cylinder quickly becomes the arc of a circle as surface tension drives fluid from one location to another. The radius of this circle is $R \sim A^{1/2}$ where $A$ is the constant cross-section area and the time required to bead up into a cylinder is determined by the balance between the capillary force $\sigma/R$ and the viscous resistance in $\mu$, $T_{cap} \sim \frac{\mu R}{\sigma}$. We have assumed that, during this interfacial adjustment, the contact lines remain motionless. This is true if the characteristic time of the contact line motion, $T_{spread} = L/U$ where U is the unkown characteristic velocity of the contact line motion, is much longer than $T_{cap}$. This then imposes the condition $(T_{cap}/T_{spread}) \sim \frac{\mu U}{\sigma} = Ca \ll 1$. We are hence restricted to small Ca spreading

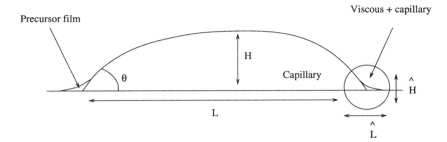

Figure 2: An example of the bulk static interface during capillary/molecular spreading with length scales $H$ and $L$. The Bretherton matching region has length scales $\hat{H}$ and $\hat{L}$ related by $\hat{H}/\hat{L} \sim Ca^{1/3}$.

and the dynamics subsequent to the beading into the cylinder with a circular arc is quasi-steady. This can be made more explicit by using $R$ as the characteristic length scale and $T_{spread}$ as the characteristic time scale to nondimenionalize (1),

$$3Ca\frac{\partial h}{\partial t} = -\frac{\partial}{\partial x}\left(h^3\frac{\partial^3 h}{\partial x^3}\right) - B\frac{\partial}{\partial x}\left(h^3\right). \tag{2}$$

where $B = \dfrac{\rho g' R^2}{\sigma}$ is the Bond number. Hence, if the cross-section dimension of the cylinder is small enough, $B \ll 1$ and if the contact line motion is slow, $Ca \ll 1$ and the bulk region during capillary/molecular spreading is quasi-static and the bulk region evolves through the members of the family of solutions of the long-wave Laplace-Young equation without gravity. These quasi-static bulk solutions intersect the solid plane at a finite angle as shown in Fig. 2.

This is the apparent dynamic contact angle $\theta$ and the relationship between $\theta$ and the spreading speed we seek essentially determines how fast the bulk shape evolves through the family of static solutions. We can also show that the small region of normal and tangential dimensions $\hat{H}$ and $\hat{L}$, near the contact line and where the macroscopic solution matches the precursor film solution, is also quasisteady for $\hat{L}/L \ll 1$. It is also clear that as one shrinks from $L$ to $\hat{L}$ and $H$ to $\hat{H}$, the buoyancy term becomes increasingly smaller relative to the capillary term and hence, at a sufficiently small neighborhood, one gets a balance between the viscous and the capillary term to yield the Bretherton equation $3Ca\dfrac{\partial H}{\partial X} = \dfrac{\partial}{\partial X}\left(H^3\dfrac{\partial^3 H}{\partial X^3}\right)$ where the typical scaling $\theta = -H_X \sim Ca^{1/3}$ holds. We note that the exact scales $\hat{H}$ and $\hat{L}$ of this Bretherton region have not been specified yet. The above analysis applies when $B \ll 1$ and the spreading is driven by gravity and intermolecular forces. When $B \gg 1$, gravity drives the spreading in the bulk and quickly flattens the cylinder into a thin film such that $H \ll L$ and the curvature in the bulk is of order $(H/L^2)$ instead of $(1/R)$ when gravity is absent. To derive the proper bulk region, we note that the characteristic tangential velocity driven by gravity is $(H^2 \rho g'/\mu)$ and hence the gravity-drainage time is $(\mu L/H^2 \rho g')$ which is now the characteristic spreading time. The dimensionless version of (1) now contains a modified Bond number $\bar{B} = (\rho g' L^3/\sigma H)$ and $\bar{B} \gg 1$ is now the appropriate condition for gravity spreading. The criteria $B \ll 1$ for capillary/molecular spreading and $\bar{B} \gg 1$ for gravity spreading apply for all spreading processes. In sessile drop spreading, for example, if one begins with a drop of small volume such that $B \ll 1$, the initial spreading is by capillary/molecular forces. But as the drop becomes a pancake, however, gravity spreading begins to dominate when $\bar{B}$ increases beyond unity as $L$ increases and $H$ decreases. This transition from one mechanism to another was dramatically demonstrated experimentally by Cazabat and Cohen-Stuart[15]. For a sessile drop, $HL^2$ scales linearly with respect to the volume $V$ and the transition pancake radius $R_c$ can be easily shown from $\bar{B} = 1$ to scale as

$$R_c \sim (\sigma V/\rho g)^{1/3}. \tag{3}$$

This simple prediction is shown to be consistent with the experimental data of Cazabat and Cohen-Stuart.

## 3. PRECURSOR FILM MODEL FOR THE APPARENT CONTACT ANGLE

To determine the lengthscales of the Bretherton regime we introduce the intermolecular terms into the Bretherton equation

$$3Ca H_X = \frac{\partial}{\partial X}\left[H^3 H_{XXX} - H^3 \frac{\partial}{\partial X}\left(\frac{\alpha}{6\pi\sigma}H^{-3} - \frac{\beta}{\sigma}H^{-3}{H_X}^4\right)\right]. \tag{4}$$

where the disjoining pressure is given by $\pi = \alpha/H^3 - \beta {H_X}^4/H^3$. The parameter $\alpha$ is negative for completely wetting fluids and positive for partially wetting ones while the parameter $\beta$ for non-parallel effects is positive for both fluids. These two terms arise from long-range intermolecular interaction as first derived by Miller & Ruckenstein[16]. A static limit exists for a partially wetting fluid yielding the static contact angle $-H_X = \tan\theta_e = \left(\dfrac{\alpha}{6\pi\beta}\right)^{1/4}$. The precursor film dimension is quite distinct between a completely wetting fluid and a partially wetting one. Following Kalliadasis and Chang[14], a completely wetting fluid exists when the Hamaker molecular term matches the viscous term exactly. The non-parallel term is negligible except it stipulates that a tip with a finite contact angle does not exist for a wetting fluid. Since the fat end of the precursor film must match into the Bretherton equation, the wetting precursor film lengthscales $\hat{H}_w$ and $\hat{L}_w$ must obey the Bretherton's scaling yielding $\hat{H}_w = \left(\dfrac{-\alpha}{6\pi\sigma}\right)^{1/2} Ca^{-1/3}$ and $\hat{L}_w = \left(\dfrac{-\alpha}{6\pi\sigma}\right)^{1/2} Ca^{-2/3}$. The matching of the asymptotic behavior of (4) away from the solid boundary occurs in the Bretherton region and this matching yields the dynamic contact angle condition which for a completely wetting fluid yields to leading order

$$tan\theta = -H_X \sim (-9\log\eta)^{1/3} Ca^{1/3}. \tag{5}$$

It is important to note, that although we use the asymptotic behavior of the Bretherton scaling with $H_X \sim Ca^{1/3}$ scaling, it is the specification of the precursor film dimensions due to a balance between intermolecular and viscous forces that allows us to specify the coefficient $(\log\eta)^{1/3}$. An intriguing observation is that this coefficient is determined from simple scaling arguments without having to resolve the precursor film. The apparent dynamic contact angle can be estimated just from simple scaling arguments. However, like all scaling arguments, the exact coefficient is not available, and we are left here with an ambiguous dependence of $\eta$ on the bulk scale $L$. Fortunately, the coefficient in front of $Ca^{1/3}$ scales as $\left[\log\left(\sqrt{|\alpha|}/L\right)\right]^{1/3}$ and it is not sensitive to the precise value of L. It is also not sensitive to the value of $\alpha$. We shall use the standard value $|\alpha| = 10^{-14}$ ergs for all our spreading models.

We can derive the apparent contact-angle condition for partially wetting liquids in a similar fashion. The result is

$$tan\theta = -H_X \sim \left[tan^3\theta_e + 9 \mid \log\eta \mid\right]^{1/3}. \tag{6}$$

where $\eta$ is now defined with a positive $\alpha$. Equation (6) indicates that away from the static equilibrium, a partially wetting fluid behaves like a completely wetting one.

While we have derived (5) and (6) by stipulating the presence of a precursor film within which intermolecular forces are important, (5) and (6) can also be derived by using slip models for both completely wetting and partially wetting fluids. In addition, several authors like for example Troian et al.[17], assume a constant-thickness precursor film for completely wetting fluids in front of the bulk. The unkown slip coefficients and film thicknesses are then empirical representation of the Hamaker constant or its dimensionless counterpart $\eta$. The most common slip condition is $3U = \gamma_i X^{-i}\left(\partial U/\partial X\right)$ at the contact line where $i$ is either taken to be zero or unity and it is usually imposed in conjuction with the static contact angle condition $H_X = -tan\theta_e$. Our asymptotic analysis

(Kalliadasis[18]) with these slip models shows that $\gamma_i \sim \eta^{1+i}$ for both completely and partially wetting fluids. For the flat precursor film model with thickness $\delta$, our analysis shows that this model can never describe partial wetting and can only describe perfect wetting if $\delta/H_0 \sim \eta$ where $H_0$ is the characteristic height of the bulk region and it is slow function of time. Since $\eta$ is constant in time, this implies the thickness of the flat-film model must vary with time. However, as long as the variation is not large, the weak logarithmic dependence in (5) on $H_0$ allows a constant thickness film to be imposed without significant error. Of course, if the solid is prewetted with a thick film, the constant-thickness model is more appropriate. These results show that, due to the Bretherton scaling at the matching region, the coefficient in front of $Ca^{1/3}$ for the contact angle condition is of order unity and is insensitive to the conditions near the contact line. Hence, all models provide the same apparent dynamic contact angle condition for the bulk. It is also interesting that (5) and (6) can be derived without detailed resolution of the precursor film. Only its characteristic dimensions are required.

## 4. APPLICATION TO SPREADING PROBLEMS

In this paper we will examine several classical capillary/molecular spreading problems by using our contact angle conditions. We begin with Blake et al.[19]'s experiment on how a meniscus climbs the inside of a vertical capillary against gravity. The Bond number $B$ is necessarily small here otherwise the meniscus will not rise. At vanishingly small $B$, the shape of the outer static meniscus is simply a sphere which represents the solution to the Laplace-Young equation. The capillary pressure generated by this curvature must then balance viscous dissipation due to the rise of liquid in the capillary and hydrostatic head. The quasi-steady balance yields the Washburn[20] equation. The classical analysis assumes that the contact angle remains constant during the spreading. However, from (5) and (6), it is clear that $\theta$ actually varies with the speed $Ca$. We then show that a perfectly wetting fluid approaches the equilibrium algebraiccally in time in contrast with the exponential approach predicted by Washburn. This algebraic approach to equilibrium is consistent with the experimental data by Blake et al.[19]. We should point out that the equilibrium of a perfectly wetting fluid is only macroscopic since the precursor film keeps on growing. A partially wetting fluid is shown to approach equilibrium exponentially in time, however, with an exponent different than Washburn's.

The next two examples concern capillary/molecular spreading of sessile and pendant drops. The gravity spreading of sessile drops does not require the contact-angle condition and has been well studied (Nakaya[21]). For the case of a sessile drop in a zero gravity environment the bulk region is described by the long-wave Laplace-Young equation without gravity, the solution of which subject to a constant volume condition is easily obtained and the slope of the bulk is set equal to the contact angle given by (5) and (6). The rate of spreading can now be obtained easily and the results for a wetting fluid are found to be in good agreement with the experimental data by Ausserre et al.[6]. For a hanging drop, we include the gravitational term in the long-wave Laplace-Young equation. The solution for the bulk region reveals the presence of an equilibrium radius for a wetting fluid which in dimensional form is $R_c = 3.832\sqrt{\sigma/\rho g}$. We found only one literature data of $R_c$ reported by Levinson et al.[22] who found an equilibrium radius of 6.0mm for a silicone oil which is in good agreement with 5.8mm predicted by our theory. Applying the apparent contact-angle condition (5) to the slope of the bulk region and integrating the resulting equation we show that a perfectly wetting fluid approaches the equilibrium algebraically in time.

## 5. SUMMARY

Since contact-line dynamics does not contribute to the spreading dynamics in gravity spreading, partially wetting fluid spreads in the same manner as wetting fluid under this mechanism. The contact-line dynamics, however, will be crucial to the instabilities that develop at the contact line and the fingers that develop from a wetting fluid are expected to be different from a partially wetting fluid. The results obtained here for capillary/molecular spreading have already revealed a difference in the bulk dynamics of the two fluids. The added retardation due to a positive disjoining pressure of a partially wetting fluid dictates that the approach to the equilibrium position, which always exists for a partially wetting fluid, must be exponential.

## 6. REFERENCES

1. Huh, C. and Scriven, L. E. Hydrodynamic Model of Steady Movement of a Solid/Liquid/ Fluid Contact Line, *J. Colloid Interface Sci.*, **35**, 85 (1971).

2. Dussan, V. E. B. and Davis, S. H. On the Motion of a Fluid-Fluid Interface Along a Solid Surface, *J. Fluid Mech.*, **65**, 71 (1974).

3. Dussan, V. E. B. On the Spreading of Liquids on Solid Surfaces: Static and Dynamic Contact Lines, *Annual Rev. Fluid Mech.*, **11**, 371 (1979).

4. de Gennes, P. G. Wetting: Statics and Dynamics, *Rev. of Modern Physics*, **57**, 827 (1985).

5. Teletzke, G. F., David, H. T. and Scriven, L. E. Wetting Hydrodynamics, *Revue Phys. Appl.*, **23**, 989 (1988).

6. Ausserre, D., Picard, A. M. and Leger, L. Existence and Role of the Precursor film in the Spreading of Polymer Liquids, *Phys. Rev. Lett.*, **57**, 2671 (1986).

7. Bretherton, F. P. The Motion of Long Bubbles in Tubes, *J. Fluid. Mech*, **10**, 166, (1961).

8. Tanner, L. The Spreading of Silicone Oil Drops on Horizontal Surfaces, *J. Phys.*, **D12**, 1473, (1979).

9. Hoffman, R. L. A Study of the Advancing Interface. I. Interface Shape in Liquid-Gas Systems, *J. Colloid Interface Sci.*, **50**, 228, (1975).

10. Joanny, J. F. Dynamics of Wetting: Interface Profile of a Spreading Liquid, *J. of Theoretical and Applied Mechanics*, **271**, 249, (1986).

11. Hocking, L. M. Moving Fluid Interface. Part 2. The Removal of the Force Singularity by a Slip Flow, *J. Fluid Mech.*, **79**, 209, (1977).

12. Ngan, C. G. and Dussan, V. E. B. On the Dynamics of Liquid Spreading on Solid Surfaces, *J. Fluid Mech.*, **209**, 191 (1989).

13. Hocking, L. M. The Spreading of Drops with Intermolecular Forces, *Phys. Fluids*, **6**, 3224, (1994).

14. Kalliadasis, S. and Chang, H.-C. Apparent Dynamic Contact Angle of an Advancing Gas-Liquid Interface, *Phys. Fluids*, **6**, 12, (1994).

15. Cazabat, A. M. anb Cohen-Stuart, M. A. Dynamics of Wetting: Effects of Surface Roughness, *J. Phys. Chem.*, **90**, 5845, (1986).

16. Miller, C. A. and Ruckenstein, E. The Origin of Flow During Wetting of Solids, *J. Colloid Interface Sci.*, **28**, 368, (1974).

17. Troian, S. M., Herbolzheimer, E., Safran, S. A. and Joanny, J. F. Fingering Instabilities of Driven Spreading Flows, *Europhys. Lett.*, **10**, 25, (1989).

18. Kalliadasis, S. *Self-Similar Interfacial and Wetting Dynamics* PhD thesis, University of Notre Dame, Notre Dame, (1994).

19. Blake, T. D., Everett, D. H. and Haynes, J. M. *Some Basic Considerations Considering the Kinetics of Wetting Processes in Capillary Systems* in Wetting, SCI Monograph 25, Staples Printers Limited, (1967).

20. Washburn, F. W. The Dynamics of Capillary Flow, *Phys. Rev.*, **17**, 273, (1921).

21. Nakaya, C. Spreading of Fluid Drops over a Horizontal Plane, *J. Phys. Soc. Japan*, **37**, 539, (1974).

22. Levinson, P., Cazabat, A. M., Cohen-Stuart, M. A., Heslot F. and Nicolet, S., The Spreading of Macroscopic Droplets, *Revue Phys. Appl.*, **23**, 1009, (1988).

# SURFACE TENSION DRIVEN THIN FILM FLOWS

T.G. Myers[†]

[†]OCIAM, Dept. of Mathematics
Oxford University, Oxford, UK OX1 3LB.

## SUMMARY

This paper reviews thin film flows where surface tension plays an important role. The aim is to highlight the wide variety of applications of surface tension driven flow and indicate where relevant work may be found. The governing equations for a thin film on an inclined plane are given in the first section. Modifications and extensions are also discussed. In the second section physical processes where these equations arise are described.

## 1. INTRODUCTION

Thin film flows have been studied for many years and for many reasons. This paper reviews just one particular aspect of thin film flows, namely that where surface tension plays an important role. The aim is to highlight the wide variety of physical applications of surface tension driven flow and indicate where relevant work may be found. A more comprehensive review dealing with both physical aspects and mathematical techniques for dealing with the relevant problems may be found in [1].

## 2. GOVERNING EQUATIONS

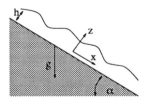

Figure 1: A thin fluid film on an inclined plane.

To derive the governing equation for a surface tension driven thin film flow the standard lubrication, or reduced Reynolds number, approximation to the Navier-Stokes equation must be used. With the configuration as shown in figure 1 the relevant boundary conditions are as follows; on the substrate $z = 0$, there is no slip so

$$u(x, 0, t) = w(x, 0, t) = 0 , \qquad (1)$$

where $u$ and $w$ are the velocity components in the $x$ and $z$ directions respectively; on the free surface, $z = h$

$$\frac{\partial u}{\partial z}(x, h, t) = 0 \qquad\qquad w(x, h, t) = \frac{\partial h}{\partial t} + u\frac{\partial h}{\partial x} \qquad (2)$$

$$p(x, h, t) = -C\frac{\partial^2 h}{\partial x^2} , \qquad (3)$$

where $p$ is the fluid pressure and $C$ is the capillary number relating surface tension to viscous forces. Conditions (2) show that shear stress is zero (for the moment) whilst the kinematic condition indicates fluid particles do not pass through the free surface. The stress condition (3) is the Laplace-Young equation stating that the normal stress is the product of surface tension and curvature. Applying these conditions to the reduced Navier-Stokes equations leads to

$$\frac{\partial h}{\partial t} + \frac{\partial}{\partial x}\left[h^3\left(C\frac{\partial^3 h}{\partial x^3} - \delta B\frac{\partial h}{\partial x}\cos\alpha + B\sin\alpha\right)\right] = 0 , \qquad (4)$$

where $\delta$ is the aspect ratio and the Bond number $B$ is the ratio of gravity to viscous forces. This is a fourth-order nonlinear degenerate parabolic equation. Note (4) is of higher order than obtained in standard lubrication theory due to the normal stress condition. A similar sixth-order equation is obtained in the work of King [2] when dealing with an elastic plate on top of a thin layer.

A generalised form of (4) is

$$\frac{\partial h}{\partial t} + \nabla \cdot \left[h^3(C\nabla\nabla^2 h - \delta B\nabla h \cos\alpha + B\vec{r}\sin\alpha)\right] = 0 . \qquad (5)$$

Most physical situations discussed in this paper can be modelled by equations of the form (5).

## 2.1. Modifications to the governing equation

Various papers deal with modifications and extensions of the above equations. A number of these are listed below. Henceforward subscripts will denote differentiation with respect to that variable.

i) Steady state
Steady forms of (4) and (5) are of considerable importance. For example a particular form derived to determine the thickness of the coating when cine-film is pulled out of a bath at constant velocity is

$$h^3 h_{xxx} - h = -1 . \qquad (6)$$

This is known as the Landau-Levich equation [3]. It neglects gravity effects whilst the no-slip condition on a moving substrate introduces the $-h$ term. It has subsequently been derived to describe the thickness of a soap-film [4], the wetting layer when a bubble moves through a capillary tube [5], the height of a tear film around a contact lens [6] and flow at the interface between a strip under tension and a coating roller [7].

Alternatively a travelling wave substitution $h(x,t) = H(x-t)$ will reduce (4) to (6) when $B$ is set to zero.

Other autonomous ordinary differential equations, related to draining flow down a vertical wall are studied by Tuck & Schwartz [8]. These include gravity and viscous shearing effects and also models for a precursor layer.

ii) Models with slip

Situations where the film thickness reaches zero, such as at the front of a moving drop, introduce the added complication of contact angles and a moving contact line. Difficulties may occur since the value of the static contact angle, usually determined from dynamic measurements, exhibits contact angle hysteresis, a non-uniqueness, on all but the smoothest surfaces. When modelling the dynamic problem the standard no-slip condition requires an infinite stress to move the contact line; any slip condition (*i.e* $u = f(h)u_z$) will alleviate this problem [9]. In practice the most common slip laws used are

$$\text{(a)} \quad u = \frac{\alpha}{3h}u_z \quad \text{or} \quad \text{(b)} \quad u = \beta u_z , \tag{7}$$

where $\alpha$ and $\beta$ are constant slip coefficients. This will modify the coefficient $h^3$ in (4) and (5) to $(h^3 + ah^m)$ where $m$ takes the value 1 or 2 for models (a) and (b) respectively. The coefficient $a \ll 1$, so that this extra term is negligible everywhere except for in the vicinity of the contact line where $h^{3-m}$ is $O(a)$. References [10,9,11,12] provide a wealth of information on the subject of static and dynamic contact lines.

iii) Surface tension gradients

Equations (4) and (5) are derived on the assumption that surface tension is constant. This is not always the case, for example when fluid evaporates (see §3.4), when there are thermal gradients (Bénards convection cells were actually driven by surface tension gradients (see Prandtl [13])) or when a film is ultra-thin (see v). In the presence of surface tension gradients the zero shear stress boundary condition becomes

$$u_z = \frac{1}{\delta^2}\sigma_x \tag{8}$$

where $\sigma$ is the surface tension. The governing equation, on a horizontal surface is

$$h_t + \left[ h^3 \left( C h_{xxx} + M \sigma_x h_{xx} - B h_x \right) + \frac{h^2}{2\delta^2} \sigma_x \right]_x = 0 \ , \tag{9}$$

where the Marangoni number $M$ is the ratio of surface tension gradients to viscous forces.

iv) <u>Curved substrate</u>
A lubrication model for surface tension driven flow on a curved substrate is developed by Schwartz & Weidner [14]. This gives rise to a variant of (4) in the form

$$h_t + \left( h^3 \left( C h_{sss} + \kappa_s \right) \right)_s = 0 \ , \tag{10}$$

where $s$ is the co-ordinate tangential to the substrate and $\kappa$ is the curvature.

v) <u>Long range molecular forces</u>
In reality a thinning film will often reach a characteristic critical thickness, typically a few hundred angstroms, and then rupture on a very short time-scale. Over such small length-scales long range molecular forces (in particular the van der Waals force) become significant and these have been proposed as a mechanism for this rapid breakdown of the film [15,16]. To model this force a potential energy function/unit liquid volume, $\phi$, was introduced in [17]. This introduces a body force $F = -\nabla \phi$ in the Navier-Stokes equations. Taking $\phi \sim \phi_0 + A'/h^3$ leads to

$$h_t + \left( C h^3 h_{xxx} + A \frac{h_x}{h} \right)_x = 0 \ , \tag{11}$$

where $A(\ll C)$ is proportional to the Hamaker constant. Numerical solutions to this equation have been obtained [18] and show an initial disturbance slowly draining until a trough develops which quickly leads to rupture.

vi) <u>Extensional flows</u>
When the fluid has two free surfaces, as in a soap film, the shear flow models (4), (5) no longer describe the problem. Now the flow is termed "extensional". Because the fluid velocity is unknown on both surfaces a higher order approximation in $\delta$ is generally required to close the leading order problem. This leads to

$$\begin{aligned} h_t + (uh)_x &= 0 \\ (4hu_x)_x + hh_{xxx} + Bh \sin \alpha &= 0 \ . \end{aligned} \tag{12}$$

This system is derived in [19,20]. To arrive at a simpler problem many authors assume one surface to be loaded with surfactant and therefore inextensible. The fluid velocity there may then be specified (and is generally set to zero) reducing the problem to (4). This is discussed further in §3.6.

vi) Fluid models with evaporation
If the thin layer comprises two components (such as a resin and solvent), one of which is volatile then a coupled system arises

$$h_t + Q_x = -E \tag{13}$$
$$(sh)_t + (sQ)_x = -E + D(hs_x)_x \tag{14}$$

where $s$ is the concentration of the evaporating fluid, $E$ the evaporation rate, $D$ the diffusion coefficient of $s$ and the flux $Q$ is given by

$$Q = \frac{h^3}{\mu}(Ch_{xx} - \delta Bh)_x - \frac{h^2}{2\mu}Ms_x \ . \tag{15}$$

This has been proposed as a model for paint drying. It is discussed in §3.4.

## 3. PHYSICAL PROCESSES

### 3.1. Coating flows

Coating flows involve covering a surface with one or more thin layers of fluid. They range from rain running down a window to manufacturing processes such as the production of video-tapes. In many situations surface tension plays an important role, for example in determining the film thickness in dip coating, levelling surface disturbances such as ribbing or brush-marks and helping to retain (or remove) imperfections such as bubbles.

Further discussion on coating would be preaching to the converted. More information on coating methods and unresolved problems can be found in reviews by Friedman [21], Ruschak [22] and Benjamin & Scriven [23].

### 3.2. Condensate motion on heat exchangers

Film condensation is a method of heat transfer used in chemical plants, refrigerators, air conditioners and other cooling devices. The method of cooling involves passing a vapour over metal fins which are kept below the condensation temperature. Condensate builds up on the fins, flows into the channel between them and subsequently drains away. At low vapour flow rates the transfer of heat to the fins has been found to be much enhanced compared to measurements at higher flow rates. Surface tension aided drainage, producing locally thin films and thus reducing thermal resistance, has been proposed as the mechanism responsible for this discrepancy [24].

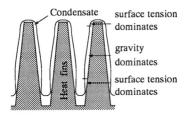

Figure 2: Fluid flow on cooling fins.

At the top of the fin large stresses resulting from the high curvature act to reduce film thickness. This region has been investigated in order to calculate the optimum radius of curvature for a fin (see Hirasawa *et al* [24]). In the trough high curvature produces a suction effect which acts to pull fluid down the side wall. These two effects, acting to reduce the film thickness and hence thermal resistance can improve heat transfer by up to three times that predicted by Nusselt theory [25].

### 3.3. Contact lenses

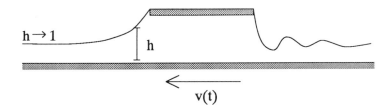

Figure 3: A typical contact lens configuration.

Despite the extensive use of contact lenses over the last few decades their motion over the eyeball is still not yet fully understood, particularly the mechanism by which a lens centres itself over the cornea. An understanding of lens behaviour is important in their design since a poorly fitting lens may lead to corneal problems; also the fluid motion under the lens controls oxygen diffusion into the eye. Moriarty & Terrill[6] develop a model with the specific aim of describing the motion of a hard contact lens on a tear film, although the model is applicable to the motion of any rigid object on the surface of a thin film. In two dimensions they arrive at the Landau-Levich equation (6) to model tear film thickness away from the lens; they also derive a three-dimensional version. The simplifying assumptions made in this analysis, in particular that the eyeball is flat and the lens has constant velocity, mean that the model cannot explain centration, but it does provide a starting point for more complex models and geometries.

## 3.4. Paint drying

A newly painted surface normally has an uneven profile, the most familiar example being "brush-marks". In order for the film to level and produce a smooth finish it must be mobile but this allows sagging (a downward flow due to gravity) to take place on an inclined surface.

Orchard *et al* [26,27] show that for certain paints surface tension may drive the levelling process. Their work, however, is unable to explain "reversal", a process by which peaks in the initial film may become troughs and vice-versa. This problem was addressed by Overdiep [28] and also Wilson [29]. In this case the paint is considered to consist of a resin dissolved in a volatile solvent. The governing equations are (13) and (14). They include surface tension, surface tension gradients and gravity. Although certain terms may be discarded depending upon the situation. An estimate of the relevant time-scales indicates which mechanisms must be considered.

## 3.5. Marangoni effects

If a free liquid film is to survive for any length of time in practice it must contain surface active agents or surfactants. Hence pure water will not foam whilst contaminated water (containing detergent for example) will.

Surfactants are molecules that preferentially accumulate at a fluid surface. It is energetically favourable for them to occupy the surface rather than the bulk fluid so they reduce surface energy or surface tension. Surfactants give stability to films by two processes [30]. To understand these processes consider a finite volume of liquid. If the surface expands then, in equilibrium, the surface concentration of surfactant must have decreased and the surface tension will therefore be higher. This is known as the Gibbs effect or Gibbs elasticity. Before reaching equilibrium stretching the surface will lead to an even lower concentration of surfactant molecules at the surface. The concentration will increase as molecules move to the surface through the bulk fluid but until equilibrium is reached the surface tension will be even higher than expected from Gibbs elasticity. This is known as the Marangoni effect. Hence both the Gibbs effect and the Marangoni effect act to return the film to its previous state thus providing stability.

The Marangoni effect has recently been exploited in a process known as Marangoni drying. This is a method of drying which allows hydrophilic surfaces to be withdrawn from water virtually dry [31,32].

Marangoni drying relies on an air-borne vapour (such as alcohol) being water soluble and lowering the surface tension of the water. The water layer thickness is position-dependent and, since the thickness is greatest close to the bath, the surface tension is also greatest there. Thus a surface tension gradient is set up which acts to pull the water back into the bath. The basic mechanism can be demonstrated experimentally by placing a piece of cotton wool soaked in alcohol above a thin water film and inducing a drying flow. This process has been modelled in [33].

The Marangoni effect can also be used to enhance spin-drying. In the presence of alcohol the film may break up into 'contracted' droplets (with a contact angle $> 90°$) which can be spun off.

The obvious advantage of Marangoni drying is the processing speed. In the final stages of conventional spin drying the film thickness decays exponentially, so the final layer must be removed by evaporation. An added advantage of this process is the fact that it leaves a clean surface. If the bath water is not scrupulously cleaned contaminants are left on the surface after evaporation but when the water is pulled back into the bath the contaminants go with it.

## 3.6. Foams and free films

A dry foam is a two-phase fluid consisting of gas separated by a thin continuous liquid film, where the volume fraction of the liquid phase is small. The majority of the liquid is contained in the film junctions or Plateau borders; the films have negligible thickness and liquid content.

As previously mentioned the leading order problem is closed by considering a higher order approximation than in standard lubrication theory. A common way of avoiding this is to assume the free surface is loaded with surfactant and therefore inextensible. Physically this may occur when a 'scum' forms on a surface but in general it is an unrealistic assumption.

Schwartz & Princen [34] use the inextensibility assumption to obtain the Landau-Levich equation as a model for the film height in the transition region between the Plateau border and the thin film. Work on foams and soap-films are also applicable in the study of liquid-liquid emulsions where two droplets coalesce after the fluid separating the droplets has drained and ruptured in the same manner as might a soap film. This work is also relevant to bubble bursting since a bubble about to break through a surface will be separated from it by a thin liquid film [35].

1. Myers T.G. Thin films with high surface tension. Submitted to SIAM review (1995).

2. King J.R. The isolation oxidation of silicon: the reaction controlled case, *SIAM J. Appl. Math.* **49**(4) pp1064-1080 (1989).

3. Landau L. & Levich B. Dragging of a liquid by a moving plate. *Acta Physicochim. URSS* **17** 42–54 (1942).

4. Mysels K.J., Shinoda K. & Frankel S. Soap films: A study of their thinning and a bibliography. Pergamon Press (1959).

5. Bretherton F.P. The motion of long bubbles in tubes. *JFM* **10** pp166-188 (1961).

6. Moriarty J.A. & Terrill E.L. Mathematical modelling of the motion of hard contact lenses. To appear *EJAM* (1994).

7. Rees S. An experimental and theoretical investigation of gravure roll coating. Ph.D. Thesis, University of Leeds (1995).

8. Tuck E.O. & Schwartz L.W. A numerical and asymptotic study of some third-order ordinary differential equations relevant to draining and coating flows. *SIAM Review* **32** pp453-469 (1990).

9. Dussan V., E.B. & Davis S.H. On the motion of a fluid-fluid interface along a solid surface. *JFM* **65(1)** pp71-95 (1974).

10. Dussan V., E.B. On the spreading of liquids on solid surfaces: static and dynamic contact lines. *Ann. Rev. Fl. Mech.* **11** pp371-400 (1979).

11. Goldstein S. Modern developments in fluid dynamics. Dover, New York pp676-680 (1965).

12. Leger L. & Joanny J.F. Liquid spreading. *Rep. Prog. Phys.* pp431-486 (1992).

13. Prandtl L. Essentials of fluid dynamics. Blackie, London (1952).

14. Schwartz L.W. & Weidner D.E. Modelling of coating flows on curved surfaces. *J. Engng Math.* **29** pp91-103 (1995).

15. Scheludko A. Thin liquid films. *Adv. Coll. Interf. Sci.* **1** pp391- (1966).

16. Vrij A. Possible mechanism for the spontaneous rupture of thin free liquid films. *Disc. Farad. Soc.* **42** pp23-33 (1966).

17. Ruckenstein E. & Jain R.K. Spontaneous rupture of thin liquid films. *Chem. Soc. Faraday Trans.* **2**, 70 pp132-147.

18. Williams M.B. & Davis S.H. Nonlinear theory of film rupture. *J. Coll. Interf. Sci.* **90(1)** pp220-228 (1982).

19. Erneux T. & Davis S.H. Nonlinear rupture of free films. *Phys. Fl. A* **5(5)** pp1117-1122 (1993).

20. Howell P.D. Models for thin viscous sheets. To appear *EJAM*.

21. Friedman, A. Unresolved mathematical issues in coating flow mechanics. *Mathematics in industrial problems, IMA volumes in Maths and its applications* **16**, Springer- Verlag, New York (1988).

22. Ruschak K.J. Coating flows. *Ann. Rev. Fl. Mech.* **17** pp65-89 (1985).

23. Benjamin D.F. & Scriven L.E. Coating flows: form and function. *Ind. Coat. Res.* **1** pp1-37 (1991).

24. Hirasawa, S., Hijikata K., Mori Y & Nakayama W. Effect of surface tension on condensate motion in laminar film condensation. (Study of liquid film in a small trough) *Int. J. Heat Mass Trans.* **23** pp1471-1478 (1980).

25. Chun K.R. & Seban R.A. Heat transfer to evaporating liquid films. *Trans. ASME J. Heat transfer* **93(4)** pp391-396.

26. Orchard S.E. Surface levelling in viscous liquids and gels *Appl. Sci. Res. A* **11** pp451-464 (1962).

27. Smith N.P.D., Orchard S.E. & Rhind-Tutt A.J. The physics of brush marks. *J. Oil Col. Chem. Assoc.* **44** pp618- (1961).

28. Overdiep W.S. The levelling of paints. *Prog. in Org. Coat.* **14** pp159-175 (1986).

29. Wilson S.K. The levelling of paint films. *IMA J. appl. Math.* **50(2)** pp149-166 (1993).

30. Aubert J.H., Kraynik A.M. & Rand P.B. Aqueous foams. it Sci. Am. **254(5)** pp58-66.

31. Leenaars A.F.M., Huethorst J.A.M. & van Oekel J.J. Marangoni drying: a new extremely clean process. *Langmuir* **7** pp2748- (1991).

32. Marra J. & Huethorst J.A.M. Physical principles of Marangoni drying. *Langmuir* **7** pp2748-2755 (1991).

33. O'Brien S.B.G.M. On Marangoni drying: nonlinear kinematic waves in a thin film. *JFM* **254** pp649-670 (1993).

34. Schwartz L.W. & Princen H.M. A theory of extensional viscosity for flowing foams and concentrated emulsions. *J. Coll. Interf. Sci.* **118(1)** pp201-211 (1987).

35. Chen J-D., Hahn P-S. & Slattery J.C. Coalescence time for a small drop or bubble at a fluid-fluid interface. *AIChE J.* **30(4)** pp622-630 (1984).

# APPLICATION PROPERTIES OF DECORATIVE PAINTS

W.S. Overdiep
Akzo Nobel Central Research
Physical Chemistry Department
P.O. Box 9600, 6800 Arnhem, The Netherlands

## SUMMARY

Starting from the requirements of a professional painter with regard to application behaviour of decorative coatings, the bounds of the relevant basic properties of a decorative paint satisfying these requirements are derived. The analysis shows that conventional low solids alkyd paints satisfy these requirements to the fullest measure, and the root of unsatisfactory behaviour of alternative concepts becomes clear.

## INTRODUCTION

The last decades the research effort spent on the fundamentals of organic film coating is growing. However, this only applies to industrial coating processes where one has good control of the process parameters. In contrast relatively little attention has been paid to painting by craft using brushes and rollers. The main reason for this is the lack of control of the relevant parameters, such as air drag, -temperature, -humidity, the complexity of the object to be painted, and, last but not least, the skill of the painter.

Nevertheless, it seems possible to establish a window of paint properties that allows the demands of the average painter with respect to application and film formation to be met. The most important demands a painter makes concerning application properties relate to brushability, open time, sagging on vertical planes and the levelling of brushmarks. Particularly for high gloss paints the highest demands are made on this last aspect.

In this paper the application properties mentioned above are discussed. As the wishes of the painter are taken as the starting point for the discussion, a specification of the paint properties is created as they should be and not necessarily as they are at present. From the discussion it will appear that the conventional low solids high gloss paints provide the best opportunity to fulfill the painter's demands. The problems that arise when we search for alternatives wil become clear.

## BRUSHABILITY AND OPEN TIME.

Practical experience reveals that for easy application of a decorative paint film of about 50 μm wet thickness by means of a brush a paint viscosity of about 1 Pas is required. Generally some paint strokes are applied first and subsequently the strokes are spread over the surface until a film of uniform thickness is obtained. This paint distribution over the surface takes time, and already within this period a slight change of the composition of the paint occurs causing a considerable increase of the viscosity. A reasonable upper limit as experienced by the painter seems to be about 2.5 Pas. If we assume that the painter needs about 5 minutes to paint a large object, e.g. a door, we arrive at the following limitation for the viscosity increase of the film:

$$\frac{\eta(300)}{\eta(0)} < 2.5 \tag{1}$$

## SAGGING

Every freshly applied paint film tends to sag to some extent on vertical surfaces. As a rule of thumb it holds that to prevent excessive sagging, which leads to the formation of curtains and tears at lower edges, the free surface of a 50 μm film should sag less than about 2 mm:

$$\int_{t_A}^{\infty} \int_0^h v_y \, dy \, dt = \rho g \int_{t_A}^{\infty} \frac{1}{\eta} \int_0^h (h-y) \, dy \, dt = \frac{\rho g h^2}{2} \int_{t_A}^{\infty} \frac{dt}{\eta} < 2.10^{-3} \, m \tag{2}$$

or

$$\int_{t_A}^{\infty} \frac{dt}{\eta} < 150 \, Pa^{-1} \tag{3}$$

where          $t_A$ = the time when the paint application is complete
The film thickness is assumed to be constant, which is of course not true if the solvent evaporates. Therefore the criterion in Eq. 3 is only a rough estimate. Regarding the time dependence of the viscosity, however, we can draw two conclusions from this criterion. For large times the fluidity $(1/\eta)$ should vanish and the integral should exist.

## LEVELLING

Particularly for glossy paint surfaces disturbances of the flatness have a detrimental effect on the coating appearance. So during film formation the brushmarks have to level completely. For the description of this levelling process we seek solutions of the partial differential equation

$$h_t + \phi_x + E = 0 \tag{4}$$

where            h = film thickness
                     E = rate of solvent evaporation (m/s)
                     $\phi$ = the volume flux in the film.

For thin films with slow motions and small gradiënts in thickness the stress field in the film is given by

$$\tau_{xx,x} + \tau_{xy,y} = 0 \tag{5}$$

subject to the boundary condition

$$\tau_{xy}\big|_h = \sigma_x \tag{6}$$

(Levich-Aris condition)
where            $\sigma$ = surface tension
Furthermore

$$\tau_{xx} = -P \approx \sigma h_{xx} \tag{7}$$

The shear stress at the boundary $\sigma_x$ follows from

$$\sigma_x = \sigma_c \, c_x \tag{8}$$

where the (volume) fraction of the resin c is given by the resin balance:

$$(ch)_t + (c\phi)_x = 0 \tag{9}$$

Given the rheological equation of state, Eqs 4-9 can be solved numerically. For the newtonian case and an initial profile

$$h(0,x) = h(0) + a(0) \sin \frac{2\pi x}{\lambda} \tag{10}$$

we get upon linearization the ordinary differential equation for the amplitude of the surface wave (ref.1):

$$a_t = -\frac{C_1}{\eta(t)} \left( a + C_2 \int_0^t a(s)\, ds \right) \tag{11}$$

with the constants

$$C_1 = \frac{1}{3} v^4 \frac{\sigma_A}{\eta_A K}$$

$$C_2 = \frac{3}{2} v^{-2} \frac{\Gamma}{\sigma_A} S_A (1-S_A) \tag{12}$$

$S_A$ = volume fraction of solvent

(after application)

Here K is the rate constant of solvent evaporation defined by

$$E = K\, S|_h \tag{13}$$

and $\Gamma$ is the slope of surface tension vs. resin concentration

$$\Gamma = \sigma_c \tag{14}$$

Finally

$$v = \frac{2\pi h}{\lambda} \tag{15}$$

$\lambda$ = wavelength of the surface ripple

To get an impression of the effect of $C_1$, $C_2$ and $\eta(t)$ on levelling we can use van der Hout's solution (as given in ref.1) of Eq.11 for large times assuming an exponential increase in viscosity versus time

$$\eta = \eta_A \, e^{\frac{t}{t_k}} \tag{16}$$

$$\frac{a(\infty)}{a(0)} = 1 + \sum_{k=1}^{\infty} \frac{(-C_1 \tilde{t}_k)^k}{k!} \prod_{j=1}^{k} \left( 1 + \frac{C_2 \tilde{t}_k}{j} \right) \tag{17}$$

where

$$\tilde{t}_k = t_k \frac{K}{h_A} \tag{18}$$

For any value of $C_1\tilde{t}_k$ a value of $C_2\tilde{t}_k$ can be found for which $a(\infty) = 0$. See Fig.1.

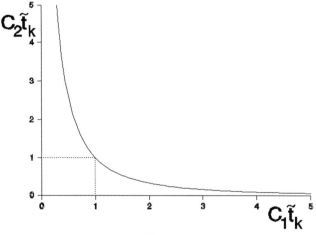

Figure 1

For the special case $C_2 = 0$ Eq.17 reduces to

$$\frac{a(\infty)}{a(0)} = e^{-C_1\tilde{t}_k} = e^{-\frac{1}{3} v^4 \frac{\sigma_A}{h_A} \frac{t_k}{\eta_A}} \tag{19}$$

which is the expression for levelling derived by Smith et. al. (ref.2) in 1961. Remembering that according to Eq.16

$$\int_{t_A}^{\infty} \frac{dt}{\eta} = \frac{t_k}{\eta_A} \tag{20}$$

equation 19 predicts perfect levelling if the fluidity integral diverges, independent of the value of $C_1$.

If $C_2 \neq 0$ the fluidity integral must exist to obtain perfect levelling, but at the expence of only one specific value of $C_1$ (Fig.1). In practical situations brushmarks represent a spectrum of wavelengths. In order to obtain optimal levelling the properties should be such that the effects of wavelength variations are as small as possible. See Fig.2. From Eq.17 the derivative of $a(\infty)$ vs $\ln(v)$ has been calculated and plotted vs $C_1 \tilde{t}_k$. It is evident that for optimal levelling $C_1 \tilde{t}_k$ should be as large as possible.

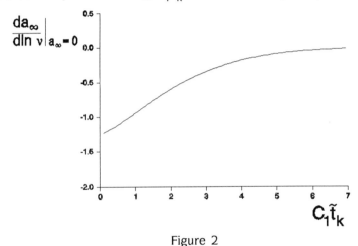

Figure 2

DISCUSSION

If we assume an exponential time dependence of the film viscosity as in Eq.16, then from Eq.1 it follows that $t_k$ should be

$$t_k > \frac{300}{\ln 2.5} \approx 325 \ s. \tag{21}$$

to avoid a too short open time. From Eqs.1,3 and 20 it follows that

$$t_k < 2.5*150 \approx 375 \; s. \tag{22}$$

to avoid unacceptable sagging. Therefore the acceptable range for $t_k$ is rather narrow:

$$325 < t_k < 375 \quad s \tag{23}$$

The largest wavelength of brushmarks is about 4 mm. For a 50 $\mu$m film we then calculate for $v$ (Eq.15) about 0.1. From Eqs.12, 18 and 22 we obtain

$$C_1 \tilde{t}_k \approx 1 \tag{24}$$

if the surface tension $\sigma_A \approx 30$ mN/m. Then, in order to achieve a levelled film, we see from Fig.1 that

$$C_2 \tilde{t}_k \approx 1 \tag{25}$$

or, from Eqs 12, 18 and 25 that

$$K \, \Gamma \, S_A \, ( 1 - S_A ) \approx 10^{-11} \; N/s \tag{26}$$

The rate constant of evaporation K for water based (ref.3) as well as for solvent based decorative coatings (ref.4) is of the order of $10^{-8}$ m/s. So for paints with a solvent content $S_A \approx 0.5$ the sensitivity of the surface tension on a change in solvent content $\Gamma$ (Eq.14) must be about $4.10^{-3}$ N/m. Alkyd resins dissolved in white spirit satisfy this requirement reasonably well, which allows us to conclude that alkyd paints satisfy all the painter's application demands if (Eqs.21,22) $t_k \approx 350$ s. for a 50 $\mu$m film.

The evolution of the fluidity integral (Eq.3) can be measured by means of the Sagging Balance. A film is applied on a test panel by means of a knife. This panel is placed on two sharp pins in an inclined position and hangs at one end on a sensitive electronic balance. The sagging is measured by the change of weight recorded by the balance. From the geometry of the film, the paint density and the angle of inclination of the panel the fluidity integral can be calculated as a function of time from the recorded curve. See figure 3. The Sagging Balance is placed in a heating stove (but it will be clear that measurements on decorative paints are carried out at room temperatere). Analysis of the curve obtained reveals that the initial viscosity $\eta(0) \approx 1.3$ Pas and $t_k \approx 775$ s. The thickness of the film h was 65 $\mu$m and the angle of inclination $20^\circ$. Separate evaporation measurements on the paint under the same conditions revealed that $K \approx 9.\,10^{-9}$ m/s.

276

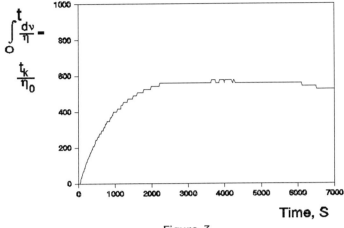

**Time, S**

Figure 3

From the solvent balance

$$(h\ S)_t = -K\ S = h_t\ S + h\ S_t = -K\ S^2 + h\ S_t \tag{27}$$

and the resin balance

$$h\ (1-S) = h(0)\ (1-S(0)) \tag{28}$$

we find for the rate constant of viscosity increase $t_k^{-1}$

$$t_k^{-1} = (\ln\ \eta)_t = -(\ln\ \eta)_s\ \frac{K}{h(0)}\ \frac{S\ (1-S)}{1-S(0)} \approx -\ln(\eta)_s\ \frac{K}{h(0)}\ S(0) \tag{29}$$

So, according to Eq.27 $(\ln\ \eta)_S \approx 20$. This order of magnitude is quite acceptable for alkyd paints. For a 50 μm film $t_k$ is about 600 s. This value is too large, but under more practical conditions the magnitude of K will easily be enhanced by a factor of 2 ($\approx 2.10^{-8}$ m/s) by air drag and free convection effects on vertical surfaces and thus $t_k$ will be expected to have the right magnitude in a more open environment. Accordingly, the measured asymptote of the fluidity integral of about 600 Pa$^{-1}$ (fig.3) is expected to be half as large in practice, which is still in conflict with the sagging requirement Eq.3. But the initial viscosity $\eta(0) \approx 1.3$ Pas is that of the fresh paint and not the maximum allowable application viscosity $\eta_A \approx 2.5$. So in practice the fluidity integral is expected to be about 150 Pa$^{-1}$, which is in agreement with Eq.3.

The comparison presented above of the painter's application requirements and the relevant properties of conventional decorative alkyd paints justifies their use as a reference for the development of alternative paint concepts.

The quest for higher solids decorative coating formulations is on the one hand faced with two fundamental problems regarding the application requirements and on the other hand offers a solution for the restrictions imposed by one of them. To maintain the application viscosity at the desired level, the molecular weight of the resin (and so also its viscosity) has to be reduced as the solvent content decreases. Consequently, the slope of the logarithmic viscosity-solvent content curve $(\ln \eta)_S$ decreases, resulting in too large magnitudes of $t_k$ (Eq.30) regarding sagging (Eq.22), but very large open times (Eq.21) as well. Thixotropic additives building a weak network in the paint at relatively low stresses reduce $t_k$ to the desired level with respect to sagging, without affecting the favourably large open time by the reversibility of the thixotropic effect at high stresses. The other fundamental problem is the reduction of the product $S_A(1-S_A)$ (Eq.12), giving rise to lower values of $C_2$. As sagging requires $C_1 t_k \leq 1$, the only solution to maintain the levelling properties of high solids formulations at the level of low solids paints is adjustment of $\Gamma$ by a proper choice of the resin-solvent combination.

As mentioned above, thixotropy introduces shear stress dependence of the viscosity. This means that instead of using Eq.17 as a solution of the levelling Eq.11, the function $\eta(\tau,t)$ should be modelled, and the development of the surface profile of an initially rippled film surface should be studied numerically using finite element techniques. Although far more complicated, a thorough analysis of thixotropic paints can be performed in essentially the same way as discussed in this paper on non-thixtropic paints.

The same holds for water based decorative dispersion paints, which are thixotropic by their formulation. The first problem met with water based dispersions is their low high shear viscosity, which approximates the viscosity of their dispersing medium, water. Thickeners are applied to lift this viscosity up to the desired application level. Then however, the low shear stress viscosity and its rate of build-up $t_k^{-1}$ is, in spite of improvements made by recent developments of so-called associative thickeners, are very large. This is demonstrated by a Sagging Balance measurement on a water based dispersion paint. See Fig.4. This curve shows that there are two mechanisms of visocosity build-up; a fast rate by thixotropy and a slow rate by evaporation of water. The initial viscosity is abt. 6 Pas. During handling of the panel with the paint film before initiation of the measurement, the low shear viscosity has increased already to this level. It will be evident, that in this case sagging drops out as a restrictive requirement, but levelling will be very poor. Even more so, as emulsifiers for the emulsion polymerisa

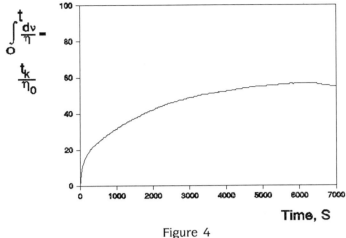

$$\int_0^t \frac{dv}{\eta} = $$

$$\frac{t_k}{\eta_0}$$

Figure 4

tion of the resin, and defoamers in the formulation will prevent the formation of surface tension gradients ($C_2$ = 0).

## CONCLUSIONS

It has been shown that the craft operation of painting is suitable for an analysis, leading to a set of physical properties to be quantified by the formulator to satisfy the application demands of the painter. Paints developed empirically without constraints regarding quality and quantity of solvent appear to satisfy these requirements best. The analysis of thixotropic paints is, however more laborous, essentially possible.

The difficulties met with application properties of alternatives such as water borne dispersion paints and high solids paints can be traced back to lack of control of thixotropy and surface tension gradients.

## REFERENCES

1. Overdiep, W.S.; POC 14,1(1986)
2. Smith, N.P.D., S.E. Orchard, A.J. Rhind-tutt; JOCCA 44,618(1961)
3. Croll, S.G.; J.Coat.Tech. 58,41(1986)
4. Gardner, G.S.; Ind.Engng.Chem. 32,226(1940)

# SPREADING OF LIQUIDS ON DRY AND PRE-WET SOLID SURFACES

Y.D. Shikhmurzaev
Institute of Mechanics
Moscow University, 119899 Moscow, Russia

A unified mathematical model for the spreading of liquids over dry and microscopically pre-wet solid surfaces is developed. For both cases the model eliminates the shear-stress singularity inherent in the classical solution of the moving contact-line problem and adequately describes available experimental data.

## 1. INTRODUCTION

One of the fundamental problems in mathematical modeling of wetting phenomena is how to describe the spreading of liquids on dry and pre-wet solid surfaces by means of a unified mathematical model formulated in the framework of continuum mechanics. Such a model would open a self-consistent way of investigating the influence of a microscopic liquid films ahead of the contact line on the dynamic contact angle and other macroscopic characteristics of the coating flow.

In general, from the mathematical point of view the term 'pre-wet' means that the solid surface ahead of the contact line is covered by a liquid film which cannot be described by the equations used for the bulk flow modeling and therefore must be incorporated into the boundary conditions on the solid surface. However, physical interpretation of this term as well as of the term 'dynamic contact angle' depends on the level of description, while the term 'dry solid surface' is understood quite uniquely.

Even in the framework of continuum mechanics, one can distinguish the following three main levels of modeling with their specific concepts, scopes and difficulties. One limiting case is the subject of *macrohydrodynamics*[1,2] which considers flows over rough and/or chemically inhomogeneous solid surfaces on characteristic length scales $\mathcal{L}$ large compared with dimensions of the solid surface inhomogeneities. The scope of macrohydrodynamics is to reduce the actual problem to considering the flow over some 'effective' *smooth* boundary (Fig. 1a) with some 'effective' boundary conditions on it and an 'effective' contact angle $\theta_{\text{eff}}$ provided that the model describing the actual wetting is known. In macrohydrodynamics, the term 'pre-wet' means that the actual liquid film ahead of an 'effective' contact line is either discontinuous (though an averaged 'effective' film will be, of course, continuous) or its averaged thickness is negligible compared with $\mathcal{L}$, though in principle this film may be thick enough to be described by the classical fluid mechanics equations.

The opposite limiting case is considered by *microhydrodynamics*[3] which deals with length scales comparable with the thickness of the interfacial layers (see Fig. 1c) so that the diffuse nature of interfaces as well as the long-range intermolecular forces must be explicitly taken into account in the bulk equations. The dynamic contact angle $\theta_{\text{act}}$ used in microhydrodynamics may be called *actual* or *microscopic*. For microhydrodynamic length scales the term 'pre-wet solid surface' means that the thickness of the liquid film ahead of the contact line is so small that it cannot be described in the framework of the

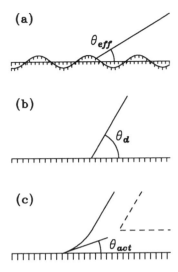

Fig. 1. An explanatory scheme for the classification of the levels of mathematical modeling of wetting phenomena: macrohydrodynamics (a), hydrodynamics (b), microhydrodynamics (c). The dashed line corresponds to the liquid-facing boundary of the interfaces.

continuum mechanics approach, which is applicable for the bulk flow.

Between the above-described limiting cases there is the area of classical *hydrodynamics* which considers the flow over the actual (not 'effective') solid surface on the length scales large compared with the interfacial layer thickness so that the interfaces may be regarded as geometrical surfaces with their specific *surface* properties, and the bulk equations will not include any intermolecular forces of the non-hydrodynamic nature. In wetting hydrodynamics, one deals with the dynamic contact angle $\theta_d$ (Fig. 1b), which may be called *macroscopic\**, and the term 'pre-wet solid surface' implies that the thickness of the liquid film ahead of the contact line is comparable with the interfacial layer thickness (Fig. 2).

---

\* Just this angle is used as a geometrical boundary condition for the capillarity equation, which determines the free-surface shape in classical fluid mechanics.

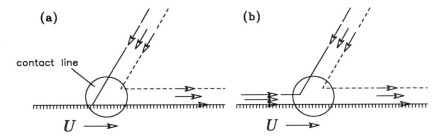

Fig. 2. The scheme of the flow near the moving contact line for dry (a) and pre-wet (b) solid surfaces. From the point of view of wetting *hydrodynamics* the three-phase-interaction zone (the circled area) is regarded as a one-dimensional structureless 'contact line'.

It should be pointed out that in this case there is no contact line from the point of view of microhydrodynamics since it treats this film as the 'bulk'.

The fundamental difficulty of wetting *hydrodynamics* is how to eliminate the non-integrable shear-stress singularity at the moving contact line inherent in the classical formulation of the moving contact-line problem[4]. It should be emphasized that the presence of a microscopic liquid film ahead of the contact line must not be used as a remedy for the singularity since the desired mathematical model should be applicable to the spreading of liquids on both dry and pre-wet solid surfaces. The scope of the present work is to propose such a model and investigate some of its properties.

## 2. PHYSICAL BACKGROUND

The present work continues a series of papers[5−7] devoted to the moving contact-line problem in gas/liquid/solid systems. An approach used in those papers is formulated on the basis of the following physical ideas:

— Experiments[8] show that near the moving contact line the liquid undergoes the so-called *rolling* motion so that liquid elements which initially belong to the liquid-gas interface traverse the three-phase-interaction zone (the 'contact line') and become elements of the liquid-solid interface in a finite time. Thus their *surface* properties (first of all, the surface tension) have to relax to the new equilibrium values;

— Experiments[9] show that the surface tension relaxation time is a macroscopic quantity so that relaxation occurs not within the three-phase-interaction zone but along the liquid-solid interface and gives rise to the surface tension gradient along the interface;

— Since in the vicinity of the moving contact line the liquid-solid interface is not in equilibrium and its properties are velocity-dependent, the dynamic contact angle is not equal to the static one and depends on the contact-line speed;

— The surface tension gradients along the liquid-solid interface, which is caused by the flow, will have a reverse influence upon the bulk flow (this feedback may be called the flow-induced Marangoni effect).

The simplest mathematical model[5,6] developed in the framework of this approach for the spreading of liquids on dry surfaces eliminates the shear-stress singularity and is in good agreement with experimental data.

If the solid surface is pre-wet, then there will be an additional surface mass flux into the three-phase-interaction zone (Fig. 2b), and in general, the tangential projection of the force acting on the contact line from the liquid-gas interface $p_{res}^s$ will differ from that in the film-free situation $p_{SG}^s$. Obviously, $p_{res}^s \to p_{SG}^s$ as the surface density of the microscopic film ahead of the contact line, $\rho_{res}^s$, tends to zero.

## 3. BASIC EQUATIONS

We will consider the flow of a Newtonian liquid in the neighbourhood of a moving contact line at small Reynolds number.

The essence of the model[5,6] is given by the generalized Navier boundary condition

$$\mathbf{n} \cdot \mathbf{P} \cdot (\mathbf{I} - \mathbf{nn}) - \tfrac{1}{2}\nabla p^s = \beta(\mathbf{u} - \mathbf{U}) \cdot (\mathbf{I} - \mathbf{nn}) \tag{1}$$

($\mathbf{I}$ is a metric tensor; a unit vector $\mathbf{n}$ normal to the interface points from the solid to the liquid) and equations which describe distributions of the *surface* parameters along the liquid-solid interface

$$p^s = \gamma(\rho^s - \rho_0^s) \tag{2}$$

$$\frac{\partial \rho^s}{\partial t} + \nabla \cdot (\rho^s \mathbf{v}^s) = -\frac{\rho^s - \rho_{2e}^s}{\tau} \tag{3}$$

$$\mathbf{v}^s \cdot (\mathbf{I} - \mathbf{nn}) = \tfrac{1}{2}(\mathbf{u} + \mathbf{U}) \cdot (\mathbf{I} - \mathbf{nn}) - \alpha\nabla p^s \tag{4}$$

$$(\mathbf{v}^s - \mathbf{U}) \cdot \mathbf{n} = 0$$

Here $\mathbf{n} \cdot \mathbf{P} \cdot (\mathbf{I} - \mathbf{nn})$ is the shear stress acting on the interface ($\mathbf{P}$ is the stress tensor in the bulk); $p^s$ is the *surface* pressure in the liquid-solid interface related with the *surface* density by the equation of state (2), which is given in the simplest form applicable for isothermal and some barotropic processes in the surface phase; $\beta$ is the coefficient of sliding friction[10]; $\mathbf{u}$ and $\mathbf{U}$ are the velocities of the liquid and the solid on the opposite sides of the liquid-solid interface. It is necessary to emphasize that $p^s$ is *not* the surface tension of the solid; it is associated with the two-dimensional pressure in a thin layer of the liquid adjacent to the solid surface (Fig. 2), and this quantity may be positive or negative. The right-hand side of the surface mass balance equation (3) describes the relaxation of the surface mass density due to the mass exchange between the interface and the bulk ($\rho_{2e}^s$ and $\tau$ are the equilibrium surface density and the relaxation time, respectively). Obviously, this mass exchange may be neglected in the boundary conditions for the bulk velocity, and therefore on the interfaces one has

$$(\mathbf{u} - \mathbf{v}^s) \cdot \mathbf{n} = 0. \tag{5}$$

Equations (4) relate the components of the surface velocity $\mathbf{v}^s$ with the components of the bulk phase velocities on the opposite sides of the interface; $\alpha$ is a phenomenological coefficient describing in an integral form the influence of the surface pressure gradient on the velocity distribution across the interface.

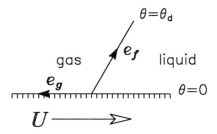

Fig. 3. A definition sketch for the flow near the contact line (the hydrodynamic level).

Boundary conditions on the free surface derived in the same way as (1)–(5) take the form

$$(\boldsymbol{u} - \boldsymbol{v}^s) \cdot \boldsymbol{n} = 0 \tag{6}$$

$$\boldsymbol{n} \cdot \boldsymbol{P} \cdot \boldsymbol{n} + p_g + p^s \kappa = 0$$

$$(\boldsymbol{I} - \boldsymbol{n}\boldsymbol{n}) \cdot \boldsymbol{P} \cdot \boldsymbol{n} + \nabla p^s = 0$$

$$p^s = \gamma(\rho^s - \rho_0^s)$$

$$\frac{\partial \rho^s}{\partial t} + \nabla \cdot (\rho^s \boldsymbol{v}^s) = -\frac{\rho^s - \rho_{1e}^s}{\tau}$$

$$(1 + 4\alpha\beta)\nabla p^s = 4\beta(\boldsymbol{u} - \boldsymbol{v}^s) \cdot (\boldsymbol{I} - \boldsymbol{n}\boldsymbol{n})$$

Here $p_g$ is the pressure in the gas, which is assumed to be inviscid; $\kappa$ is the curvature of the free surface; $\rho_{1e}^s$ is the equilibrium surface density in the liquid-gas interface; the unit normal vector $\boldsymbol{n}$ points from the gas to the liquid.

The distributions of the surface parameters along the interfaces are linked by the boundary conditions at the contact line. The mass balance conditions at an arbitrary point $\boldsymbol{r}_0$ of the contact line in a reference frame fixed with respect to the contact line (Fig. 3) takes the form:

$$(\rho^s \boldsymbol{v}^s)_{|\boldsymbol{r} \to \boldsymbol{r}_0, \, \boldsymbol{r} \in \Sigma} \cdot \boldsymbol{e}_f = (\rho^s \boldsymbol{v}^s)_{|\boldsymbol{r} \to \boldsymbol{r}_0, \, \boldsymbol{r} \in S} \cdot \boldsymbol{e}_g + \rho_{\text{res}}^s \boldsymbol{U} \cdot \boldsymbol{e}_g. \tag{7}$$

Here $\rho_{\text{res}}^s$ is the surface density of the microscopic film ahead of the contact line (Fig. 2b); $\Sigma$ and $S$ denote the liquid-gas and liquid-solid interface, respectively; the notation $\boldsymbol{r} \to \boldsymbol{r}_0$, $\boldsymbol{r} \in \Sigma(S)$ is used to denote the limit of a function as $\boldsymbol{r}$ tends to $\boldsymbol{r}_0$ along $\Sigma(S)$; $\boldsymbol{e}_f$ and $\boldsymbol{e}_g$ are the unit vectors normal to the contact line and directed along the free surface and the liquid-solid interface, respectively.

The tangential momentum balance condition takes the form of the Young equation

$$p^s_{|\boldsymbol{r} \to \boldsymbol{r}_0, \, \boldsymbol{r} \in \Sigma} \boldsymbol{e}_f \cdot \boldsymbol{e}_g = p^s_{|\boldsymbol{r} \to \boldsymbol{r}_0, \, \boldsymbol{r} \in S} - p_{\text{res}}^s \tag{8}$$

$$(\cos \theta_d = -\boldsymbol{e}_f \cdot \boldsymbol{e}_g)$$

Boundary conditions (1)–(8) must be completed by some conditions far away from the contact line for determining the outer flow. Obviously, far away from the contact line, where the surface tension gradients vanish, boundary conditions (1)–(6) degenerate into the classical boundary conditions on free and solid surfaces.

## 4. RESULTS

For small capillary numbers and relaxation lengths the flow in the neighbourhood of the contact line may be considered independently of the details of the outer flow, and we have a *local* moving contact-line problem. Under the same assumptions as in the case of the spreading of a viscous liquid on a dry solid surface[5,6], we may obtain an algebraical equation which relates the dynamic contact angle with non-dimensional parameters of the problem

$$(\cos \theta_0 - \cos \theta_d) \left[ 1 + \frac{\sqrt{V^2 + 1} - V}{2V} \right] =$$
$$\frac{1 + \rho_{1e}^s u_{r(0)}(\theta_d) - \rho_{res}^s}{1 - \rho_{1e}^s} + \cos \theta_0 + p_{res}^s \qquad (9)$$

Here $V$ is the contact-line speed scaled with $\sqrt{\sigma(1 + 4\alpha\beta)/((1 - \rho_{1e}^s)\tau\beta)}$ ($\sigma$ is the equilibrium surface tension of the liquid-gas interface); the surface densities are scaled with $\rho_0^s$; $\theta_0$ is the static contact angle on the pre-wet solid surface (its relation with the static contact angle on the dry solid surface, $\theta_s$, is discussed elsewhere[7]); $u_{r(0)}(\theta_d)$ is the inner limit of the outer solution for the radial velocity of the free surface in the reference frame fixed with respect to the contact line (Fig. 3) scaled with the contact-line speed. If the viscosity of the displaced gas is neglected, then

$$u_{r(0)}(\theta_d) = \frac{\sin \theta_d - \theta_d \cos \theta_d}{\sin \theta_d \cos \theta_d - \theta_d}$$

Fig. 4 shows the numerically obtained dependence of the advancing contact angle $\theta_d$ on $V$ for dry (curves 1–3) and wet (4–7) solid surfaces. In the case of a dry solid surface, the value and the sign of $p_{SG}^s$ qualitatively influence the flow characteristics. If $p_{SG}^s > 0$, the system has a maximum speed of wetting $V_*$: if $V > V_*$, then the solution of the moving contact-line problem with boundary conditions (1)–(8) fails to exist. This fact was interpreted as the transition from the rectilinear contact line to a 'sawtooth' one[5]. The relation between the maximum speed of wetting and the onset of air entrainment was demonstrated experimentally[11]. If $p_{SG}^s < 0$, then $\theta_d$ asymptoties to a certain value $\theta_{max} < 180^0$ as $V$ increases. However, for reasonable values of $p_{SG}^s$ the deviation of $\theta_d$ from $180^0$ is relatively small[5,6] though in experiments[12,13] for some systems $\theta_{max} = 115^0$. Thus, it is interesting to see if variations of any other parameter can lead to $\theta_{max}$ considerably less than $180^0$.

In Fig. 4, it is shown that even a very thin microscopic film ($\rho_{res}^s = 0.1$) strongly decreases $\theta_d$ even for positive $p_{res}^s$. In complete agreement with

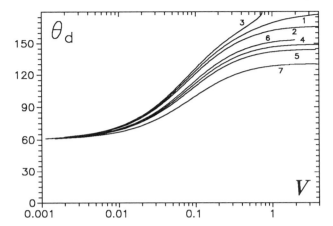

Fig. 4. The advancing dynamic contact angle versus the dimension-less contact-line speed. Curves 1–3 correspond to a dry solid surface $(\rho^s_{res} = 0)$ and $\rho^s_{1e} = 0.95$, $\theta_s = 60^0$:1 — $p^s_{SG} = 0$; 2 — $p^s_{SG} = -0.5$; 3 — $p^s_{SG} = 0.5$. Curves 4–7 were obtained for a wet surface and $\rho^s_{1e} = 0.95$, $\theta_0 = 60^0$: 4 — $\rho^s_{res} = 0.1$, $p^s_{res} = 0$; 5 — $\rho^s_{res} = 0.1$, $p^s_{res} = -0.5$; 6 — $\rho^s_{res} = 0.1$, $p^s_{res} = 0.5$; 7 — $\rho^s_{res} = 0.2$, $p^s_{res} = 0$.

experimental observations, the dynamic advancing contact angle associated with the flow over a pre-wet solid surface is lower than that obtained for the dry solid surface[14,15], and, if the static contact angle in both cases is the same[14], $\theta_d$ increases not so rapidly in the case of a pre-wet solid surface as it does if the surface is dry. Quantitative comparison with the above-cited experiments is difficult due to the lack of information about some parameters involved in the theory.

Boundary conditions (6) can be easily reformulated to take into account the gas viscosity. Then equation (9) remains valid if $u_{r(0)}(\theta_d)$ is replaced by $u_{r(0)}(\theta_d, k_\mu)$, where $k_\mu$ is the viscosity ratio[16,17]. The corresponding curves for the dry solid surface are shown in Fig. 5. As is clear from this figure, if the viscosity of the displaced has is taken into account, then the maximum speed of wetting (and therefore the air entrainment) corresponds to $\theta_d$ considerably less than $180^\circ$ just as it was observed in some experiments[15].

It should be emphasized that, strictly speaking, if the gas viscosity is neglected, this case physically corresponds to a *vacuum*/liquid/solid system. When $\theta_d$ approaches $180^\circ$, the viscosity of the displaced gas becomes impor-tant since the gap between the advancing liquid and the solid surface dimin-ishes.

In Fig. 6 the present theory is compared with recent experiments[18] using the procedure described in the early papers[5,6].

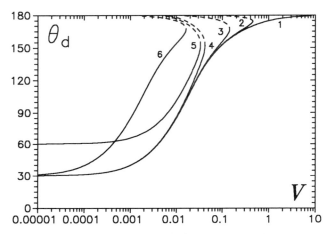

Fig. 5. The velocity dependence of the macroscopic contact angle formed by an interface between a viscous gas and a liquid with a solid surface for different viscosity ratios. Curves 1–4 correspond to $k_\mu = 0$, $10^{-4}$, $10^{-3}$, $10^{-2}$, respectively ($\theta_s = 30°$, $\rho^s_{1e} = 0.99$, $p^s_{SG} = 0$); 5 — $\theta_s = 60°$, $k_\mu = 10^{-2}$; 6 — $\rho^s_{1e} = 0.999$, $k_\mu = 10^{-3}$. Dashed lines correspond to the branches which have no physical meaning.

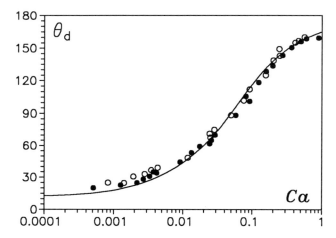

Fig. 6. Comparison of the theory with experiments[18] for a silicone–air interface in a capillary tube.

# REFERENCES

1. Hocking, L. M. A Moving Fluid Interface on a Rough Surface. *J. Fluid Mech.* **76,** pt 4, 801–817 (1976).
2. Jansons, K. M. Moving Contact Lines at Non-Zero Capillary Numbers. *J. Fluid Mech.* **167,** 393–407 (1986).
3. de Gennes, P. G. Wetting: Statics and Dynamics. *Rev. Mod. Phys.* **57,** 827–863 (1985).
4. Dussan V., E. B., and Davis, S. H. On the Motion of a Fluid-Fluid Interface along a Solid Surface. *J. Fluid Mech.* **65,** 71–95 (1974).
5. Shikhmurzaev, Y. D. The Moving Contact Line on a Smooth Solid Surface. *Int. J. Multiphase Flow* **19,** 589–610 (1993).
6. Shikhmurzaev, Y. D. Mathematical Modeling of Wetting Hydrodynamics. *Fluid Dyn. Res.* **13,** 45–64 (1994).
7. Shikhmurzaev, Y. D. Dynamic Contact Angles and the Flow in the Vicinity of a Moving Contact Line. *AIChE J.* (accepted) (1995).
8. Dussan V., E. B. On the Spreading of Liquids on Solid Surfaces: Static and Dynamic Contact Lines. *Ann. Rev. Fluid Mech.* **11,** 371–400 (1979).
9. Kochurova, N. N., Noskov, B. A. and Rusanov, A. I. Taking Account of the Velocity Profile in the Determination of the Surface Tension by Means of an Oscillating Jet. *Colloid J.* (USSR) **36,** 559–561 (1974).
10. Bedeaux, D., A. M. Albano, and Mazur, P. Boundary Conditions and Non-Equilibrium Thermodynamics. *Physica* **A82,** 438–462 (1976).
11. Blake, T. D., and Ruschak, K. J. A Maximum Speed of Wetting. *Nature* (London) **282,** 489–491 (1979).
12. Elliott, G. E. P., and Riddiford, A. C. Dynamic Contact Angles I. The Effect of Impressed Motion. *J. Colloid Interf. Sci.* **23,** 389–398 (1967).
13. Inverarity, G. Dynamic Wetting of Glass Fibre and Polymer Fibre. *Br. Polym. J.* **1,** 245–251 (1969).
14. Rillaerts, E., and Joos, P. The Dynamic Contact Angle. *Chem. Eng. Sci.* **35,** 883–887 (1980).
15. Ghannam, M. T., and Esmail, M. N. The Effect of Pre-Wetting on Dynamic Contact Angles. *Can. J. Chem. Eng.* **70,** 408–412 (1992).
16. Cox, R. G. The Dynamics of the Spreading of Liquids on a Solid Surface. Part 1. Viscous Flow. *J. Fluid Mech.* **168,** 169–194 (1986).
17. Shikhmurzaev, Y. D. Moving Contact Lines in Liquid/Liquid/Solid Systems. *J. Fluid Mech.* (submitted) (1995).
18. Fermigier, M. and Jenffer, P. An Experimental Investigation of the Dynamic Contact Angle in Liquid-Liquid Systems. *J. Colloid Interf. Sci.* **146,** 226–241 (1991).

# THE DYNAMICS OF PLANAR AND AXISYMMETRIC
# HOLES IN THIN FLUID LAYERS

S. K. Wilson

*Department of Mathematics, University of Strathclyde,
Livingstone Tower, 26 Richmond Street, Glasgow G1 1XH*

and

E. L. Terrill

*Faculty of Mathematical Studies, University of Southampton,
Southampton S09 5NH*

## SUMMARY

In this paper we consider the dynamics of planar and axisymmetric holes in thin
fluid layers lying on a solid horizontal substrate and bounded laterally by solid
vertical walls. Employing the lubrication approximation the thickness of the
layer is governed by a non-linear partial differential equation representing the
unsteady balance between capillary and gravity effects. We model the dynamics
of the contact line by assuming a functional relationship between its speed and
the contact angle which generalises the models of earlier authors and leads to an
ordinary differential equation for the motion of the contact line. We solve the
resulting coupled problem in the quasi-steady limit of small Capillary number
and determine the stability of the equilibrium solutions.

## 1. INTRODUCTION

In the coating industry it is frequently important to ensure that any holes
which may occur in a fluid layer applied to a substrate will close during the
coating process so that the final product is free of bare patches. It is, therefore,
rather surprising that while the mechanisms for hole *formation* have been extensively studied both experimentally and theoretically (see, for example, the work
of Kornum and Raaschou Nielsen [1] and Burelbach *et al.* [2] and the references
therein), the *dynamics* of a hole once it has been formed have not received such
widespread attention.

Dombrowski and Fraser's [3] experimental observations of thin fluid sheets in
free fall clearly show that large holes rapidly expand into larger holes, but leave
the fate of small holes unresolved. Inspired by this work Taylor and Michael [4]
investigated the somewhat simpler problem of the stability of axisymmetric
holes in fluid layers of infinite extent lying on a solid horizontal substrate under
the action of capillary and gravity effects. In this situation a single equilibrium configuration exists if the layer is sufficiently thin and by considering the
potential energy of the fluid Taylor and Michael [4] showed this equilibrium to
be unstable. Taylor and Michael [4] conjectured that holes with radius smaller
than that of this unstable equilibrium hole would close while those with larger
radius would open. To test their hypothesis Taylor and Michael [4] conducted a

series of experiments in which holes were made in a horizontal layer of mercury standing on a glass disc with a series of cylindrical probes with different radii. All the holes either opened or closed, and the division between the two kinds of behaviour was in good agreement with the theoretically calculated critical radius.

Recently Moriarty and Schwartz [5] developed a numerical algorithm to solve the unsteady equation (derived using the lubrication approximation) governing the dynamics of an axisymmetric hole in a thin fluid layer of finite extent lying on a horizontal substrate. For sufficiently small volumes of fluid they found that there are two possible equilibrium configurations and, after conducting a series of numerical experiments, concluded that the hole with smaller radius was "highly" unstable while the hole with larger radius was stable. Moriarty and Schwartz [5] also conducted a series of numerical experiments using "realistic" non-equilibrium holes created by imposing a non-uniform pressure distribution at the free surface of the fluid. They concluded that realistic holes with radius somewhat larger than that of the smaller (unstable) equilibrium may either open or close depending on the amount of slip permitted at the solid/fluid interface. The algorithm Moriarty and Schwartz [5] used to move the contact line was as follows; if the numerically calculated contact angle was greater than the advancing contact angle then the contact line was advanced one spatial grid point in one time step, while if it was less than the retreating contact angle then it was moved back one grid point. Their results must therefore be qualified with the remark that they correspond to one particular rule for the motion of the contact line in which the speed of the contact line depends on the numerical grid spacing used.

Evidently understanding the dynamics of the moving three-phase contact line is of vital importance to the dynamics of holes, just as it is to the spreading of droplets. The fundamental problem of a moving contact line, reviewed by Dussan V. [6], has been the subject of much debate in recent years. The key issue is the determination of the relationship between the experimentally measured contact angle (inferred from global properties or measured some distance from the contact line) and the speed of the moving contact line. One approach is to determine this relationship empirically and then *prescribe* this behaviour as part of the specification of the problem. For example, Greenspan [7] assumed that the contact angle was linearly dependent on the contact line speed and subsequently Ehrhard and Davis [8] generalised this to a power law dependence. The major appeal of this approach is that it circumvents the need for a detailed consideration of the flow in the vicinity of the contact line. A conceptually superior approach is to examine this flow in detail and hence *determine* this relationship theoretically. Several different physical mechanisms have been proposed to relieve the unacceptable force singularity which would otherwise occur

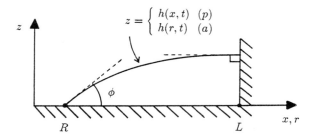

Figure 1: Geometry of the problem.

at the contact line. For example, Hocking [9] analysed the effect of slip and subsequently [10] incorporated intermolecular forces. In the present preliminary study we shall adopt the former approach; work employing the latter approach is currently underway.

## 2. PROBLEM FORMULATION

Consider a volume of fluid with constant viscosity $\mu$ and density $\rho$ lying on a solid horizontal planar boundary under the influence of constant surface tension $\sigma$ and acceleration due to gravity $g$. Both planar (denoted by $p$) and axisymmetric (denoted by $a$) situations will be investigated for which we shall employ Cartesian $(x, z)$ and cylindrical polar $(r, z)$ coordinates respectively. The thickness of the fluid layer is denoted by $z = h(x, t)$ $(p)$ or $z = h(r, t)$ $(a)$ where $t$ denotes time. In both cases the fluid is laterally bounded by a solid vertical wall at $x, r = L$ and surrounds a hole of radius $R(t) < L$ as shown in Fig. 1.

We assume that the speed of the contact line, $R_t$, is related to the contact angle, $\phi(t)$, by the equation

$$R_t = \kappa(\phi_0 - \phi)|\phi_0 - \phi|^n, \tag{1}$$

where $\kappa > 0$ and $n$ are empirically determined constants and $\phi_0$ is the constant static contact angle. If $\phi > \phi_0$ then Eq. (1) is equivalent to the corresponding expression used by Ehrhard and Davis [8], which itself reduces to that used by Greenspan [7] in the case $n = 0$. The experimental evidence cited by Ehrhard and Davis [8] suggests that in many situations the value $n = 2$ gives good agreement with experimental data. For simplicity we take the contact angle at the outer wall to be $\pi/2$. In order to relieve the force singularity at the contact line, we introduce a slip condition $u = \lambda u_z$, where $u$ is the horizontal component of velocity and the constant $\lambda$ is the slip length, on $z = 0$.

We non-dimensionalise the problem using $L$ as the characteristic horizontal lengthscale and a typical value for the contact line speed, $\kappa\phi_0^{n+1}$, as the characteristic horizontal velocity scale. Thus the non-dimensional version of Eq. (1) is

$$R_t = (1 - \phi)|1 - \phi|^n. \tag{2}$$

Hereafter all quantities will be dimensionless unless stated otherwise.

We assume that the fluid layer is sufficiently slender and that the flow is sufficiently slow that we can make the familiar lubrication approximation to the governing Navier-Stokes equations and boundary conditions. It is easily shown (see, for example, Ehrhard and Davis [8]) that using this approximation the equation for $h$ is given by

$$\begin{cases} Ch_t + \left[ h^2 \left( \dfrac{h}{3} + \beta \right) (h_{xx} - G^2 h)_x \right]_x = 0, & (p) \\ Ch_t + \dfrac{1}{r} \left[ rh^2 \left( \dfrac{h}{3} + \beta \right) \left( \dfrac{1}{r}(rh_r)_r - G^2 h \right)_r \right]_r = 0, & (a) \end{cases} \tag{3}$$

where we have introduced a non-dimensional Capillary number, $C = \mu\kappa\phi_0^{n-2}/\sigma$, Bond number, $G^2 = \rho g L^2/\sigma$, and slip coefficient, $\beta = \lambda/\phi_0 L$. The appropriate boundary conditions for Eq. (3) are

$$h(R, t) = 0, \tag{4}$$

$$\begin{cases} h_x(R, t) = \phi, & (p) \\ h_r(R, t) = \phi, & (a) \end{cases} \tag{5}$$

$$\begin{cases} h_x(1, t) = 0, & (p) \\ h_r(1, t) = 0, & (a) \end{cases} \tag{6}$$

$$\begin{cases} h_{xxx}(1, t) = 0, & (p) \\ (rh_{rr})_r(1, t) = 0, & (a) \end{cases} \tag{7}$$

where $\phi$ is related to the speed of the contact line through Eq. (2). In addition, we prescribe the initial conditions

$$R(0) = R_I, \tag{8}$$

and

$$\begin{cases} h(x, 0) = h_I(x), & (p) \\ h(r, 0) = h_I(r). & (a) \end{cases} \tag{9}$$

The volume of fluid (or volume per unit width in the planar case), denoted by $2V$ $(p)$ or $2\pi V$ $(a)$, remains constant throughout the motion. In the unsteady case the initial condition for $h$ determines the value of this constant. If, however, we do not impose the initial condition for $h$ (because the motion is steady or quasi-steady) then to close the system we require the additional condition

$$V = \begin{cases} \int_R^1 h(x,t)\,dx, & (p) \\ \int_R^1 rh(r,t)\,dr. & (a) \end{cases} \tag{10}$$

We note that in a laterally bounded planar (but not, of course, an axisymmetric) geometry the problem describing the dynamics of a hole is equivalent to that describing the dynamics of a symmetric droplet.

## 3. EQUILIBRIUM SOLUTIONS

In equilibrium $\phi = 1$ and Eq. (3) can be readily solved subject to the boundary conditions given in Eq. (4), (5), (6) and (7) to yield

$$\begin{cases} h(x,t) = f(x,R), & (p) \\ h(r,t) = f(r,R), & (a) \end{cases} \tag{11}$$

where

$$\begin{cases} f(x,R) = \dfrac{\cosh(G(1-R)) - \cosh(G(1-x))}{G\sinh(G(1-R))}, & (p) \\ f(r,R) = \dfrac{K_1(G)(I_0(Gr) - I_0(GR)) + I_1(G)(K_0(Gr) - K_0(GR))}{G(K_1(G)I_1(GR) - K_1(GR)I_1(G))}. & (a) \end{cases} \tag{12}$$

We remark that these solutions are not new. The solution for a planar hole can easily be obtained from the familiar solution for a planar droplet, while the solution for an axisymmetric hole was first given by Moriarty & Schwartz [5]. The value of $R$ in equilibrium is determined by the constant volume condition given by Eq. (10) which implies that $R$ must satisfy

$$V = g(R) \tag{13}$$

where

$$g(R) = \begin{cases} \dfrac{1}{G^2}[G(1-R)\coth(G(1-R)) - 1], & (p) \\ \dfrac{1}{G^2}\left[\dfrac{G(1-R^2)(K_1(G)I_0(GR) + I_1(G)K_0(GR))}{2(K_1(GR)I_1(G) - K_1(G)I_1(GR))} - R\right]. & (a) \end{cases} \tag{14}$$

In the special case of zero Bond number, $G = 0$, the functions $f(x,R)$ $(p)$, $f(r,R)$ $(a)$ and $g(R)$ are given by

$$\begin{cases} f(x,R) = \dfrac{(1-R)^2 - (1-x)^2}{2(1-R)}, & (p) \\ f(r,R) = \dfrac{R}{2(1-R^2)}\left[R^2 - r^2 + 2\log\left(\dfrac{r}{R}\right)\right], & (a) \end{cases} \tag{15}$$

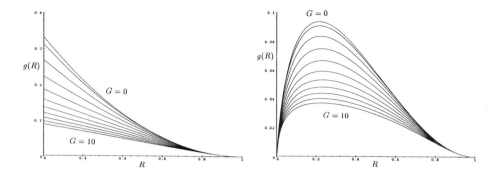

Figure 2: The function $g(R)$ plotted as a function of $R$ for a range of values of $G$ in (a) planar and (b) axisymmetric cases.

and

$$g(R) = \begin{cases} \dfrac{1}{3}(1 - R)^2, & (p) \\[2ex] -\dfrac{R}{8(1 - R^2)}\left[(3 - R^2)(1 - R^2) + 4\log R\right]. & (a) \end{cases} \tag{16}$$

Fig. 2 shows $g(R)$, the volume of the equilibrium hole with radius $R$, plotted as a function of $R$ in both planar and axisymmetric cases for a range of values of $G$. Evidently in the planar case there is a single equilibrium hole with radius $0 < R < 1$ provided that $V < V_c$ where $V_c = g(0) = (G\coth G - 1)/G^2$, and if $V > V_c$ then no equilibrium hole is possible. In the axisymmetric case there are two equilibrium holes with radii $R_1$ and $R_2$ such that $0 < R_1 < R_2 < 1$ provided that $V < V_c$ where $V_c$ denotes the maximum value of $V(R)$ which occurs at $R = R_c$, and if $V > V_c$ then no equilibrium hole is possible. Fig. 3 shows examples of the layer thickness profiles of these equilibrium holes, all of which have the same volume, for a range of values of $G$. Numerically calculated values of $R_c$ and $V_c$ in the axisymmetric case are plotted as functions of $G$ in Fig. 4. These results show that $V_c$, the maximum volume of fluid which can adopt an equilibrium hole configuration, is a monotonically decreasing function of $G$ and that the radius of this hole, $R_c$, exhibits a global maximum with respect to $G$.

## 4. THE LIMIT $C \to 0$

In the remainder of the present work we shall adopt the approach taken by both Greenspan [7] and Ehrhard and Davis [8] for the problem of a spreading droplet and seek the leading order solution in the limit $C \to 0$. In this limit the

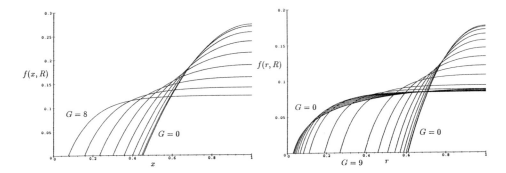

Figure 3: Typical equilibrium holes for a range of values of $G$ in (a) planar ($V = 0.1$) and (b) axisymmetric ($V = 0.04$) cases.

contact line moves *slowly* relative to the bulk of the fluid and so the dynamics of the motion are controlled by those of the contact line. At leading order in $C \ll 1$ we drop the unsteady term in Eq. (3) and the initial condition for $h$, but impose the volume condition and the initial condition for $R$. Eq. (3) can then be solved to give

$$\begin{cases} h(x,t) &= \phi f(x, R), \quad (p) \\ h(r,t) &= \phi f(r, R). \quad (a) \end{cases} \tag{17}$$

Note that because the motion is quasi-steady these solutions are independent of the slip coefficient $\beta$. The volume condition given by Eq. (10) requires that

$$V = \phi g(R). \tag{18}$$

Substituting for $\phi$ from Eq. (18) into Eq. (2) yields a first-order non-linear ordinary differential equation for $R$,

$$R_t = \left(1 - \frac{V}{g(R)}\right)\left|1 - \frac{V}{g(R)}\right|^n, \tag{19}$$

subject to the initial condition given by Eq. (8).

## 5. STABILITY ANALYSIS

We analyse the effect of small perturbations about an equilibrium hole with radius $R = R_0$ by writing $R(t) = R_0 + R_1(t)$ where $R_1(0) = \bar{R} \ll R_0$. At leading order in $R_1$ Eq. (19) yields

$$R_{1t} = (cR_1)|cR_1|^n \tag{20}$$

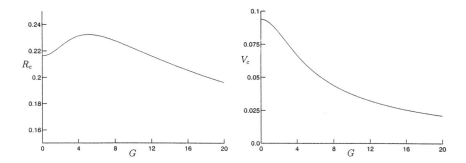

Figure 4: Typical values of (a) $R_c$ and (b) $V_c$ plotted as functions of $G$.

where the constant $c$ is given by $c = g'(R_0)/g(R_0)$, and so

$$R_1 = \begin{cases} \bar{R}e^{ct} & \text{if } n = 0, \\ \bar{R}(1 - nc|c\bar{R}|^n t)^{-1/n} & \text{if } n > 0. \end{cases} \tag{21}$$

Thus if $c < 0$ then the equilibrium $R = R_0$ is stable and small perturbations to the equilibrium state decay exponentially when $n = 0$ and algebraically when $n > 0$. Conversely, if $c > 0$ then the equilibrium $R = R_0$ is unstable and small perturbations to the equilibrium state grow exponentially when $n = 0$ and algebraically when $n > 0$. In the planar case $c$ is a negative, monotonically decreasing function of $R_0$ given by

$$c = \frac{G(a - \cosh a \sinh a)}{1 + (a \sinh a - \cosh a) \cosh a} \tag{22}$$

where $a = G(1 - R_0)$, while in the axisymmetric case $c$ is again a monotonically decreasing function of $R_0$, but now taking the value zero at $R = R_c$. We can deduce at once that the single equilibrium hole in the planar case and the larger hole in the axisymmetric case are unconditionally stable while the smaller hole in the axisymmetric case is unconditionally unstable. Note that since $c$ has the same sign as $g'(R_0)$ the stability of the equilibrium holes follows directly from the sign of the gradient of the curves plotted in Fig. 2.

## 6. NUMERICAL CALCULATIONS

In order to determine the evolution of a general quasi-steady initial hole profile, Eq. (19) was solved numerically using a Runge-Kutta-Merson technique implemented by using NAG routine D02BBF running under UNIX on a SUN

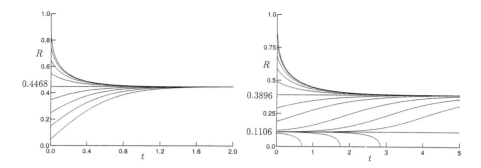

Figure 5: Typical values of $R$ plotted as a function of $t$ in (a) planar ($V = 0.1$, $G = 1$, $R_0 = 0.4468$) and (b) axisymmetric ($V = 0.04$, $G = 8$, $R_1 = 0.1106$, $R_2 = 0.3896$) cases when $n = 0$.

SPARCstation 10. Fig. 5 depicts typical results of these calculations and shows $R$ plotted as a function of $t$ for a number of different initial values. These results confirm the conclusions of the stability calculation presented above. In the planar case all holes evolve towards the single stable equilibrium hole, while in the axisymmetric case all holes initially smaller than $R_1$ become monotonically smaller as $t$ increases and eventually close altogether, while all holes initially larger than $R_1$ monotonically approach the stable equilibrium hole with radius $R_2$. Note that Greenspan's [7] Fig. 1 gives the corresponding results for a spreading axisymmetric droplet in the case $n = G = 0$.

## 7. CONCLUSIONS

In this paper we have obtained a complete description of the dynamics of planar and axisymmetric holes in laterally bounded fluid layers in the limit of small Capillary number. In particular, we have demonstrated analytically that the single equilibrium hole possible in a planar geometry is stable, while in an axisymmetric geometry, where two equilibrium holes are possible, the hole with smaller radius $R_1$ is unstable while that with larger radius $R_2$ is stable. Typical evolutions were presented which show that in the planar case all holes evolve monotonically towards the single stable equilibrium hole; in the axisymmetric case all holes with radius initially smaller than $R_1$ become monotonically smaller as $t$ increases and eventually close, while all holes initially larger than $R_1$ evolve monotonically towards the stable equilibrium hole with radius $R_2$.

In the limiting case of a laterally unbounded layer the present analysis confirms Taylor and Michael's [4] conclusion that the single axisymmetric equilibrium hole possible in a sufficiently thin layer is always unstable and shows that holes with an initially smaller radius will close while those with an initially larger radius will open.

Finally, we note that although the results presented here are for one particular choice of the relationship relating the contact angle to the speed of the contact line, the linear stability analysis indicates that we should expect qualitatively similar behaviour whenever $R_t$ is a monotonically decreasing function of $\phi - \phi_0$.

## 8. ACKNOWLEDGEMENT

SKW greatfully acknowledges the financial support of the Nuffield Foundation through grant SCI/180/92/143/G which made it possible for him to visit ELT in Southampton. SKW also acknowledges valuable discussions with Dr B. R. Duffy (University of Strathclyde).

## 9. REFERENCES

1. Kornum, L. O. and Raaschou Nielsen H. K. Surface defects in drying paint films, *Prog. Org. Coat.* **8**, 275-324 (1980).

2. Burelbach, J. P., Bankoff, S. G. and Davis, S. H. Nonlinear stability of evaporating/condensing liquid films, *J. Fluid Mech.* **195**, 463-494 (1988).

3. Dombrowski, N. and Fraser, R. P. A photographic investigation into the disintegration of liquid sheets, *Proc. R. Soc. Lond.* **247**, 101-130 (1954).

4. Taylor, G. I. and Michael, D. H. On making holes in a sheet of fluid, *J. Fluid Mech.* **58**, 625-639 (1973).

5. Moriarty, J. A. and Schwartz, L. W. Dynamic considerations in the closing and opening of holes in thin liquid films, *J. Coll. Int. Sci.* **161**, 335-342 (1993).

6. Dussan V., E. B. On the spreading of liquids on solid surfaces: static and dynamic contact lines, *Ann. Rev. Fluid Mech.* **11**, 371-400 (1979).

7. Greenspan, H. P. On the motion of a small viscous droplet that wets a surface, *J. Fluid Mech.* **84**, 125-143 (1978).

8. Ehrhard, P. and Davis, S. H. Non-isothermal spreading of liquid drops on horizontal plates, *J. Fluid Mech.* **229**, 365-388 (1991).

9. Hocking, L. M. The spreading of a thin drop by gravity and capillarity, *Q. Jl Mech. appl. Math.* **36**, 55-69 (1983).

10. Hocking, L. M. The spreading of drops with intermolecular forces, *Phys. Fluids* **6**, 3224-3228 (1995).

# SECTION 6

# Rheological Effects in

# Coating Processes

# A REVIEW OF THE ROLE OF RHEOLOGY IN COATING PROCESSES

D C-H Cheng
Consultant in Industrial Applied Rheology
191 Icknield Way
Letchworth, Herts, UK SG6 4TT

## SUMMARY

This review provides a guide to the literature, points to research results useful in practical industrial application and notes problem areas, needs and opportunities for research. A summary of rheological properties is given, covering viscosity or flow curve, normal stresses, extensional viscosity, linear viscoelasticity, thixotropy, non-linear viscoelasticity and tensorial constitutive equations; and listing idealised rheological properties or fluid models used in coating research. The coating processes considered are forward and reverse roll coating, bead coating, blade coating, pick-up onto roller and slide coating. The topics discussed are velocity and stress fields, flow rate or flux, film thickness, drag on roller and roll-separating force, and forces on blades; ribbing, tracking, spatter, fly-off and other flow instabilities and film imperfections. Rheological evaluation of coating fluids to correlate with the more complicated fluid behaviour is also included.

## 1. INTRODUCTION AND SCOPE

There is no need to describe the importance of coating processes in industrial production operations at this symposium. But there are a number of non-coating operations that make use of the basic flow processes. Such as: in the production of powdered baby food by drying over heated drums, refining and flaking of chocolate, and the manufacture of cornflakes. Also, in the manufacture of twisted plastic ropes and pelletising of hot-melt adhesives. Not to mention the mixing of pigments, inks and paints in roll-mills, and calendering and mastication of rubbers, elastomers and mastics. The materials being processed in these processes are highly non-Newtonian and these non-coating applications lend an added dimension to coating research and this review.

In order to limit the scope, the review is restricted to selected coating processes, namely two-roll coating in either forward or reverse rolling mode, bead or meniscus coating, blade coating, pick-up onto roller and slide coating. Dip coating and extrusion coating are excluded; also curtain, slot and air-knife coating.

The bulk of the literature, especially the theoretical, deals with Newtonian liquids. This allows the basic fluid mechanics of coating flows to be explored and the mathematical and experimental techniques of investigation to be developed without the complication of the non-Newtonian characteristics of coating fluids. This elucidates the effect of the magnitude of viscosity. Non-Newtonian characteristics cover a wide range of properties, from variable (non-Newtonian) viscosity to normal stresses and non-Trouton extensional viscosities, and from steady-state straining to time-dependent behaviour. A brief summary of these properties is given before the coating processes are discussed one by one.

The review provides a guide to the literature, identifies research results that are useful in practical industrial application and points out problem areas, needs and opportunities for further research.

## 2. SUMMARY OF RHEOLOGICAL PROPERTIES

Rheology is the science of the deformation and flow of matter. It is concerned with the constitutive relationship between stresses (force per unit area) acting on a material element and strains that the element undergoes, time and temperature. The idea is that the constitutive equation can be solved in conjunction with the equations of motion and continuity to predict the velocity and stress fields of a flowing fluid, from which we can derive macroscopic quantities of technical significance, such as flow rates, forces, pressures and power consumption. Or, they are used to correlate with experimental data. This section briefly reviews the different rheological properties that exist and names the fluid models that have been used in coating research. More details may be found in standard texts, such as references 1 and 2.

### 2.1 Viscosity

A fluid undergoing steady-state shear develops shear stress ($\tau$). Its ratio to shear rate ($\dot{\gamma}$, velocity gradient) is the viscosity:

$$\eta = \frac{\tau}{\dot{\gamma}}$$

(1)

For liquids and solutions of low molecular weight at constant temperature, the viscosity is constant with respect to shear rate. They are described as Newtonian. For solutions of macromolecules, suspensions of fine particles, emulsions and other dispersions of relatively low concentrations, the viscosity decreases with shear rate. The variable or non-Newtonian viscosity is described by the $(\tau, \dot{\gamma})$ flow curve or the viscosity function, $\eta(\dot{\gamma})$. Many mathematical expressions or non-Newtonian viscosity models exit in the literature. The following are some that have been used in coating research.

Many coating fluids are shear thinning; ie, the viscosity decreases with shear rate. A very popular model is the power law [3-8]:

$$\tau = k\dot{\gamma}^n, \eta = k\dot{\gamma}^{n-1}, n < 1$$

(2)

It is widely used in material characterisation and flow modelling, including coating flows. But it is not realistic because it gives an infinite viscosity at zero shear rate and zero viscosity for infinite shear rate. More realistic models are also used. For fluids that are slightly non-Newtonian, a simple model is the Sisko:

$$\eta = \eta_0 (1 - \alpha \dot{\gamma}^m)$$

(3)

where it can be assumed that $m = 2$ [9]. The model cannot be used for high shear rates because it leads to negative viscosities. This can be overcome by using [3,10]:

$$\eta = \frac{\eta_0}{1 + (\theta\dot{\gamma})^m}$$

(4)

where $\theta$ can be interpreted as a relaxation time. A more realistic model in which the viscosity does not decrease to zero as shear rate tends to infinity is the Carreau [11,12]:

$$\eta = \frac{\eta_0 - \eta_\infty}{1 + (\theta\dot{\gamma})^m} + \eta_\infty$$

(5)

where $\eta_\infty$ is the upper Newtonian viscosity.

Alternatively, the viscosity can be expressed as a function of shear stress. A popular one is the Ellis model [13]:

$$\eta = \frac{\eta_0}{1 + \beta \tau^m}$$

(6)

A specific version has $m = 2$ [5,8,14].

Suspensions, particularly of hard particles or when the solids concentration is high, can show dilatancy, in which viscosity increases with increasing shear rate. The power law equation, if employed with $n > 1$, is a model that can be used. But in practice, dilatancy is found when the shear rate exceeds some critical value. At lower shear rates, the fluid is shear thinning. At an even higher shear rate, the viscosity can show a more-or-less sudden jump to an even higher value, followed by strong shear thinning if the shear rate is further increased. Similar dilatant behaviour and step increase in viscosity has also been observed with polymer solutions. In the case of suspensions, the mechanism is packing structural change due to particle migration or phase separation and particle frictional interaction. With polymer solutions, it is molecular agglomeration.

In the extreme, suspensions or dispersions can show chaotic behaviour due to break up or loss of integrity in narrow gaps and/or at high shear rates, which causes problems in viscometers and in coating processes. Such behaviour cannot be described by the simple concept of flow curves or non-Newtonian viscosity.

Polymeric gels, suspensions, dispersions and pastes can posses yield stress and do not flow unless the applied stress is greater. In shearing flow, a simple model is the Bingham plastic and it has been used in coating research [15,16].

## 2.2 Normal Stresses

In addition to shear stress (from which viscosity is derived), a shearing fluid can also develop (deviatoric) normal stresses (NS): $\tau_{xx}, \tau_{yy}, \tau_{zz}$. For incompressible fluids, the significant quantities are the first and second NS differences, $N_1 = \tau_{xx} - \tau_{yy}$ and $N_2 = \tau_{yy} - \tau_{zz}$. Their ratio to $\dot{\gamma}^2$ are the NS coefficients:

$$\Psi_i = \frac{N_i}{\dot{\gamma}^2}, i = 1,2$$

(7)

At low shear rates, the zero-shear NS coefficients $\Psi_{io}$ are constants. Fluids that have constant viscosity and constant NS coefficients are known as second-order fluids [1,2,5]. At higher shear rates, the NS coefficients become dependent on shear rate.

Normal stresses have been observed mainly in polymer solutions and melts. The first NS difference is the largest. The second NS difference is much smaller, typically 10-20% of the first, and negative in sign. The shear rate dependence is shear thinning. They may be described by equations similar to the viscosity ones above, such as the power law [17] and the Ellis [18]. It is also calculated from shear stress: either $N_1 = \tau_{xy}^2/G$ [17] or $\tau_{yy} = -\frac{1}{2}(\theta/\eta_o)\tau_{xy}^2$ [3,10]. Normal stresses are also measured on suspensions or dispersions in polymeric fluids. Measurements of suspensions in Newtonian liquids are not common.

## 2.3 Extensional Viscosity

Fluids are not only sheared during flow, but undergo extensional deformation. Such as in the streamline moving away from a stagnation point, along which the velocity increases, or in a stretching filament or sheet. The ratio of the extensional stress to the extensional or stretch rate is the (uniaxial) extensional viscosity:

$$\eta_E = \frac{\tau_{xx} - \tau_{rr}}{\dot{\varepsilon}}$$

$$(8)$$

For low molecular weight Newtonian liquids, there is a simple relationship between the shear and the extensional viscosities, the Trouton relationship:

$$\eta_E = 3\eta$$

$$(9)$$

This means that once one viscosity is determined, the other viscosity is fixed also; there is no need to measure both separately. For polymeric systems, however, this is not the case. Trouton ratios, $\eta_E/\eta$, as large as $10^3$ or $10^4$ have been measured. More importantly, it means that in general $\eta_E$ and $\eta$ are independent quantities and have to be separately measured. Furthermore, the dependence of $\eta_E$ on stretch rate $\dot{\varepsilon}$ is more complex. At low $\dot{\varepsilon}$, $\eta_E$ is constant (related to the zero-shear viscosity by a Trouton constant of 3). But as $\dot{\varepsilon}$ is increase, $\eta_E$ initially increases (showing tension-thickening), then it decreases. At even higher stretch rates, it can show further tension-thickening behaviour. This complex non-Newtonian extensional viscosity property is the subject of active current research [19,20]. Extensional viscosity has also been measured on suspensions of fibres in Newtonian liquid.

## 2.4 Linear Viscoelasticity

The description of viscosity, normal stresses and extensional viscosity implies steady-state shearing or stretching and the fluid models described apply strictly only to such conditions. The fluid behaviour during start up or under conditions where there are changes in strain rates is described by a separate set of rheological properties, the transient properties. This is because non-Newtonian fluids are not just viscous, but their behaviour is tempered by elastic property. A simple way to illustrate the resulting viscoelastic behaviour is to use mechanical models composed of viscous dashpots and elastic springs in different combinations. (This approach gives rise to constitutive equations in differential form. Equivalent formulation has been made also in the integral form, involving the use of the memory functions [1,2]. These are sometimes used in coating research, in the form of the Lodge Rubber-like Liquid [21] or a K-BKZ model [12]).

The simplest is the Maxwell model made up of a dashpot $(\eta_o)$ and spring (G) in series. The behaviour under simple shear is given by the equation:

$$\tau + \theta \frac{d\tau}{dt} = \eta_o \dot{\gamma}, \theta = \frac{\eta_o}{G}$$

(10)

When a shear rate is suddenly imposed and held constant, the shear stress responds by an exponential growth at the characteristic time $\theta$ (the relaxation time). The time-dependent stress gives the stress growth coefficient, $\eta^+(t) = \tau(t)/\dot{\gamma}$. Under steady shearing, the model behaves as a Newtonian liquid with viscosity $\eta_o$.

The next in complication is the Jeffreys model, which is made up of a dashpot $(\eta_o)$ in series with a Kelvin model. (The Kelvin model is composed of a dashpot $(\eta_K)$ and a spring (G) in parallel. If it is subjected to a suddenly imposed constant shear stress (the creep test), the shear strain response is an exponential growth at the characteristic time (the retardation time) of $\theta_K = \eta_K/G$.) The equation for the Jeffreys model is:

$$\tau + \theta \frac{d\tau}{dt} = \eta_0 (\dot{\gamma} + \theta_K \frac{d\dot{\gamma}}{dt})$$

(11)

The general transient behaviour is governed by two characteristic times, the retardation time $\theta_K$ and a relaxation time $\theta$ ($= \eta_0/G$). Under steady-state shearing, it is again Newtonian with viscosity $\eta_0$.

The Maxwell and the Jeffreys are models for fluids that show elastic property during unsteady shear. Other models can be constructed with a range of discrete relaxation and retardation times; the governing equation is a general linear differential equation of arbitrary order [1,2]. (An integral model has a spectrum of relaxation times.) They all show Newtonian behaviour under steady-state shearing. They are models of linear viscoelasticity and describe the behaviour of real fluids under conditions of small shear stress or shear rate; generally over short times or when the total strain suffered by the fluid is small. The critical limits vary from fluid to fluid, and depends on the material, polymeric or dispersion, and associated parameters such as molecular weight, particle size, strength of colloidal forces and concentration. Many suspensions and dispersions have very low limits that their linear viscoelasticity is not detectable in conventional rheometers.

## 2.5 Thixotropy

The non-Newtonian viscosity described in Section 2.1 is related to linear viscoelasticity (Section 2.4) by having zero relaxation and retardation times. The model response to a sudden change in shear is instantaneous, and under constant shear rate, the shear stress and viscosity are time independent. However, the viscosity depends on shear rate. Such behaviour is described as purely viscous or inelastic.

An inelastic fluid can show time-dependent behaviour due to a second mechanism, namely structural changes at the microscopic and molecular levels. It is described as thixotropy [22]. When a constant shear rate is suddenly applied on the fluid, previously undisturbed, the shear stress responses instantaneous and then decreases with time; the viscosity correspondingly decreases with time. Mathema-tical thixotropic models can be constructed using the idea that viscosity is a function of a structural parameter ($\lambda$), in addition to being shear rate dependent: $\eta = \eta(\dot{\gamma}, \lambda)$. The ($\tau, \dot{\gamma}, \lambda$) relationship is described as the equation of state. The structure is further considered to be shear sensitive, breaking down under high shear and recovering at low shear or at rest. The changes are governed by a rate equation which is a function of shear rate: $d\lambda/dt =$

$g(\lambda, \dot{\gamma})$. The pair of equations of state and rate made up the constitutive relationship for thixotropy and can be used in flow modelling.

On prolonged application of the shear rate, a steady-state is reached where there is dynamic equilibrium between structural build-up and break-down. The equilibrium viscosity can be found by setting the rate $g(\lambda, \dot{\gamma}) = 0$ and solving for the equilibrium value $\lambda_e(\dot{\gamma})$ as a function of shear rate. On substituting into the equation of state, the equilibrium viscosity can be derived: $\eta_e = \eta_e(\dot{\gamma}, \lambda_e(\dot{\gamma}))$. Many of the viscosity functions given in Section 2.1 can be obtained in this way.

In early research, many ad hoc tests have been devised to study thixotropy. Empirical parameters derived from the test results are used to relate to fluid behaviour or performance in coating processes. A popular test is to determine the hysteresis loop obtained when the shear rate is cycled, ie ramped up to a high shear rate and then down again. The area of the loop is taken to be a measure of the degree of thixotropy. We now know how to refine these tests and interpret the results to construct model equations of state and rate. But because the proper testing and model building is a complex procedure, industrial practitioners still favour the ad hoc tests to assess the rheology of coating fluids [23,24].

### 2.6 Non-linear Viscoelasticity

Linear viscoelastic fluids with non-zero relaxation and retardation times show stress growth response on the sudden imposition of shear rate. A further type of response is possible, described as stress overshoot. There is initially a gradual increase in shear stress and viscosity, but on reaching a maximum, they then decrease. Crudely speaking, the stress growth is described as due to viscoelastic property of the fluid, while the later stress decay is due to thixotropic structural break-down. The viscosity level and the rates of growth and decay depend on the shear rate. The behaviour is described as non-linear viscoelasticity or viscoelastic thixotropic depending on one's starting point. Simple mathematical models can be constructed to illustrate the underlying physics. This starts with a mechanical model, such as the Maxwell model, and allows the component viscosity ($\eta_o$) and modulus (G) to dependent on the structural parameter ($\lambda$). Then, assuming a rate equation for $d\lambda/dt$ completes the model. It will predict the stress overshoot [25].

In real fluids, the three types of responses to sudden imposition of shear rate are not exclusive but form a hierarchy of material behaviour. At low shear rates, over the time scale of the experiment and subject to the sensitivity of stress measurement, stress growth is observed. At very low shear rates, the behaviour may be linear, ie the stress growth coefficient, $\eta^+(t)$, is independent of shear rate. At higher shear rates, it depends on shear rate. At medium shear rates, stress overshoot is observed. As the shear rate is increased, the time to reach the peak becomes shorter and the peak proportionally higher. At very high shear rates, the time to reach the peak may become so short that only the stress decay is observed. If an experiment produces only this result, one may think that the fluid is purely-viscous thixotropic and does not posses elasticity. The hierarchy of the three types of behaviour is easily predicted by the simple mechanical model just described.

Normal stresses also show the overshoot behaviour. Although not common in experiments, this is predicted by tensorial fluid models (Section 2.7). Extensional viscosity also shows the full range of viscoelasticity, from linear to non-linear behaviour, including overshoot. The mechanical models described for shear above can be directly translated to apply to extensional straining or stretching if extensional stress ($\sigma$), stretch rate ($\dot{\varepsilon}$), extensional viscosity ($\eta_E$) and Young's modulus (E) are invoked. If structure is allowed to vary, the model will predict extensional stress overshoot. In fact, with mobile fluids that have low viscosities, because it is extremely difficult to achieve steady-state stretching flow in practice, it has not been possible to measure the steady-state extensional viscosity $\eta_E(\dot{\varepsilon})$. What have been measured are transient extensional viscosities which depend on the particular stretching history generated in the extensional viscometer used [20].

## 2.7 Tensorial Constitutive Equations

In introducing the wide range of rheological properties in the previous sections, the flow has been assumed to be simple shear or uniaxial stretching, which are one dimensional. The constitutive equations given are scalar, relating to either shear stress and shear rate, or the extensional counterparts only. However, flow fields are two or three dimensional and there are, in general, three shear and three normal or extensional stresses and corresponding strain rates to contend with. The scalar constitutive equations have therefore to be generalised into tensorial forms if they are to be used in general flow modelling. This is a vast and highly mathematical subject [1,2]. In this section, I try to summarise some of the salient points and name some examples used in coating research.

### 2.7.1 Models from generalisation of Newtonian viscosity
The first group of tensorial constitutive relationships are derived by generalisation of the Newtonian model, giving inelastic non-Newtonian fluids. The most general is the Reiner-Rivlin equation:

$$\tau_{ij} = -p\,\delta_{ij} + 2\eta(I_2, I_3)\,d_{ij} + 4\zeta(I_2, I_3)\,d_{ij}d_{ij}, i, j = x, y, z$$

$$(12)$$

where $\tau_{ij}$ is the stress tensor, p hydrostatic pressure, $\delta_{ij}$ Kronecker delta, $d_{ij}$ the rate-of-strain tensor, $I_2$ and $I_3$ invariants of $d_{ij}$, and $\eta$ and $\zeta$ are material functions. The rate-of-strain tensor is made up of shear and stretch rates:

$$d_{xy} = \frac{1}{2}\left(\frac{\partial u_x}{\partial y} + \frac{\partial u_y}{\partial x}\right), \text{etc}$$

$$d_{xx} = \frac{\partial u_x}{\partial x}, \text{etc}$$

$$(13)$$

in which $(u_x, u_y, u_z)$ is the velocity. The model predicts normal stresses to be $N_1 = 0$ and $N_2 \neq 0$, which is not realistic [2] (or at least not so far observed in experiments) and the full equation is not now used in flow modelling.

However, the simplified version, the Stokesian model, given by:

$$\tau_{ij} = -p\delta_{ij} + 2\eta(I_2)d_{ij}$$

$$(14)$$

is often used for flow modelling [26]. The dependence of viscosity on $I_3$ is neglected because it is not determined in practice. Any of the non-Newtonian viscosity models given in Section 2.1 can be used.

A second general constitutive relationship is the Rivlin-Ericksen equation:

$$\begin{aligned}
\tau_{ij} + p\delta_{ij} =\ & \mu A_1 + \alpha_1 A_2 + \alpha_2 A_1^2 \\
& + \beta_1 A_3 + \beta_2(A_2 A_1 + A_1 A_2) + \beta_3(\mathrm{tr} A_2) A_1 \\
& + \gamma_1 A_4 + \gamma_2(A_3 A_1 + A_1 A_3) + \gamma_3 A_2^2 \\
& + \gamma_4(A_2 A_1^2 + A_1^2 A_2) + \gamma_5(\mathrm{tr} A_2) A_2 \\
& + \gamma_6(\mathrm{tr} A_2) A_1^2 + \gamma_7(\mathrm{tr} A_3) A_1 + \gamma_8(\mathrm{tr} A_2 A_1) A_1 \\
& + \ldots\ldots
\end{aligned}$$

$$(15)$$

where $A_1 = 2d_{ij}$, and $A_n$ are the higher order time derivatives of $A_1$ or the rate-of-strain tensor, and the 14 parameters $\mu, \alpha_n, \beta_n, \gamma_n$ are material constants. Although derived from the Coleman-Noll "simple fluid", which is an memory fluid showing elasticity, it is the limiting case when the various relaxation moduli have all decayed to zero [27]. It can therefore be taken to describe inelastic behaviour. At low strain rates, the behaviour is Newtonian with constant viscosity $\eta_o = \mu$ and Trouton extensional viscosity $\eta_E = 3\mu$. As strain rate increases, the model predicts normal stresses (given by the two $\alpha_n$ parameters; second-order behaviour) and then non-Newtonian viscosity (given by the $\beta_n$ parameters; third order). The stretch rate dependent extensional viscosity involves all these parameters. The $\gamma_n$ parameters give fourth-order behaviour. The second-order effect has been studied in forward rolling [5,28].

## 2.7.2 Models from generalisation of Maxwell and Jeffreys models

The second group of tensor constitutive relationships are obtained by generalisation of the Maxwell and Jeffreys viscoelastic models (Section 2.4). A naive way to generalise the Maxwell model is to write:

$$\tau_{ij} + \theta \frac{\partial}{\partial t} \tau_{ij} = 2\eta_o d_{ij}$$

$$(16)$$

(The generalised Jeffreys model is obtained similarly.) But this refers to changes at a particular point in space, whereas what are required are changes in a fluid element. This is achieved by replacing the partial derivative $\partial/\partial t$ with the convected time derivative (also known by way of Jaumann [28], Lagrangean [2] and Stokesian [29]):

$$\frac{D}{Dt} = \frac{\partial}{\partial t} + u_i \frac{\partial}{\partial x_i}$$

$$(17)$$

with the usual summation convention over repeated suffix. The convected derivative refers to changes in a fluid element, which is tracked as it moves in a flow field. The resulting model is non-linear and predicts a shear thinning viscosity (Eq (4) with m = 2); although the extensional viscosity is constant, $\eta_E = 3\eta_o$ (Eq (9)). The model has been used in coating research [3,28,29]. ---- Such generalisations allow steady and dynamic shear properties to be related, and so the calculation of G and $\theta$ from measurements of $\eta$ and $N_1$ [30].

The use of D/Dt is valid for small deformation and vorticity. For large deformations and vorticities that are found in lubrication and coating flows, the absolute time derivative has been used instead [29]:

$$\left(\frac{D\,a_{ij}}{Dt}\right)_A = \frac{D\,a_{ij}}{Dt} - v_{ix}\,a_{xj} + v_{xj}\,a_{ix}$$

(18)

where $a_{ij}$ is a tensor and the vorticity is:

$$v_{ij} = \frac{1}{2}\left(\frac{\partial u_i}{\partial x_j} - \frac{\partial u_j}{\partial x_i}\right)$$

(19)

A more thorough generalisation of the Maxwell model is to use a convected coordinate system embedded in the fluid and deforming continuously with it. The strain rate is defined in terms of $\gamma_{ij}$, the metric tensor, which measures the distance between parts of a fluid element:

$$\tau_{ij} + \theta\frac{D}{Dt}\tau_{ij} = \eta_o\frac{D}{Dt}\gamma_{ij}$$

(20)

As it stands, Eq (20) refers to a convected coordinate system. It needs to be transformed into the fixed laboratory coordinate system for practical use. This is done [2] by replacing D/Dt in Eq (20) with the "codeformational derivative", $\Delta/\Delta t$. There are two expressions for $\Delta/\Delta t$, depending on whether the symmetric tensor is covariant or contravarient, and referred to respectively as the lower or upper convected derivative. Constitutive equations may be formulated with either the lower or the upper convected derivative, or with the two being mixed in the same equation. It is not at all clear from a casual reading of the literature, what the difference is in physics when different derivatives are used, or mixed. But, the fact that the different constitutive equations make different predictions for the different rheological properties is well understood.

One example of such constitutive equations that has been used in coating research is the upper-convected Oldroyd-B model [3,31] which is a generalisation of the Jeffreys model, Eq (11). It has a constant viscosity and one each of relaxation and retardation time. Another version is also used [32].

### 2.7.3 Structural models

The third group of tensorial constitutive equations are derived by consideration of structural changes, and are more commonly referred to as network models. They described non-linear viscoelasticity or viscoelastic thixotropy. They predict stress overshoot for transient response and strain-rate dependent viscosity, normal stresses and extensional viscosity.

One example is the Phan Thien-Tanner model [33], in which the fluid is assumed to contain entangled macromolecules. The structure is envisaged as made up of a distribution of network strands spanning entanglement junction points, and is represented by the vectors between the ends of the strands. The stress tensor is derived from the distribution of the vectors, specifically their deformed lengths under flow. During deformation and flow, there is also slip at the junction points leading to destruction and creation of strands. Rate equations relating the evolution of the vector distribution function to the strain rates are proposed. This leads to the final tensorial constitutive equations. The form used in coating research [33] is:

$$\tau = \tau_1 + \tau_2$$

$$\text{where } \tau_1 = 2\,\eta_1\,d_{ij}$$

$$\text{and } \exp(\frac{bc}{\eta_2}\,\text{tr}\,\tau_2)\,\tau_2 + b[(1-\frac{a}{2})\tau- +\frac{a}{2}\tau-] = 2\,\eta_2\,d_{ij}$$

$$(21)$$

with five material parameters: $a, b, c, \eta_1, \eta_2$; $\tau-, \tau-$ are the upper and lower convected derivatives of $\tau$. It contains terms that do not appear in the generalised Maxwell and

Jeffreys equations. The model predicts shear thinning and elastic behaviour, and a finite uniaxial extensional viscosity.

## 3. FORWARD ROLL COATING

This section considers two rollers running in the forward or direct rolling mode, the peripheral velocities of the roller surfaces in the nip being in the same direction. In the practical coater, the coating fluid is taken into the nip on one roller and after passing through the nip, the stream splits into two, giving one film on each roller. One of these actually coats the substrate which is draped over that roller. There are therefore two fluid/air menisci or cavities up- and down-stream of the nip, their positions and conditions determine the boundary conditions for the theoretical modelling of the nip flow. Before such splitting flow is considered, the situation where the two rollers are fully immersed in the fluid is reviewed, because this is where tensorial constitutive equations are used in a rigorous way. Although film splitting is not considered, the results give some insight into the effect of rheology.

### 3.1 Fully Immersed Rollers

Analytical or numerical solutions of the equation of motion for the <u>Newtonian liquid</u> have been obtained by many workers [13,26,28,34-37]. Between them they considered equal diameter rollers and unequal (including the roller-plane case), in an infinite sea or contained in a vessel, running at equal or unequal speeds; and covered details of velocity (which is compared with photographs) and stress fields, volume flow rate, torque (which is compared with experiments), pressure distribution, roll-separation force and power consumption. The effect of temperature-dependence of viscosity is also considered in terms of viscous heating in the nip. For <u>inelastic non-Newtonian fluids</u>, finite elements (FE) computational solutions have been obtained for the power law and Carreau models for the velocity field [26].

<u>Elastic fluids</u> have been studied experimentally using polymer solutions. Theoretical modelling has also been attempted [26,28,31,33], and computational solutions obtained for a simplified Oldroyd-B model (with constant viscosity and one each of relaxation and retardation time) and for the Phan Thien-Tanner model (which is both shear thinning and elastic).

<u>Practical significance and research needs</u>. Most of this research appears to be concerned with the development of the mathematical computational technique, but the results show up differences in the flow fields for different types of rheological properties and different models. Qualitative conclusions are drawn by the authors, but it is not sure how universally valid they are. There is no comprehensive study of the

effect of varying material parameters for each fluid model, nor detailed comparison of the differences between different fluid models. There is practically no comparison with experiment and the significance in practical coating situation has not been considered. ---- That the theoretical results are dependent on the fluid models assumed is an important conclusion. It warns that the fluid model is not to be casually chosen if flow modelling predictions are to be compared with experiment.

## 3.2 Flow With Film Splitting

The main concern here is with the prediction of fluid film thickness on the rollers. In the case of Newtonian liquids, many papers have appeared by the early 1980s and reviewed [38-40]. In this early work, theoretical solution was sought by treating the flow as one-dimensional and solving the corresponding Reynolds lubrication equation analytically for the velocity and pressure fields. It is usual to assume that it is fully flooded up-stream. The problem turns out to be the determination of the location and/or conditions of the down-stream meniscus, as boundary conditions for closure of the lubrication equation. Many assumptions are made, mostly based on reasonable physical considerations, giving different solutions. The assumptions and the corresponding solutions generally refer to different flow regimes, some with recirculation at the cavity, others without. Appeal is made to experiments to decide between them [38]. But it turns out that neither the total flux through the nip nor the film thicknesses are sufficiently sensitive to the assumptions under industrial operating conditions (more of which below when the role of a modified capillary number, Eq (22), is discussed). The limitation of the lubrication approach is clear [39].

For more accurate prediction, the two-dimensional nature of the flow in the region of the cavity has to be accounted for. At least a local two-dimensional analysis has to be carried out. A number of attempts have been made but with simplifications and assumptions [39]. A rigorous matched asymptotic expansion (MAE) method has been used for the limit $(Ho/R)^{1/2} \rightarrow 0$ [39,41]; R is the average roller radius and Ho is the nip width. The two-dimensional splitting flow is that confined between two parallel planes and it is solved by the Galerkin FE method. The matching involves using the rectilinear lubrication velocity for the flow far up-stream of the meniscus. The flux and film thickness ratio are calculated as functions of speed ratio and capillary number, $Ca = \eta U/\sigma$, where U is roller velocity, $\eta$ viscosity and $\sigma$ surface tension.

The entire flow for the nip, cavity and down-stream of the split can be calculated by solving the two-dimensional Navier-Stokes equation using the Galerkin FE method [39,40] without being confined to the limit $(Ho/R)^{1/2} \rightarrow 0$. The effects of liquid inertia (in the form of Reynolds number) and gravity (as Stokes number) can be included. The method gives a detailed description of the recirculation flow and its relation to film splitting. The predictions of flux and film thickness are well supported by experimental data, but we know that these are not sensitive quantities to verify the method. Instead, the location of the meniscus is used as a better quantity for testing the theory.

Industrial application of Newtonian results. The Galerkin FE method gives the most accurate prediction and can cover inertia and gravity effects. Commercial computer software packages are now available and that will make it more accessible for industrial application. However, it requires dedicated staffing and makes calculations only case by case. In the papers describing the method [39,40], only illustrative calculations are given. It is generally beyond the capability of the general run of industry. What is needed therefore is for calculations to be done giving a comprehensive coverage of the parameter space relevant to industrial operations. The numerical results can be fitted with equations and design charts constructed. In this respect, the matched asymptotic expansion (MAE) method should be quite accurate for most industrial applications [39]. A typical coater operating condition is 25 µm gap with 250 mm diameter rollers, giving Ho/R = 0.0001. For Ho/R = 0.001, the FE element calculation for the entire flow agrees well with the MAE prediction when Ca > 0.1; for Ho/R = 0.01, Ca > 1. For some operating situations, it may be necessary to account for the effect of Ca.

In the lubrication theory, the effect of rheology appears as the parameter, a modified capillary number:

$$\beta = \frac{1}{12Ca}\left(\frac{Ho}{R}\right)^{\frac{1}{2}}$$

(22)

Typical industrial conditions are: R > 25 mm, Ho < 250 µm (making Ho/R < 0.01) , U ≈ 1 m/s, η ≈ 1 Pa.s and σ < 75 mN/m (Ca ≈ 10). The β parameter is therefore of the order of $10^{-3}$. This turns out to be negligible [38], which means that under practical situations Newtonian velocity is independent of viscosity. The stresses and pressures are of course proportional to viscosity. For the prediction of film thickness, a convenient design equation can be chosen from among the analytical results. This is done in reference 42 which also gives some examples of application in a number of industrial coaters.

For inelastic Non-Newtonian fluids, the one-dimensional lubrication theory has been extended to the power law model (Eq (2)) [3-7]. Between the different references, they considered equal and unequal rollers, running at equal or unequal speeds; velocity and pressure distributions are solved; and volume flow rate or flux

and film thickness are calculated, as are the roller drag and roll-separation force. But, because different assumptions are made for the conditions of the down-stream cavity where film splitting takes place, the numerical predictions differ from paper to paper. However, a consensus is reached on the qualitative effect of non-Newtonian viscosity. In general, the velocity depends on a modified capillary number:

$$\beta = \left(\frac{\sigma}{\eta_a U_1}\right) \left(\frac{Ho}{R}\right)^{\frac{1}{2}}, \eta_a = k \left(\frac{U_1}{Ho}\right)^{n-1}$$

(22A)

But as in the Newtonian case, for practical industrial purposes, the limiting results for $\beta = 0$ is sufficiently accurate. The assumptions made in some of the references are equivalent to this. All of them show that the effect of non-Newtonian characteristic given by the power law index is an increase in the flux and film thickness as n is decreased. But the magnitude of the increase is small. The largest increase in flux predicted is about 5% at n = 0.5, and 17% when n = 0.2. The reduced film thickness $\lambda_1 = h_1/Ho$ is related to the reduced flux $\lambda$ by a perturbation equation [4]:

$$\lambda_1(n) = \frac{\lambda(n)}{2}\left[1 + \frac{Rs-1}{2}\frac{n}{n+1}\right]$$

(23)

where Rs is the speed ratio.

The roller drag and roll-separation force are proportional to $k(U/Ho)^n$ and vary with respective functions that depend on index n. These functions do change significantly with n, increasing by roughly an order of magnitude as n decreases to 0.2; the exact amounts depend on the assumptions made [5,6].

Analytical solutions of the one-dimensional lubrication equation has also been obtained for the non-Newtonian flow curve given by Eq (4) [3,29] under conditions of small Deborah number (De, ratio of characteristic material time parameter to characteristic process time). Although the work had been carried out to investigate the effect of viscoelasticity, the solutions may be treated as examples of the generalised Newtonian or purely-viscous model Eq (14). Tanner [29] showed that normal stress effects are negligible, while Greener and Middleman [3,10] assumed decoupling of shear and normal stresses, the latter having an effect on roll-separation force only.

The FE method has also been used in conjunction with the power law model [7]. Results are given for the film thickness ratio $\lambda_1/\lambda_2 = Rs^{n/(n+1)}$. Finite element calculations have also been carried out using the Carreau model (Eq (5); some results of velocity have been published [11].

For viscoelastic fluids, analysis of the flow of a Maxwell fluid generalised using the absolute time derivative [29] (Section 2.7.2) shows that when $\theta U/Ho$ (De, a Deborah number) is small compared with unity ( 0.3 at the most), the one-dimensional lubrication equation applies with the steady shear viscosity Eq (4). In the case of unequal rollers but equal speed rolling, Tanner [29] obtained solutions for pressure and shear stress at the roller surface, from which the drag (D) and roll-separating force (F) are calculated. While D is increased by a factor that varies with $De^2$; F is reduced.

Greener and Middleman [3] also used the shear-thinning fluid model Eq (4). They decoupled the shear from the normal stresses and calculated $\tau_{yy}$ from $\tau_{xy}$ (Section 2.2). The NS $\tau_{yy}$ is added to pressure to calculate F. All the coating process variables depend on the index m and also the second rheological parameter, De = $(\theta U/Ho)$. Computed results show the flux to be greater than Newtonian, but the magnitude of the increase is small as the modified capillary number $\beta$ tends to zero. The pressure and roll-separation force are reduced. (There seems to be some error in the analysis which casts doubt on the numerical results [4].)

Comparison of theory has been made with experiment for non-Newtonian film thickness [42,43], but this cannot be said to be really rigorous. Although the flow curves of experimental fluids were measured, their viscoelastic properties were not determined. Generally, the measured film thickness is higher than the Newtonian over a wide range of speed ratio, in qualitative agreement with flow modelling predictions. It is quite insensitive to the nature of the fluid for many examples of industrial polymer solutions and a suspension containing solid filler particles. An anomaly was observed with a CMC solution. It was tempting to attribute this to elasticity [42], except that CMC solutions are not usually considered to be highly elastic. ---- Experimental measurement of drag and roll-separating force have been made [44,45], but there is little comparison with theory.

Industrial application and research needs. The theoretical modelling of non-Newtonian forward roll coating has been done for purely viscous and slightly-elastic fluids. The general consensus seems to be that shear thinning viscosity increases the flux and film thickness compared with Newtonian, but only marginally. But experimental increases may be larger, but still not large. The non-Newtonian data appear to parallel the Newtonian design equation quite well. For practical purposes, this design equation might be used to predict trends in non-Newtonian film thickness when operating conditions are altered in the factory. It can be used as the basis for the establishment of calibration curves for specific coating fluids.

As regards further research, the case of strongly elastic fluids remains to be tackled, but this is not an easy problem [28].

## 3.3 On-set Of Ribbing

At low running speeds, the films produced by splitting are smooth, but on exceed a critical speed they become uneven. This takes the form of ridges, tram-lines or ribs, running at right angles to the line of the nip. They originate at the cavity, the fluid growing into buttresses or webs which lead to the ribs. In most industries, the aim is to produce smooth films, but at practical economic coating speeds, ribbing is always present as a general rule; one then depends on levelling to give a smooth film. In some industries, such as rope making, ribbing is needed. In which case one wants to control the rib properties. In lubrication, ribbing is preceded by cavitation and formation of gas bubbles due to low local pressure [46-53]. But the phenomenon has a straight forward hydrodynamic origin and is readily predicted by stability analysis. The literature has been reviewed [40,54-58].

Newtonian liquids. The flow between rollers is too complex for an analytical solution to be obtained by classical stability analysis rigorously [59]. The early attempts have used a simplified stability criterion which considered the effect of perturbing the balance of pressure and surface tension force at the meniscus. The disturbance either grows with time, or decays. The neutral conditions defines the critical point for on-set of ribbing. The detailed solution depends on the undisturbed base flow considered in Section 3.2. Depending on the assumptions made about the conditions of the cavity, different base flows are predicted. This leads to different predictions for the critical condition for on-set of ribbing in equal speed rolling: $Ca^* = A(Ho/R)^B$ in which the constants A,B vary from model to model. Comparing these with experimental results, it became clear that the models refer to different sections of the overall $(Ca^*,Ho/R)$ curve and have limited range of applicability. The index B increases from 1/2 to 2 as Ho/R increases from zero; the actual $(Ca^*,Ho/R)$ curve shows the whole gradation of variation. A more detailed description of the flow field, taken together with various physical assumptions and mathematical approximations, allowed the $(Ca^*,Ho/R)$ curve to be predicted covering a wide range of Ho/R [60]. The agreement with experiment is good, given that practical determination of the start of ribbing is not a clear-cut matter.

This approach has been extended for rollers running at different speeds [61,62]; the condition of on-set of ribbing is now dependent on speed ratio. Comparison with experiments showed that the model over predicts $Ca_2^*$ for Ho < ca. 600 µm and under predicts for larger Ho; the cross-over value of Ho being approximately the same for different roller radii.

A rigorous theory for ribbing on-set depends on an accurate description of the base flow. This is now possible by means of finite element methods (Section 3.2), as is the attendant stability analysis. The predicted critical $(Ca^*,Ho/R)$ curve agrees with experiment [40,58]. The capillary number is defined in terms of the average speed $U = (U_1+U_2)/2$ and the curve applies to equal and unequal speed running. It is not very sensitive to speed ratio.

The case of a roller next to a stationary plate can be treated as the special case of a roller of infinite radius and zero speed, but the situation is complicated if the plate is dry and there is a contact line where the meniscus touches the plate [40,58]. The critical ribbing condition is very sensitive to the conditions of the contact line. Experiments suggest that $Ca^*$ is lowered significantly. If the plate is wetted to eliminate the contact line, the two-roller $(Ca^*,Ho/R)$ curve applies.

Industrial significance of the Newtonian results. It seems that the on-set of Newtonian ribbing is well understood and the $(Ca^*,Ho/R)$ curve can be used in industry to assess and control the phenomenon. Changes may be made on the coating fluid, the viscosity and/or the surface tension, to try to eliminate or promote ribbing. It is not easy to very the roller diameters in practice, but the nip gap can be used for control. In unequal speed rolling, although $Ca^*$ defined in terms ofU  is not sensitive to speed ratio (Rs), the separate $Ca_1^*$ and $Ca_2^*$ do show significant variation with Rs and so some control may be possible through varying the relative speed. ---- Some practical ways of combating ribbing have been reviewed [63].

Non-Newtonian fluids. An analytical theory has been attempted using the inelastic Sisko model (Eq (3)) with m = 2 and small values for the parameter $\alpha$ [9]. The result suggests that increasing $\alpha$ has a slight stabilising effect, increasing $Ca^*$ somewhat. Another theoretical modelling study has also been carried out for shear-thinning fluids, presumably by the FE method, but the results are not generally available as yet [58]. The predictions indicate that if Ca is defined in terms of the viscosity at a "process" shear rate ofU /Ho, then the critical condition for on-set of Newtonian ribs applies.

Ribbing has been observed in a wide range of experimental non-Newtonian fluids and industrial materials such as butter, condensed milk, grease, paints, paper coatings and printing inks [53,58,62-65],some of which show much more prominent webs and ribs, while others give unusual or irregular patterns; but the critical conditions for on-set were not recorded. The observations are rather ad hoc. A 2% bentonite slurry, which is expected to be inelastic, gave the same results as Newtonian liquids in unequal speed rolling [62]. Solutions of high MW polyacrylamide at 10 and 100 ppm concentration reduced $Ca^*$ by factors up to 30, even though there was no detectable increase in viscosity nor normal stresses [21]. Dramatic reduction of $Ca^*$ was found for a range of fluids having the same viscosity but different dynamic uniaxial extensional viscosity [66].

There is as yet no proper theoretical study of the role of viscoelasticity in ribbing on-set, but a preliminary analysis has been attempted [21]. Elasticity gives rise to normal stresses and non-Trouton extensional viscosity. An estimate can be made of the magnitude of normal stress in the roller nip, but this provided no insight into likely effect on ribbing on-set. The nip flow has an extensional component and the extensional stresses contribute to the local pressure driving the flow instability. An estimate of the effect, using the Lodge Rubber-like Liquid, with an exponential memory function with a single relaxation time (Section 2.4), shows that the pressure is reduced. This is interpreted to mean that elasticity destabilises the flow. Clearly, much thorough investigation remains to be done.

Industrial significance and research needs. The on-set of ribbing in non-Newtonian fluids has not been studied as thoroughly as for Newtonian. But it seems that non-Newtonian viscosity as such, in inelastic fluids, does not introduce undue complications. Provided a viscosity value corresponding to a process shear rate of U /Ho is used, the critical condition $(Ca^*,Ho/R)$ for on-set of Newtonian ribs applies. Elasticity in a fluid, in the guise of strongly non-Trouton extensional viscosity, appears to be the dominant factor to promote instability and reduce $Ca^*$, as well as affecting the rib shape and amplitude stability. Theoretical modelling of viscoelastic effects is the next challenge, but it would not be easy.

## 3.4 Rib Number

Newtonian liquids. When the ribs first form, the amplitude is small and they appear to be sinusoidal in profile. But at large Ca and amplitude, the meniscus becomes distinctly non-sinusoidal [50,57]. Ad hoc experimental data on the wave length of ribs or rib number exit in the literature [30,51,59,65], together with an extensive set of data [67]. This has been supplemented by measurements carried out under the Warren

Spring Laboratory RRS/TLP project and all the data, covering equal and unequal roll radii, have been correlated empirically [68]. The rib number (expressed as number of ribs over a length Ho, ie NHo) falls into two regimes, one capillary number dependent, the other not. The existence of the two regimes is found in later work also [58]. The correlation equations are respectively:

$$NHo = 0.062 + 0.035 \log Ca$$

(24)

$$NHo = 1.62 \left( \frac{Ho}{R} \right)^{\frac{1}{2}}$$

(25)

The spread of the data about either equations is $\pm 20\%$.

The theoretical prediction of rib number forms part of the story of the modelling of ribbing on-set (Section 3.3). The analysis assumes a sinusoidal curvature for the cavity with initially unknown frequency N. Solution of the stability equation determines the value of N for the conditions where ribs are established. The complication is again that the result depends on the base flow and different results are found [57,61,69]. The second complication is that the models predict N as bi-valued functions of Ca and Ho/R. Although both the two roller and roll-plate arrangement have been modelled, numerical results have been published only for the roll-plate case. Comparisons with experiments [50,59] show that measurements in general fall in between the two theoretical values, but some data at low Ca agree with the higher of the two values. The ribbing equation reduces on linearisation to the form [57]:

$$(2\pi NHo)^2 = A(Ca - Ca_o) + \text{terms of } O(Ca - Ca_o)^2$$

(26)

where $Ca_o$ is the value for $N = 0$. This is to be compared with the empirical correlation Eq (24) for the Ca-dependent regime.

Non-Newtonian fluids. Measurements of rib spacing or numbers have been made on a wide range of materials [30,63,65,70-73], whose rheological properties were characterised by experiments to varying degrees. The rib number results were plotted in a variety of ways, either versus rheological parameters or rolling conditions.

One group of materials have their viscosity measured over the shear rate range of about $1\text{-}10^4$ s$^{-1}$. The CMC solutions and latex suspensions showed distinct zero-shear viscosity before becoming shear thinning. Solutions of polyethylene oxide and Xanthan gum were strongly shear-thinning. The rib number for these fluids falls into the two regimes as for Newtonian liquids. The Ca-independent data appear to be bunched about the Newtonian correlation, Eq (25), but there is much scatter. An attempt was made [73] to relate the Ca-dependent non-Newtonian data to the Newtonian correlation, Eq (24), in order to establish the apparent shear rate for the calculation of Ca$^*$. This was unsuccessful, except to show that the apparent shear rate is well beyond the upper limit of the range of the viscometer used.

The viscosity of butter and printing inks decreased during rolling due to either viscous heating or thixotropic break-down [30]. The butter data was plotted against viscosity. The ink data, against either Ho or speed. Empirical equations were fitted. These results showed that the rib number increases with decrease in viscosity, which is the opposite to that found for Newtonian liquids.

The elastic property of silicone oils was modified by irradiation to induce crosslinking. Elastic modulus G and relaxation time $\theta$ were calculated from measurements of viscosity and first normal stress difference. The results showed that the rib number increased with increase in relaxation time [30].

While these results are valid for the particular fluid and roller rig, the complexity of the fluid rheological properties do not really allow for general conclusions to be drawn. Clearly a more systematic experimental investigation of the role of rheology is needed, not to mention the need for theoretical modelling.

### 3.5 Height Of Ribs

A knowledge of the rib amplitude would be useful where an industrial process requires ribs to be present. This is also important where smooth films are desired. Given that economic production has to be carried out at speeds that rib formation is inevitable, one relies on levelling of the ribs to give smooth films. The prediction of levelling rate is a separate topic not discussed in this paper, but we need to know the initial rib amplitude to calculate levelling time. Despite these needs, there is very little information on rib height. One paper [30] studied rib height of silicone oils whose relaxation time is changed by irradiation. The rib height is relatively insensitive to relaxation time until a critical value, then it increases rather shapely to a maximum and then decreases. The rate of rise and the maximum height increase with increasing speed.

## 3.6 Filamentation, Fly/Misting And Spatter

These phenomena are prominent in two industries. In high speed printing [74-77], the splitting of the ink film in the nip forms filaments which eventually break up. Tiny droplets are thrown off as fly which leads to misting. The strands still attached to the films retract to give the film a mottled or orange-peel surface. In hand-roll application of paints, the filaments can become very long, trailing behind or wrapping round the roller. On breaking, they give rise to spatter or fly-off, while the remnants form irregular pattern on the paint films called tracking [66,78-80]. This behaviour clearly defies theoretical modelling, but much experimental research has been done to relate it to the rheological properties of inks and paints. Attempts have been made to obtain correlations with one or other of the properties, but without much success. A lot of the work was qualitative speculation albeit accompanied, in the more recent research, by better understanding and measurement of rheological properties, especially extensional viscosity. While it is generally accepted that the non-Trouton extensional viscosity of polymeric fluids is very important, as shown by the work of Glass and co-workers, very little hard conclusion has been drawn. There always seems to be some exceptions, or contrary observations are made by different workers. It may be simply that the measurements of rheological properties have not been carried out under the right conditions of strain rate and strain history to be relevant to filament formation and break-up.

For the record, an attempt has been made to map out the boundaries of different degrees of misting (detachment of fine droplets), fly (of larger droplets) and spraying in terms of roller speeds and nip width [70,71,81]. Differences were observed with a mineral oil as compared with a CMC solution. But the data are not sufficiently comprehensive for generalised correlation to be attempted.

## 4. REVERSE ROLL COATING

In reverse rolling, the peripheral speeds of the two rollers are in opposite directions. A typical reverse roll coater has three rollers forming an applicator/metering roller pair and an applicator/back-up roller pair. The coating fluid is picked up onto the applicator roll and the film thickness is doctored by the metering roll. This is then transferred to the substrate draped over the back-up roll. The hydrodynamics of the two nips are differentiated in that the metering roll is wiped by a doctoring blade so that no fluid is returned to the nip by it ($\lambda_1 = 0$), while in the transfer nip both $\lambda_1$ and $\lambda_2$ (reduced film thicknesses, divided by Ho) are non-zero.

## 4.1 Film Thickness

For **the metering nip**, experimental measurement of the film thickness of Newtonian liquids [40,82-86] shows that $\lambda_m$ ($= \lambda_2$) initially decreases with increasing speed ratio Rs ($= |U_1|/U_2$), but then turns and increases. The initial decrease can be fitted to a straight line:

$$\lambda_m = K(1-Rs)$$

(27)

where the experimental value for K ranges between 1.0 to 1.28. The deviation from the line takes place at lower Rs as Ca number increases; the same for the position of the minimum $\lambda_m$.

Equation (27) is readily derived from the one-dimensional lubrication model for Newtonian liquid by taking a mass balance around the film splitting cavity and setting $U_1$ negative [86]. The theoretical value for K is trapped between 1.226-1.333, with an average of 1.28. More exact modelling has been carried out using FE computation to solve the two-dimensional Navier-Stokes equation [87]. The prediction is insensitive to the up-stream condition and the reported results show good quantitative agreement with experiments. ---- Predictions are also made for drag and roll-separating force as function of Ca and Ho/R [87].

Experiments on non-Newtonian fluids have been carried out using inelastic and elastic fluids [84,86,88]. For inelastic fluids, K = 1.38 was a little higher than Newtonian and the ($\lambda_m$,Rs) curves follow the Newtonian ones if Ca is calculated using viscosity at a "process" shear rate of $2U_2/Ho$. For the elastic fluids, the minimum was extended to higher values of Rs.

Finite element computation has also been carried out for inelastic non-Newtonian fluids using the Carreau model (Eq (5)) [88], confirming the experimental findings. However, FE modelling is not without deficiencies [87] because the effect of gravity and fluid inertia are not yet included. There is also the problem of the physics of the dynamic wetting line of which there is a separate body of literature and still to be resolved. The computation for viscoelastic fluids is a formidable challenge.

For **the transfer nip**, experiments have been carried out using a wide range of Newtonian and non-Newtonian fluids, including oils, polymer solutions and a suspension, showing varying degrees of shear thinning viscosity. The transferred film thickness $\lambda_2$ was found to be insensitive to the fluids, but strongly dependent on Rs and $\lambda_1$; and can be fitted to an empirical correlation [86]:

$$\lambda_2 = \frac{1}{2}\left(\frac{X_o + X_\infty}{1 + 1.9\lambda_1^3} - X_\infty\right) + Rs\lambda_1$$

$$X_o = 1.28 X_\infty = 1.28|1 - Rs|^{1.3}$$

(28)

Finite element modelling has been extended to the flow in the transfer nip [87], but no results appear to have been published. From the experimental insensitivity to fluid rheology, it has been suggested [83,86] that reverse rolling flow may be dominated by transient "start-up" flow, rather than steady-state creeping flow, and it is better modelled if fluid inertia is assumed to dominate. The FE method has been used to investigate the effect of small Reynolds number. It means applying it for Re tending to infinity.

Industrial significance and research needs. The FE method has proved to be useful for predicting reverse rolling flow, as for forward rolling. It is good for case by case film thickness predictions, Newtonian and purely viscous non-Newtonian. Commercial software packages are available. But this may be beyond the capability of small companies. For them, what is needed is for the Newtonian metering case to be comprehensively treated so that simple design equations may be fitted and design charts drawn up. These may then be used for inelastic non-Newtonian fluids at the "process" shear rate of $2U_2/Ho$. For the transfer case, the empirical Eq (28) may be used, although one should consider using FE calculation for more accurate predictions. For viscoelastic fluids, the situation is wide open for research.

## 4.2 On-set of Ribbing

Experimental observations of ribbing in reverse rolling have been made under a variety of conditions [40,53,70-72,82,87-89] and the results of different workers have yet to be reconciled with each other. The literature has to be consulted for a full description. A brief impression is given here.

In the metering nip, experiments with Newtonian liquids [87,89] shows that at low Ca, the film might be smooth over the entire experimental Rs range. At high Ca, there was ribbing at low Rs and a type of uneven film, described as cascade, at high Rs; this is also described as "a period cross-web disturbance", but what was observed was really quite complicated and varied. There was a span of intermediate Rs over which the film was stable and smooth. At intermediate Ca, the film may pass over regimes of being smooth/ribbing/smooth/cascade as Rs is increased. The exact picture depended on Ho and condition of the cavity up-stream of the nip; and maybe disposition of the rollers with respect to the direction of gravity.

The FE method has been used to investigate the mechanism for cascading in metering flow [87]. The cascade is linked to the cavity or the dynamic wetting line being periodically dragged past the centre (the narrowest part) of the nip, giving rise to periodic changes in film thickness in the circumferential direction. The exact details are closely tied to the time-dependent flow in the nip.

The critical (Ca,Rs) loci were also determined experimentally for non-Newtonian fluids [88]. With shear thinning but relatively inelastic algin solutions. The rib amplitude was noticeably greater than Newtonian, which was thought to be the effect of the small increase in extensional viscosity in these solutions. The on-set of cascade occurred at higher Rs than Newtonian. For strongly elastic solutions, the ribbing was irregular and time-dependent, and there was no sharp transition to cascading. Estimates of the process shear and extensional strain rates and the fluid relaxation time gave a Deborah number (ratio of material to process characteristic times) of about 100, indicating that the elastic property dominates the flow [88].

In **the transfer nip**, in one experiment with Newtonian liquids [70,71], Ho and $U_2$ were held constant, and $U_1$ increased. At low Rs, ribs were observed on the lower roller 2, but the upper roller was smooth. At high Rs, the situation was reversed. There was a region of Rs on either side of Rs = 1 where both films were smooth, the span of the critical Rs depending on doctoring of the rollers. The loci of the critical Rs forms a trumpet shape, opening out as Ho was increased. ---- When both Rs and Ho were large, there was spraying from the nip.

The trumpet-shape of the critical (Rs,Ho) curves may be explained qualitatively by an application of the simple stability criterion to the lubrication model [90].

Experiments with non-Newtonian fluids, a CMC solution [71,72], gave a trumpet-shape curve similar to the Newtonian, but the ribs formed were not as regular. They may be broken or develop into patterns similar to the cascade.

In reverse roll coating of a polythene film with a latex in an industrial kiss-coater without a back-up roll, complex herring bone pattern was obtained. The flexible substrate and the wrap tension are additional variables to the coating process.

### 4.3 Rib Number And Height Of Ribs

Some measurements have been made on rib numbers in reverse rolling for a Newtonian oil and the CMC solution [71,72]. The results appeared to fall in the Ca-independent regime and bunched about the Newtonian forward rolling correlation (Eq (25)).

Measurements were made of rib number and height of ribs for a range of latex based fluids [53]. The coater consisted of a train in which the pick-up/applicator pair was in forward rolling, while the applicator/panel pair in reverse rolling (the panel roll carried the substrate). Ribs were formed already at the forward rolling nip; it was not clear whether they were modified at the reverse rolling nip or not. The rib number and height were correlated empirically to a low shear viscosity and a rheological index (the ratio of a high shear viscosity to the low shear value), the viscosities being measured on prescribed instruments.

### 5. BEAD COATING

The term bead or meniscus coating applies to different coating arrangements [91,92]. The bead may be formed between a roll and a pan or reservoir of fluid, or the fluid is supplied by a die fountain. Or it is formed between two rollers in the forward rolling mode with the up-stream cavity severely starved [93]; the film thickness supplied to the nip is thinner than the nip width. Such coating processes are widely used in industry, having the advantage of excellent precision with thin films of low viscosity fluids. Two rollers in reverse rolling can also be used to achieve the same effect [87].

The case of <u>Newtonian</u> **bead coating with forward rolling rollers** has been studied [93] by a combination of flow visualisation experiments, one-dimensional lubrication modelling and FE calculations. Each method gave inspiration in the application and interpretation of the others. It is shown that the one-dimensional lubrication modelling, giving analytical solutions for pressure and velocity, can predict important features of the flow in the bead, which is confirmed in better precision by FE calculations. The flow field is very complex, with recirculation flow regions and, under certain conditions, a stream (described as a snake) that zigzagged along the length of the bead from cavity to cavity. This is reflected qualitatively by flow visualisation experiments; and measurement of pressure has been made in the roll-plate geometry. Predictions are made for the film thickness ratio and the conditions under which bead coating is possible. The process parameters Ho/R, Rs and Ca are not independent quantities but are prescribed for a given flow rate or flux. The lubrication model predicts their relationship to some degree, but the FE method is needed to give the precise details. This computational work has yet to be carried out.

A variety of flow instabilities are found. Single or multiple "bead break" was observed in an experimental rig [93] when $U_1$ was increased, producing a regular array of air cells along the length of the nip. This may be the origin of the characteristically shaped dry patches, "cigar misses", observed in industrial coated webs. Another form of instability is uneven film with broad bands that ran parallel to the nip [54]. This seems to be like the cascade described for reverse rolling (Section 4.2). Another common defect in bead coating is the appearance of bubble lines in the bead and on the coated web [93]. This may well be due to the sub-ambient pressure in the bead, but it is not clear whether it is vapour cavitation or de-gassing. Ultimately, if the flux is excessively reduced by reducing $U_2$, there will be "bead collapse" [93] and contact with the other roller is lost.

Newtonian **bead coating in reverse rolling** has been investigated by FE calculation and experiment at low Ca and high Rs when the wetting line is dragged past the nip [87]. Very thin metered films are produced, as low as only 5% of Ho.

## 6. BLADE COATING

Blade or knife coating is also known as doctoring or metering in view of its function. An initial film is formed on the roller or substrate by some means and the blade or knife is applied to control the thickness. A large variety of arrangements is used [10,17,91,94-104]. The blade may be set at any angle, trailing, normal or against the velocity; and the tip may be shaped variously, square edged, stepped, bevelled, rounded, hooked, etc, or worn down by use. It may be made of metal, though flexible to some extent, or made of plastic or rubber.

In developing a theoretical model, the flow under a blade is divided into four regions. The nip is designated region 2; it is immediately underneath the tip or "sole" of the blade and can be of a variety of geometries as noted. Up-stream is a pool of fluid (region 1). Immediately down stream at the "toe" is region 3, the meniscus, where surface tension is important and the visco-capillary equation applies. Further on is region 4: if the flow is horizontal, the film thickness is constant; otherwise the visco-gravitational equation applies. The flow in the four regions are generally treated separately and the solutions matched at the boundaries to give the overall solution. The one-dimensional lubrication equation is widely assumed even when the flow, especially in region 1, is distinctly two-dimensional. In recently research, the two-dimensional Navier-Stokes equation is solved by FE methods. The velocity and the pressure are derived using various physical assumptions concerning the conditions of the meniscus in region 3. The flux depends on Ho among other parameters and translates directly to the film thickness; any adjustment for gravitational drainage is small.

Of more interest is the prediction of the force acting on the blade. This is needed to predict the flexure of the blade. Together with the mechanical property of the blade, it determines the nip gap Ho and hence flux. The force is depends on normal stress in the fluid and the following review concentrates on this aspect.

The force on the blade is calculated from the pressure and stresses acting on the sole. Under the usual assumption of one-dimensional flow, this is resolved into a W-force (analogous to the roll-separating force or load capacity in lubrication) and a F-force (roller drag or friction force) per unit width:

$$W = \int_0^L pdx + \frac{\Psi_1}{4}\int_0^L \left(\frac{du}{dy}\right)^2_w dx$$

$$F = \int_0^L \tau_w dx$$

(29)

where L is the length of the sole, and the stresses are evaluated on the surface (w) of the sole. The forces W and F depend on the geometry of the sole and the theoretical method used to model the flow. The geometries considered in the literature are: (i) the blade tip is flat and placed parallel to a flat substrate to form a parallel channel (Newtonian [103-107] and non-Newtonian [5]), (ii) or is used with a roller (Newtonian [105]); (iii) the flat tip face is set inclined at a small angle to the flat substrate (the plane slider in lubrication) (Newtonian [5,10,17,56,94,101,104,108], non-Newtonian [5,10,15,17,94,101]), (iv) or at a large angle against a flat or roller surface (Newtonian [10,109,110], non-Newtonian [32]); (v) the tip is profiled, with a radiused heel and triangular toe, and the substrate is not only angled to the sole but also assumed to be compressible so that the flow channel has a complicated profile (Newtonian [104]). Geometry (iii) is described here to illustrate the role of rheology.

The flow between **flat blade tip inclined at a small angle to flat substrate** is confined within a convergent channel of angle m = (Hi-Ho)/L where L is the length, and Hi and Ho are the inlet and outlet widths. For <u>Newtonian</u> liquids, solving the Reynolds lubrication equation for region 2 and neglecting the flows in regions 1 and 3 predicts that the film thickness is [10,56]:

$$T = \frac{1+b}{2+b} Ho, b = \frac{mL}{Ho}$$

$$(30)$$

which is independent of Ca as the influence of region 3 is neglected. The forces per unit width of the blade are [5,17,56,101,104,108]:

$$W = \frac{6\eta U}{m^2}\left[\ln(1+b) - \frac{2b}{2+b}\right]$$

$$F = \frac{4\eta U}{m}\left[\ln(1+b) - \frac{3b}{2(2+b)}\right]$$

$$(31)$$

Various fluid models have been assumed in the modelling of <u>purely viscous non-Newtonian</u> effects. For the Ellis model, Eq (6) with m = 2, the forces acting on a blade with Hi = 2Ho are [5]:

$$W = 0.159\,\eta_o\,U\left(\frac{L}{Ho}\right)^2\left[1 - 1.76\beta\left(\frac{\eta_o\,U}{Ho}\right)^2\right]$$

$$F = 0.773\,\eta_o\,U\frac{L}{Ho}\left[1 - 0.828\beta\left(\frac{\eta_o\,U}{Ho}\right)^2\right]$$

$$(32)$$

Other fluid models also considered are the shear-thinning model Eq (4) [10], power law [5] and Bingham plastic fluid [15,16].

The effect of normal stress has been estimated by an approach which implies assuming the second-order fluid (Section 2.2) [5,17]. The rigorous treatment, however, is not a trivial problem and quite a number of simplifying assumptions have to be made [5]. The analysis starts with a consideration of the flow between parallel planes for which it has been shown that the velocity is the same as Newtonian, unaffected by normal stresses. The normal stress is:

$$\tau_{yy} = -p^o - \frac{\Psi_1 U}{2\eta}\frac{dp^o}{dx} - \frac{\Psi_1}{4}\left(\frac{\partial u}{\partial z}\right)^2_w$$

(33)

where $p_o$ is the Newtonian pressure and $\Psi_1$ is the first NS coefficient; the velocity gradient is evaluated on the moving plane. On substituting into Eq (29) and using the Newtonian velocity gradient:

$$W = W_o + \Psi_1 U^2\frac{L}{H_o^2}\frac{(1+m+m^2)}{(1+m)(2+m)^2}$$

(34)

For the case $Hi = 2Ho$, channel angle $m = 1$:

$$\frac{W}{Wo} = 1 + 1.05\frac{\Psi_1 U}{\eta L}$$

(35)

where Wo is the Newtonian value. The F-force is unchanged from Newtonian.

The combined effect of shear thinning and normal stress on W for a fluid with Ellis viscosity and constant NS coefficient is [5]:

$$\frac{W}{Wo} = 1 - a\beta\left(\frac{\eta_o U}{Ho}\right)^2 + b\left(\frac{Ho}{L}\right)\left(\frac{\Psi_1}{\eta_o}\right)$$

(36)

With

$$\beta = \left(\frac{\theta}{\eta_o}\right)^2, b\frac{\Psi_1}{\eta_o} \approx b'\theta, De = \frac{\theta U}{Ho},$$

$$\frac{W}{Wo} = 1 - a\,De^2 + b'\frac{Ho}{L}\,De$$

(36A)

where the constants a,b are positive and of the order of unity. This shows that at low speeds, normal stress increases W; but at high speeds, the shear-thinning effect takes over to reduce it.

The combined effect of shear thinning viscosity and normal stress where the thinning is significant has also been modelled [10] by using Eq (4) in the lubrication model to predict the velocity and the pressure. The velocity gradient is used to calculate the shear stress, which is used to calculate $\tau_{yy}$ (Section 2.2) and so the force W by:

$$W = \int_0^L [p - \tau_{yy}]dx$$

(37)

Results are obtained numerically and plotted as graphs. The W-force is reduced by increasing $De = \theta U/Ho$.

There does not appear to be any theoretical modelling for **non-linear viscoelastic fluids**. But it can be noted that geometry (iii) is akin to flow between intersecting planes one of which is moving. This situation is also akin to the pond or reservoir in roll coating [109]. The Navier-Stokes equation has been solved in two dimensions [110,111] for Newtonian flow. The flow of viscoelastic fluid has been studied using the Oldroyd-B model (Section 2.7.2) [32].

## 7. PICK-UP ONTO ROLLER

One way of introducing fluid into a train of coating rollers is by partially dipping a roller in a reservoir to pick up fluid and passing it on to the next roller by forward or reverse rolling. The pick-up flow has been investigated theoretically and experimentally for Newtonian liquids. Many theoretical models are based on developing differential equations for film thickness and solving them numerically [118-123]. For more rigorous modelling, the Navier-Stokes equation can be solved by the FE method [106,124].

The flow in the reservoir, which involves bulk recirculation, is generally neglected. The pick-up flow is divided into three regions. Region 1, close to the reservoir surface, the meniscus is assumed to be static and only surface tension effect is considered. Region 2 is the dynamic meniscus and viscous, surface tension and gravitational forces all apply. Region 3 is the steady state and has essentially constant thickness, except for a small gravity drainage effect; surface tension has no effect. At low pick-up speeds, the three forces are all that need to be considered; at high speeds, fluid inertia has to be included also. Depending on the detailed assumptions made, different models are obtained with different predictions. Comparisons with experiments showed that, while the models do describe the essential features of pick-up film thickness, the agreement between theory and experiment is not exact. We look to the FE method for improvement in prediction.

Research has yet to be carried out on non-Newtonian flow. But it should be noted that the withdrawal of a flat plate from a non-Newtonian fluid has been modelled for the power law, Ellis and Bingham plastic fluids, and studied experimentally for viscoelastic materials [8,14]. The theory for Newtonian plate withdrawal is a starting point for the modelling of Newtonian pick-up onto roller. It can be expected that the non-Newtonian plate withdrawal theory would be useful in modelling non-Newtonian pick-up flow.

Pick-up flow can give uneven films as with other coating flows. Ribbing can be expected, but it does not seem to have been described in the literature. (Ribbing of the liquid inside a horizontal glass bottle is commonly seen on the laboratory rotary mixer.) Long surface waves, travelling in a direction perpendicular to the main flow, have been observed; and also waves travelling in the direction of the main flow on flat substrate at higher Reynolds number [118]. These are phenomena still to be fully investigated.

## 8. SLIDE COATING

In slide coating [89,112,113], the fluid is fed through a slot in a die and flows down an inclined plane. The film then crosses a small gap (forming a bead or meniscus) onto the upward moving substrate which is backed up by a roller. A partial vacuum is created on the underside of the bead to promote adherence to the substrate. The die may have a number of slots to give a cascade of films in order to form a multi-layered coat. It is widely used in the manufacture of photographic and x-ray films etc in which the coating fluids are non-Newtonian, but basic research has concentrated on the flow of Newtonian liquids.

The flow can be divided into three regions. First is the feed and the establishment of steady-state film flow on the inclined plane. Second is the bead, which has two free surfaces and is subject to differential pressure, in which the fluid undergoes rapid changes in flow direction. The third region is on the substrate on which the film attains fully developed plug flow, except for a small effect of gravity. Basic research on slide coating [brief review in [112,113]] has laid much emphasis on the second of these, the bead region, with the objective to describe the complex flow pattern and to identify regimes of various flow instabilities. In the experimental work, different flow visualisation methods are used [112,114,115]. The results provide information which assist in setting the details of theoretical modelling. Finite element computation is used to predict the flow field and the free surface profiles. The two-dimensional Navier-Stokes equation for Newtonian liquids is solved, with allowance for the effects of density, viscosity, surface tension and gravity. Assumptions have to be made concerning the dynamic wetting line, as this is still a poorly understood phenomenon. Results of computation [112] allow the effect of substrate speed to be described, as are the effects of vacuum, gap width, liquid inertia, capillarity, inclination of the slide and substrate, dynamic contact angle and slip (due to an invisibly thin film of entrained air at the liquid-solid interface, which rapidly breaks down). ---- The effect of surfactant in the fluid, which is assumed Newtonian, has also been studied by the FE method [116]. The model describes convective diffusion of the surfactant molecules and predicts the concentration profile and the surfactant effect on free surface speed and profiles, wetting line location.

Slide coating is subject to many instabilities which limit practical operation [12,89,112,113]. There can be waves on the slide on the free surface and at the interfaces of a multi-layer flow [117], and neck-down or standing waves at the bead. These may not necessarily lead to uneven film on the substrate. Uneven coating can take the form of ribbing with the rib lines running along the length of the substrate. In the extreme, the flow can split into individual streams separated by dry stripes. A second type of unevenness is made up of cross-web bands described as barring or chatter. Barring is caused by oscillation of the bead and have a hydrodynamic origin, while chatter is attributed to mechanical vibration. Different authors described dry patches interspersed with coated areas or light spots; these could be the same phenomenon. There can be air entrainment. Under some conditions, there is recirculatory flow or vortices in the bead which is the source of several coating defects. The residence time is infinitely long and allows alteration of the fluid, for example by inducing agglomeration of macromolecules or particles. The pressure is low, which allows bubbles or low density particles to gather and maybe coalesce or floc. The vortices are prone to three-dimensional disturbances that cause intermittent and local discharge, giving irregular imperfections on the film, such as streaking. There could be bead flooding; or bead break-up in which case no film is formed.

These phenomena have been studied experimentally or by FE modelling to varying degrees. They are described by many process and material parameters, 10 being listed by Christodoulou and Scriven [112] when the coating liquid is Newtonian. One effect of the flow instabilities is to put a low limit to the film thickness that can be achieved without the coating quality becoming unacceptable; or a high limit to the speed. In experimental work, the minimum film thickness has been correlated empirically with these parameters [89,113]. More detail descriptions of the effects of different parameters have been done by carrying out FE calculations [112,117] as already mentioned. The results can be presented in terms of operability windows on plots of parameter space [12,89].

The effect of non-Newtonian rheology seems also to have been studied in an ad hoc manner. Experiments showed that a polyvinyl alcohol solution can be coated to a higher speed than glycerol of the same viscosity [89,113]. This is attributed to the fact that polymer solutions having high extensional viscosity at high stretch rates and can carry higher tensile stress than Newtonian liquids of the same viscosity; the high tensile stress serving to resist the disruptive forces in the bead. Other qualitative discussions of the role of viscoelasticity are made in the literature. Other experiments were performed using cellulose solutions, both the shear-thinning viscosity and tensile-thickening extensional viscosity data of which could be fitted to the Carreau equation [12]. It is reported that the FE method has been applied to study the instability of these non-Newtonian fluids and operability windows may be plotted on parameter maps [12,89], but little detail is available.

## 9. RHEOLOGICAL EVALUATION OF COATING FLUIDS

Coating processes are already complex enough in the idealised hydrodynamics and rheological properties assumed in theoretical modelling described in the preceding sections. Industrial practical situations are even more so. The roller may be made of rubber and compressible, or gravured. The substrate may be porous, as well as deformable, as in paper coating and printing on paper or woven fabric. And the coating fluids have really quite complicated rheological properties. We have already seen how viscoelastic properties have to be simplified in theoretical modelling. Industrial coating fluids are further complicated by being thixotropic or showing non-linear viscoelasticity, so that the fluids appear to change with the straining imparted by flow. In order to control coating processes under these situations, industry has looked to simple rheological measurement, parameter or index, that will correlate with performance in coating processes. The question is which of the many rheological properties is the correct one to choose. In earlier research, the measurement is dictated by the capability of existing instruments. Now-a-days, with better understanding of rheology, we are more concerned with the need that measurements should be made at strain rates and strain histories relevant to coating flows.

The <u>steady-shear viscosity or flow curve</u> is the most easy to measure. Estimates of the shear rate in the roller nip range from $10^4$ to $10^6$ s$^{-1}$ [24,125-127]. While it is generally considered necessary to measure at these shear rates, in practice, the usual run of viscometers typically achieve only $10^3$ or $10^4$ s$^{-1}$ [78,79,125,126,128-134].

An example of the use of flow curve data is in the correlation of roller fly-off in hand-painting [78,79,135]. The rheological parameters used were viscosities at $10^3$ and $10^4$ s$^{-1}$ or some other high shear rate (HSV) or the power law index at $10^3$ s$^{-1}$. Some trends were observed between fly-off ranking and other performance parameters, but the correlation coefficients were low. Also, the rheological parameters giving trends vary with the type of paint involved. In another example, coil coating of metallic strips [53], a low shear rate viscosity (LSV) is used with a "rheological index" (the ratio HSV/LSV) to correlate with rib number and height. The printing industry [136] used the plastic or slope viscosity, U, and the ratio Y/U (termed "shortness", where Y is the extrapolated yield stress) to relate to the transfer of ink between rollers and the quality of the printed image; again trends were observed. All in all, however, the results from different industries did not lead to any conclusion that can be generalised for practical quality control.

Many industrial coating fluids are <u>thixotropic</u> and measurements give hysteresis loops. More elaborate rheological parameters were derived from the straight line down curve [65,138-140]. First is an index or "coefficient of thixotropy": $M = (U_1 - U_2)/\ln(\dot{\gamma}_2/\dot{\gamma}_1)$, where the plastic viscosity is defined as $U = (\tau - Y)/\dot{\gamma}$. The other, a "levelling index": $I = M/U_2$. They were used by various authors to correlate with different aspects of the coating process. It has been suggested [141,142] that the up- and down-curves of the hysteresis loop should be applied separately to different stages of the coating process: the up-curve to roller nip flow and brush transfer, and the down-curve to flow out and levelling or sagging.

The preceding work was based on steady-shear viscosity. In coating processes, the time of passage of the fluid through the nip or residence time is extremely short. It is in fact subjected to very <u>rapid changes in shear rate</u>. Rates of shear rate change has been estimated [127] to be $10^7$ s$^{-2}$ for reverse rolling, $10^8$ s$^{-2}$ for slide coating and $10^{11}$ s$^{-2}$ for blade coating. To be strictly correct, rheological evaluation should be carried out on viscosity under such transient shear histories, which is perhaps possible only in a coating machine.

The practical alternative is to look to time-dependent dynamic properties. Many workers have measured complex modulus or viscosity [128,131,143-146], but the emphasis had been to demonstrate the complicated rheology of coating fluids and to study the effect of composition. Few had related dynamic property data to fluid performance in the coating process. The drag or tack force measured on a commercial instrument was found to correlate well with complex viscosity at 45 kHz for pigment dispersions in standoil [126]. The complex modulus of paints was found to increase with time after pre-shearing and thixotropic break-down at high shear rate [131,143]. The "G$^*$ recovery" was found to correlate generally with tracking in hand-rolling when extensional viscosity is low to medium [143,147]. Non-linear complex modulus or viscosity depends on frequency, oscillatory strain amplitude and time and a search has to be made to determine which rheological data would correlate with fluid performance. It was found that different rheological parameters correlated with different phenomena [148]. The strain-sweep G' value at 10 rad/s frequency and 25% strain amplitude correlated with spatter resistance in hand rolling, but the exact relationship depended on the type of paint involved.

Time-dependent behaviour is considered to be due to fluid elasticity and complex modulus or viscosity is a manifestation of elasticity. Normal stresses are another indication of elastic property with polymeric fluids and the easier to measure first NS difference has been investigated to relate to hand-rolling performance [79,147]. It seems that notable spatter could be selectively associated with large $N_1$ values, but if the data were taken all together, the correlation was questionable. On the other hand, tracking pattern appears to correlate generally with $N_1$ also when extensional viscosity is low to medium.

Extensional viscosity is recognised to be of paramount importance in coating processes [24,58,66,80,113,149,150]. Estimates of extensional strain rate ($\dot{\varepsilon}$) make it $10^3$ s$^{-1}$ in reverse rolling, $10^4$ s$^{-1}$ in slide coating and $10^5$ s$^{-1}$ in blade coating, and the corresponding rate of stretch rate $10^6$, $10^8$ and $10^{10}$ s$^{-2}$ [127,151]. These figures do not reflect the complicated strain rate history experienced by the fluid when passing through the nip or bead.

An extensive experimental investigation of the relevance of a DUEV (dynamic uniaxial extensional viscosity, as measured using the spin-line technique) to ribbing, spatter, tracking and misting in hand-roll coating has been carried out by Glass and co-workers [66,80,149]. Detailed comparison was made between DUEV and the roller phenomena as observed visually and photographically. High DUEV was found to give large ribs, very stable fibres (which were extremely long filaments) and prevent spatter. With decreasing viscosity, the degree of spatter was increased, while tracking, ie the flow out of surface irregularities, was improved. No quantitative correlation was obtained however. This was thought to be due to the fact that the strain rate history of the spin-line extensional viscometer does not match that in the coating process. It was also observed that in the large web-shaped precursors to the ribs, the flow is better approximated by biaxial extension and biaxial extensional viscosity would be more relevant than the uniaxial.

Glass also noted the synergistic effect of the different rheological properties. The effect of $G^*$ Recovery and $N_1$ mentioned above, is conditional to extensional viscosity being low or medium. Glass further discussed the interplay of extensional viscosities with viscoelasticity as given by relaxation time which is calculated from shear stress and the first NS difference $N_1$. All this shows that the fluid undergoes different types of straining in the coating process and so no one single rheological property can be expected to act alone. It would appear that the correlations being sought are multi-variate in terms of rheological properties.

There are additional complications with porous substrates, typically paper, in paper coating and printing, because of absorption of the liquid phase of the coating fluid. This changes the composition of the fluid and hence the rheological property. It is therefore important to determine the effect of solids concentration on rheological property, and not just confine oneself to measurement on the fluid as supplied to the reservoir of the coating machine.

Industrial relevance and research needs. The search for some simple rheological parameter or index for use in quality control of coating fluids has been carried out for various complex coating phenomena, especially those described by technical terms such as tack, transfer, ribbing, tracking, fly-off, spatter, misting, levelling and sagging. While it is clear that rheological properties are important in these phenomena, the attempt to identify one or other rheological parameter singly as the key has not really been too successful. Gross effects are observed for large changes in rheology, but quantitative correlations are more elusive. Where they are found, they are valid only for particular fluid types or have not been widely tested. There is therefore no "off the shelf" correlation equation that can be applied universally. Nevertheless, the research results that have been obtained encourages one to try to establish the correlation for your specific fluid. They also warn that it is not sufficient just to concentrate on viscosity, even though the shear rate dependence is allowed for.

The other rheological properties have to be included in the search for a multi-variate correlation. The extensional viscosities are especially important.

## 10. SUMMARISING AND CONCLUDING REMARKS

This review aims to introduce research into some selected coating processes, to identify research results that are of use in industrial application and to point out outstanding problems and future research needs. It summarises the wide range of rheological properties exhibited by non-Newtonian fluids to which coating fluids belong. For theoretical modelling and the correlation of experimental results, the rheological properties are idealised. The fluid models that have been used in coating research are listed. Two theoretical approaches have been used in the coating research. The first more traditional approach seeks analytical solutions of the one-dimensional Reynolds lubrication equation and it goes quite a long way to predict significant results. The second approach solves the basic equations of motion using the Galerkin finite element method, which allows two dimensional flow with free surface to be tackled. The flow of Newtonian liquids has been studied in all the coating processes and the modelling has been extended to purely-viscous non-Newtonian fluids in some processes. The derived velocity field is used to predict the effect of normal stress. The effect of viscoelasticity, provided the elasticity is slight, is considered in some cases. The effect of full-fledged non-linear viscoelasticity has only been studied for the fully immersed rollers. The theoretical results compare reasonably well with experimental data for Newtonian liquids and non-Newtonian fluids that are not highly elastic. Details of what has been achieved in research vary with the coating processes reviewed and are given in the body of this paper.

A third research approach is used to deal particularly with the complex flows, the so-called unstable flows, those that give rise to industrially undesirable films, defects and fluid behaviour, and other practical problems. This is to look for simple rheological parameters or indices that correlate with the fluid behaviour. The search has been made using firstly non-Newtonian viscosity, and extended to the other rheological properties. There is now much interest in the role of extensional viscosities. This is summarised in a section on rheological evaluation.

Given in each section are the results that are useful in industrial application, such as design equations and design charts. And also future research needs, which fall into two areas. Analytical results from theoretical modelling can qualify automatically as design equations and be readily turned into design charts, but FE computation is carried out case by case. What has been done so far has provided an understanding of the physics. For industrial application, the computer software is being made available commercially. But this may not be accessible to every one. The first kind of future research need, therefore, is for comprehensive computation to be carried out covering industrial ranges of process and fluid parameters. The results can then be fitted with design equations, or otherwise summarised as design charts or similar, that can be more easily and widely applied.

The theoretical modelling of coating flows of elastic fluids has yet to be tackled more thoroughly and there is a lack of experiments. These make up the second area of research needs. The theoretical investigation would not be simple, but there is a wide choice of elastic fluids for the experiments.

One last point: In conventional usage, rheology refers to the bulk properties of materials. But, it is an essence of coating processes that free liquid surfaces are formed and so surface tension is an important property that affects coating processes. While surface tension of pure liquids and solutions of low molecular weight species does not present a problem, that for solutions of surfactants and macromolecules is a separate phenomenon in its own right. This is because of the relatively slow rate of transport, whether by diffusion or convection, of the macromolecules to the newly formed surface. The effective parameter is the dynamic surface tension, which depends on the solute concentration distribution in the bulk as at the surface. Additional parameters have to be taken into consideration, namely surface tension gradient and interfacial rheological properties. The same behaviour can be expected in suspensions and dispersions. The use of surfactants to improve the performance of coating processes is well known in industry, and some fundamental research has been carried out. It is decided not to discuss the role of interfacial rheology in this review, but some references have been made here and there so that the matter is not overlooked.

## REFERENCES

1. RB Bird, RC Armstrong and O Hassager, "Dynamics of Polymeric Liquids. Vol. 1, Fluid Mechanics"; RB Bird, O Hassager, RC Armstrong and CF Curtiss, "Dynamics of Polymeric Liquids. Vol. 2, Kinetic Theory". New York: Wiley and Sons, 1977. Second edition, 1987.
2. HA Barnes, JF Hutton and K Walters, "An Introduction to Rheology". Amsterdam: Elsevier, 1989.
3. Y Greener and S Middleman, Polymer Eng Sci, 1975, 15, 1-10.
4. MD Savage, J Appl Math Phys (ZAMP), 1983, 34, 358-369.
5. RI Tanner, "Engineering Rheology", Chap. 6. Oxford: Clarendon Press, 1985.

6. MN Tekic and VO Popadic, in I Cheremisinoff (Ed), "Encyclopedia of Fluid Mechanics. Vol. 7. Rheology and Non-Newtonian Flows", Chap.35. Houston, Texas: Gulf Publishing, 1986.

7. DJ Coyle, CW Macosko and LE Scriven, AIChE Journal, 1987, 33, 741-746.

8. JA Tallmadge and C Gutfinger, Ind Eng Chem, 1967, 59 (11), 18-34; correction in a later issue, p.74.

9. MD Savage, unpublished TLP Rept No. 64. Stevenage: Warren Spring Laboratory, 1982.

10. Y Greener and S Middleman, Polym Eng Sci, 1974, 14, 791-796.

11. DJ Coyle, CW Macosko and LE Scriven, in T Provder (Ed) "Computer Applications in Applied Polymer Science", pp 251-264. ACS Symp Series 197. Washington, DC: Amer Chem Soc, 1982.

12. SK Ahuja, M Stevanovic and LE Scriven, in P Moldenaers and R Keunings (Eds), "Theoretical and Applied Rheology", pp 443-445. Amsterdam: Elsevier, 1992.

13. F Dobbels and J Mewis, AIChE Journal, 1977, 23, 224-232.

14. RE Hildebrand and JA Tallmadge, AIChE Journal, 1968, 14, 660-661.

15. AA Milne, Proc Conf Lubric and Wear, paper 102. London: Instn Mech Engrs, 1957.

16. C Dorier and J Tichy, J Non-N Fluid Mech, 1992, 45, 291-310.

17. W Windle and KM Beazley, TAPPI, 1968, 51, 340-348.

18. JV Kelkar, RA Mashelkar and J Ulbrecht, Trans Instn Chem Engrs, 1972, 50, 343-352.

19. NE Hudson and TER Jones, J Non-N Fluid Mech, 1993, 46, 69-88.

20. J Ferguson and NE Hudson, in P Moldenaers and R Keunings (Eds), "Theoretical and Applied Rheology", pp 470-471. Amsterdam: Elsevier, 1992.

21. T Bauman, T Sullivan and S Middleman, Chem Eng Commun, 1982, 14, 35-46.

22. DC-H Cheng, Intern J Cosmetic Chemists, 1987, 9, 151-191.

23. DC-H Cheng, "Viscosity measurement on a PVC plastisol as an example of dense suspensions". Rept No. LR 711 (MP/BM). Stevenage: Warren Spring Laboratory, 1989.

24. O Cohu and A Magnin, J Rheol, 1995, 39, 767-785.

25. DC-H Cheng, "Prediction of stress overshoot behaviour - a simple theoretical model approach". Rept No. LR 626 (MH). Stevenage: Warren Spring Laboratory, 1987.

26. TS Obaid and P Townsend, Rheol Acta, 1984, 23, 255-260.

27. AB Metzner, JL White and MM Denn, Chem Eng Prog, 1966, 62(12), 81-92.

28. JM Broadbent, DC Pountney and K Walters, J Non-N Fluid Mech, 1978, 3, 359-378.

29. RI Tanner, Int J Mech Sci, 1960, 1, 206-215.
30. M Fukushima, Japan J Appl Phys, 1976, 15, 525-530.
31. TS Obaid and P Townsend, Rheol Acta, 1985, 24, 260-264.
32. Y-N Huang, RK Bhatnagar and KR Rajagopal, Rheol Acta, 1993, 32, 490-498.
33. P Townsend and EOA Carew, J Non-N Fluid Mech, 1986, 20, 293-298.
34. GB Jeffery, Proc Roy Soc, 1922, A101, 169-174.
35. MGN Perera, 1982, private communication; paper accepted for publication, J Eng Mech.
36. R Takserman-Krozer, G Schenkel and G Erhmann, Rheol Acta, 1975, 14, 1066-1076.
37. R Nahme, Ing.Arch, 1940, 9, 191-209.
38. DC-H Cheng, Res Rept No. LR 506 (MH). Stevenage: Warren Spring Laboratory, 1985.
39. DJ Coyle, CW Macosko and LE Scriven, J Fluid Mech, 1986, 171, 183-207.
40. DJ Coyle, in ED Cohen and EB Gutoff (Eds), "Modern Coating and Drying Technology", Chap 3. New York: VCH Publishers, 1992.
41. KJ Ruschak, J Fluid Mech, 1982, 119, 107-120.
42. DC-H Cheng, unpublished TLP Rept No.68. Stevenage: Warren Spring Laboratory, 1984.
43. H Benkreira, MF Edwards and WL Wilkinson, Chem Eng Sci, 1981, 36, 429-434.
44. R Rautenbach, Fette Seifen Anstrichmittel, 1959, 61, 571-575; "Das Fliessverhalten von Kunststoffen im Walzspalt, untersucht am Beispiel von Polyathylen". Koln: Westdeutscher Verlag, 1961; Rheol Acta, 1961, 1, 653-656. (In German.)
45. DR Oliver, SI Bakhtiyarov and M Shahidullah, J Non-N Fluid Mech, 1983, 12, 269-282; DR Oliver, Proc. I Chem E Ann Res Mtg, April 1986, Bradford, pp. 105-110; DR Oliver and M Shahidullah, J Non-N Fluid Mech, 1986, 21, 39-50; DR Oliver, in DR Oliver (Ed), "Third European Rheology Conference", pp.377-380. London: Elsevier Applied Science, 1990.
46. WH Banks and CC Mill, Proc Roy Soc, 1954, A223, 414-419.
47. GI Taylor, J Fluid Mech, 1963, 16, 595-619.
48. HJ van de Bergh, "A study of cavitation for a cylinder sliding against a plane". MSc Expt Project Rept. University of Leeds, Dept of Mech Eng, 1974.
49. D Dowson and CM Taylor, Ann Rev Fluid Mech, 1979, 11, 35-66.
50. MD Savage, J Fluid Mech, 1977, 80, 743-755.
51. B Chalmers and WE Hoare, J Iron Steel Inst, 1941, 144, 127P-132P.
52. JC Miller and RR Myers, Trans Soc Rheol, 1958, 2, 77-93; RR Myers, JC Miller and AC Zottlemoyer, J Coll Sci, 1959, 14, 287-299.
53. T Matsuda and WH Brendley Jr, J Coatings Tech, 1979, 51, 46-60.

54. DC-H Cheng, unpublished TLP Rept No. 40. Stevenage: Warren Spring Laboratory, 1981.

55. DC-H Cheng, "Lecture 15. Roller coating and other processes involving transfer of materials to and from rollers". In Course Manual to 8th Training Course in Fluid Rheology, November 1982. Stevenage: Warren Spring Laboratory, 1982.

56. KJ Ruschak, Ann Rev Fluid Mech, 1985, 17, 65-89.

57. MD Savage, Ind Coating Res, 1992, 2, 47-58.

58. DJ Coyle, Ind Coating Res, 1992, 2, 33-45.

59. JRA Pearson, J Fluid Mech, 1960, 7, 481-500.

60. MD Savage, AIChE Journal, 1984, 30, 999-1002.

61. GC Carter and MD Savage, Math Engng Ind, 1987, 1, 83-95.

62. GC Carter, PhD Thesis. University of Leeds, 1985.

63. BR Parker, unpublished Rheological Research Service Rept No. 4, Part 1. Stevenage: Warren Spring Laboratory, 1977.

64. T Sone, M Fukushima and E Fukada, J Phys Soc Japan, 1960, 15, 1708.

65. JB Yannas and RN Gonzalez, TAPPI, 1962, 45, 156-159.

66. RH Fernando and JE Glass, J Rheol, 1988, 32, 199-213.

67. CC Mill and GR South, J Fluid Mech, 1967, 28, 523-529.

68. BR Parker, unpublished Rheological Research Service Nos 4, Part 1; 6, Part 1, 1977; Rept No. 8, Part 1, 1978. Stevenage: Warren Spring Laboratory.

69. VV Gokhale, J Rheol, 1981, 25, 421-432; AIChE Journal, 1983, 29, 865-866.

70. DC-H Cheng and AK Patel, unpublished Rheological Research Project Rept No. 12, Part 1.12B. Stevenage: Warren Spring Laboratory, 1979.

71. DC-H Cheng and AK Patel, unpublished Rheological Research Project Rept No. 13, Part 1.13.B. Stevenage: Warren Spring Laboratory, 1980.

72. GH Barton and DC-H Cheng, unpublished TLP Rept No. 52, Part II. Stevenage: Warren Spring Laboratory, 1981.

73. DC-H Cheng, unpublished Rheological Research Service Rept No. 9, Part 1. Stevenage: Warren Spring Laboratory, 1978.

74. A Voet, "Ink and Paper in the Printing Process", Chap. V. New York: Interscience, 1952.

75. FA Askew, Printing Tech, 1958, 2, 24-39.

76. FA Askew, Rheol Acta, 1965, 4, 285-286.

77. IG Thomson and FR Young, JOCCA, 1975, 58, 389-390.

78. JD Dormon and DMD Stewart, JOCCA, 1976, 59 (4), 115-126.

79. JD Glass, JOCCA, 1978, 50 (5), 53-60.

80. DA Soules, RH Fernando and JE Glass, J Rheol, 1988, 32, 181-198.

81. DC-H Cheng, unpublished Rheological Research Project Rept No. 15, Pt 1.15B. Stevenage: Warren Spring Laboratory, 1980.

82. J Greener and S Middleman, Ind Eng Chem Fundam, 1981, 20, 63-66.

83. JT Jurewicz, Proc Heat Tranf Fluid Mech Inst, 1978, 103-116.

84. H Benkreira, MF Edwards and WL Wilkinson, Chem Eng Sci, 1982, 37, 277-282.

85. Y-T Kang and T-J Liu, Chem Eng Sci, 1991, 46, 2958-2960.

86. DC-H Cheng, in H Benkreira (Ed), "Thin Film Coating", pp. 58-74. Cambridge: The Royal Society of Chemistry, 1993.

87. DJ Coyle, CW Macosko and LE Scriven, AIChE Journal, 1990, 36, 161-174.

88. DJ Coyle, CW Macosko and LE Scriven, J Rheol, 1990, 34, 615-636.

89. EB Gutoff and ED Cohen (Eds), "Coating and Drying Defects. Troubleshooting Operating Problems". New York: Wiley & Sons, 1995.

90. MD Savage, unpublished TLP Rept No. 64. Stevenage: Warren Spring Laboratory, 1982.

91. DF Benjamin, LE Scriven and Colleagues, Ind Coating Res, 1992, 2, 1-31.

92. SC Zink, in Kirk-Othmer "Encyclopedia of Chemical Technology, 3rd Ed, Vol.6", pp.386-426. New York: John Wiley & Sons, 1979.

93. PH Gaskell and MD Savage, in SF Kistler and PM Schwizer (Eds), Liquid Film Coating, Chap. 12, Self-metered coating processes. Chapman and Hall Publishers, in press.

94. E Bohmer, Svensk Papperstidning, 1964, 67, 347-355. (In English.)

95. NO Clark, W Windle and CA Restall, Paper Trade J, 19 Sept 1966, 49-54.

96. D Satas, in D Satas (Ed), "Handbook of Pressure-Sensitive Adhesive Technology", Chap. 26. New York: van Nostrand Reinhold, 1982.

97. W Windle and KM Beazley, Wochenblatt für Papierfabrikation, 10.1973, 332-338. (In German.)

98. EJ Barber, Hercules Chemist, Dec 1972, 15-20.

99. JF Hern, TAPPI, 1961, 44, 838-849.

100. H Benkreira, in H Benkreira (Ed), "Thin Film Coating", pp. 49-57. Cambridge: Royal Society of Chemistry, 1993.

101. W Windle and KM Beazley, TAPPI, 1967, 50, 1-7.

102. WJ Follette and RW Fowells, TAPPI, 1960, 43, 953-957.

103. H Benkreira, in H Benkreira (Ed), "Thin Film Coating", pp. 88-93. Cambridge: Royal Society of Chemistry, 1993.

104. WC Bliesner, TAPPI, 1971, 54, 1673-1679.

105. SS Hwang, Chem Eng Sci, 1979, 34, 181-189.

106. BG Higgins and LE Scriven, IEC Fundam, 1979, 18, 208-215; BG Higgins, IEC Fundam, 1982, 21, 168-173.

107. JA Walowit and JA Arno, "Modern Developments in Lubrication Mechanics", Section 2.3.

108. WF Hughes, "An Introduction to Viscous Flow", Chap.3. Washington, USA: Hemisphere Publ Corp, 1979.

109. CC Mill, JOCCA, 1961, 44, 596-617.

110. EO Tuck and M Bentwich, J Fluid Mech, 1983, 135, 51-69.

111. JR Jones and TS Walters, J Appl Mech, 1971, 38, 1056-1057.

112. KN Christodoulou and LE Scriven, J Fluid Mech, 1989, 208, 321-354.

113. EB Gutoff, in ED Cohen and EB Gutoff (Eds), "Modern Coating and Drying Technology", Chap 4. New York: VCH Publishers, 1992.

114. W Mues, J Hens and L Boiy, AIChE Journal, 1989, 35, 1521-1526.

115. KSA Chen and LE Scriven, Ind Coating Res, 1992, 2, 59-64.

116. R Schunk, PhD Thesis, Chap 5. Univ Minnesota, 1989.

117. SJ Weinstein, paper to AIChE Spring National Mtg, Orlando, Florida, March 1990.

118. KDP Nigam and MN Esmail, Canad J Chem Eng, 1980, 58, 564-568.

119. S Tharmalingam and WL Wilkinson, Chem Eng Sci, 1978, 33, 1481-1487.

120. JA Tallmadge, AIChE Journal, 1971, 17, 243-246.

121. MN Tekic and S Jovanovic, Chem Eng Sci, 1982, 37, 1815-1817.

122. NPC Chao, Textile Res J, 1977 May, 311-316.

123. OH Campanella and RL Cerro, Chem Eng Sci, 1984, 39, 1443-1449.

124. BG Higgins, WJ Silliman, RA Brown and LE Scriven, Ind Eng Chem Fundam, 1977, 16, 393-401.

125. W Windle and KM Beazley, World's Paper Trade Rev, 1966, 1838-1850.

126. L Hellinckx and J Mewis, Rheol Acta, 1969, 8, 519-525.

127. "13. Basic phenomena of rheology in coating". Notes for Coating Process Fundamentals Short Course, Antwerp, Belgium, March 1993. Dept of Chem Eng and Mater Sci, University of Minnesota.

128. U Daum and H Benninga, TAPPI, 1970, 53, 1710-1718.

129. E Franz, "Viscosity measurements of paper coatings up to a shear rate of $D_{max} = 4 \times 10^4$ s$^{-1}$". Tech Bull TB-806. Saddle Brook, NJ, USA: Haake Buchler Instruments, undated.

130. R Byrne, "Rheological characterisation of flexographic ink under process conditions". Tech Bull TB-850. Saddle Brook, NJ, USA: Haake Buchler Instruments, undated.

131. JE Glass, JOCCA, 1975, 58, 169-177.

132. WC Arney Jr and JE Glass, JOCCA, 1976, 59, 372-378.

133. JE Glass, RH Fernando, SK Egland-Jongewaard and RG Brown, JOCCA, 1984, 67, 256-261.

134. RH Fernando and JE Glass, JOCCA, 1984, 67, 279-283.

135. JD Glass, JOCCA, 1978, 50 (5), 61-68.

136. CO Rosted, JOCCA, 1971, 54, 520-545.

137. JW Smith, RT Trelfa and HO Ware, TAPPI, 1950, 33, 212-218.
138. JF Hern, TAPPI, 1961, 44, 838-849.
139. RA Diehm, Paper Ind, July 1961, 225-228, 238.
140. CF Stubbert Jr and RS Sterritt, TAPPI, 1962, 45, 193A-197A.
141. BK Kim, Korean J Rheol, 1992, 4, 89-98; tables and figures in English.
142. DC-H Cheng, in H Benkreira (Ed) "Thin Film Coating", pp. 3-22. Cambridge: Royal Society of Chemistry, 1993.
143. JE Glass, JOCCA, 1976, 59, 86-94.
144. PM McGenity, PAC Gane, JC Husband and MS Engley, in "TAPPI Coating Conf Proc, 1992", pp.133-146. Atlanta: TAPPI Press.
145. JC Husband and JM Adams, Coll Polym Sci, 1992, 270, 1194-1200.
146. DJ Preston, PM McGenity and JC Husband, in H Benkreira (Ed) "Thin Film Coating", pp. 23-33. Cambridge: Royal Society of Chemistry, 1993.
147. JE Glass, J Coatings Tech, 1978 (6), 50, 72-78.
148. S Tso and C Rohn, Water Borne Coatings, May 1988, 15-20.
149. JE Glass, J Coatings Tech, 1978 (6), 50, 56-71.
150. G Pangalos, JM Dealy and MB Lyne, J Rheol, 1985, 29, 471-491.
151. JJ Cai, CW Macosko, LE Scriven and RB Secor, "A comparison of extensional rheometers". Paper to 6th Intern Coating Process Sci & Tech Symp, AIChE Spring National Mtg, New Orleans, LA, USA. Mar/Apr 1992.

# DYNAMICS OF SUSPENSIONS IN COATING FLOWS

Cyrus K. Aidun and Yannan Lu
Institute of Paper Science and Technology
500 10th Street, NW
Atlanta, Georgia 30318, USA

**ABSTRACT**

The lattice Boltzmann method, a well developed and very efficient approach for modeling fluid phase, is implemented to analyze the dynamics of particles suspended in coating flows. The interaction rules between the fluid and the solid particles are developed in which the particle boundaries are treated as no-slip impermeable surfaces. For two-dimensional sedimentations, the results of lattice Boltzmann simulations agree very well with the results of the direct simulations by solving the Navier-Stokes equations. In three-dimensional cases, the relative viscosity of suspension flows is in good agreement with the experiments.

## 1. INTRODUCTION

Analysis of rheology of suspension systems is a big challenge to both theoretical and experimental researchers. Understanding the macroscopic transport behavior of particles suspended in a fluid medium is of fundamental interest. It is also important to many industries that deal with slurries, colloids, polymers, ceramics, etc. In the paper and photographic film industries, the flow of suspensions occurs in important manufacturing processes including paper formation, coating and printing applications. Effective experimental methods, such as magnetic resonance imaging, are being developed for investigation of the macroscopic behavior of suspensions. There is a great need for some theoretical approaches to analyze and predict accurately the microstructural dynamics of suspensions in many manufacturing processes.

One of the methods that has been successful for analysis of suspensions is the Stokesian Dynamics[1]. This method solves the Stokes equations for the fluid phase and the $N$-body Langevin equation to obtain the motion of $N$ particles. With this method, the computational time scales as $N^2$, if the mobility matrix is directly constructed, and $N^3$, if the hydrodynamic interactions are included. The lattice Boltzmann method solves the full Navier-Stokes equations and the computational time scales with $N$.

A two-dimensional direct simulation of the motion of sedimenting circular and elliptic particles in a channel has been carried out recently by Hu et al[2]. They used the finite element method to solve the Navier-Stokes equation for the fluid phase and applied Newtonian dynamics for the solid particles. Their method is very accurate. However, it can only handle a few particles at present[3].

348

It is also not practical to extend their method to three-dimensional cases. Considering that in coating applications, the systems are three-dimensional, the number of suspended particles is large, particles are irregularly shaped, and the particle Reynolds number is not small, other methods need to be developed for analysis of various particulate transport processes.

Recent developments of the lattice Boltzmann method[4] provide a very efficient way to simulate fluid systems dynamically. In the lattice Boltzmann simulations, the fluid phase is modeled by pseudo fluid particles which spread in a regular lattice and move along the lattice links with discrete velocities. The distribution of fluid particles evolves according to the lattice Boltzmann equation, which reduces to the Navier-Stokes equations[5]. The reason for the remarkable efficiency is the local nature of time evolution operation with the lattice Boltzmann equation. Ladd[6] first applied the lattice Boltzmann method to analysis of suspended particles. In his method, the fluid phase evolves according to lattice Boltzmann equation whereas the suspended particles move along in a continuous way according to Newtonian dynamics. The interaction between the solid and the fluid is treated by the boundary rule which takes account of momentum transfer. Ladd's boundary rule[6], however, allows small amount of mass to transfer across the surface of the solid particles, due to the conflict between the continuous motion of suspended particles and the discrete nature of fluid particles. In other words, the lattice nodes inside and outside the particles are treated in an identical manner so that the fluid occupies the whole computational domain, inside and outside the suspended particles.

We have developed a method for microdynamical analysis of particles suspended in fluid where the fluid phase can not penetrate into the solid particle. The fluid phase is simulated by using the lattice Boltzmann method in which the fluid is modeled by a group of pseudo fluid particles moving in a cubic lattice with discrete velocities. These pseudo particles fall in three categories. The first type is the fluid particle at rest. The second type of fluid particles is moving along the lattice links and the third one is moving along the diagonal directions in the coordinate planes. For the rest fluid particles, the moving direction is represented by the vector $\mathbf{e}_{01} = (0, 0, 0)$. There are 6 different moving directions for the second type fluid particles, denoting as $\mathbf{e}_{1i}$ with $i = 1, \cdots, 6$. One of them is $(1, 0, 0)$. The rest can be obtained by permuting index 1 and replacing index 1 by $-1$. For the third type of fluid particles, there are 12 moving directions which are denoted as $\mathbf{e}_{2i}$ with $i = 1, \cdots, 12$. A typical one is $(0, 1, 1)$. The others can be obtained from permutations and changing the signs.

At each time step, the moving fluid particles will arrive only at their nearest neighbor nodes. The lattice Boltzmann equation applied to the fluid particles is given by

$$f_{\sigma i}(\mathbf{x} + \mathbf{e}_{\sigma i}, t + 1) - f_{\sigma i}(\mathbf{x}, t) = -\frac{1}{\tau}[f_{\sigma i}(\mathbf{x}, t) - f_{\sigma i}^{(0)}(\mathbf{x}, t)], \qquad (1)$$

where $f_{\sigma i}(\mathbf{x}, t)$ $(\sigma = 0, i = 1; \sigma = 1, i = 1, \cdots, 6; \sigma = 2, i = 1, \cdots, 12)$ is the single-particle distribution function, $f_{\sigma i}^{(0)}(\mathbf{x}, t)$ is the equilibrium distribution at $(\mathbf{x}, t)$, and $\tau$ is the single relaxation time. In our simulations, $f_{\sigma i}^{(0)}(\mathbf{x}, t)$ is taken as

$$f_{\sigma i}^{(0)}(\mathbf{x}, t) = A_\sigma + B_\sigma(\mathbf{e}_{\sigma i} \cdot \mathbf{u}) + C_\sigma(\mathbf{e}_{\sigma i} \cdot \mathbf{u})^2 + D_\sigma u^2, \qquad (2)$$

with

$$A_0 = \frac{\rho}{4}, \quad B_0 = 0 \quad C_0 = 0, \quad D_0 = 0$$

$$A_1 = \frac{\rho}{12}, \quad B_1 = \frac{\rho}{6} \quad C_1 = \frac{\rho}{4}, \quad D_1 = -\frac{\rho}{4} \qquad (3)$$

$$A_2 = \frac{\rho}{48}, \quad B_2 = \frac{\rho}{12} \quad C_2 = \frac{\rho}{8}, \quad D_2 = 0$$

where $\rho$ is the mass density at the node. For this model, the speed of sound is $c_s = \sqrt{1/3}$, and the kinematic viscosity is $\nu = (2\tau - 1)/6$. With this equilibrium distribution, the Navier-Stokes equations can be derived using the Champan-Enskog expansion[5].

The dynamics of suspended particles is based on Newton's equation of motion. We have developed a new boundary rule for the interactions between the surface of impermeable particles and fluid[7]. The new rule treats the suspensions as solid particles and prevents mass exchange across the solid-fluid interface while taking account of the momentum exchange between the fluid and the solid particle. The particles have solid boundaries and their motion relative to the fluid phase is governed by Newton's law of motion. In our simulations, we first update distribution of fluid particles according to the lattice Boltzmann equation. While the boundary rule is imposed along the surface of the suspended particles, the force and the torque on each of them is calculated. Knowing the net force and the torque on each solid particle, the motion of the particle is recovered by solving the Newtonian equation. The suspended particles arrive at their new positions by replacing the fluid particles. A detailed discussion on the interaction and the motion of solid particles is presented in the paper by the authors[7] and will not be repeated here.

In this study, we first test the new boundary collision rule in two-dimensional cases. Single-particle sedimentations are simulated and compared with the results of Feng et al[3]. we have also simulated the dynamics of a monolayer of spheres suspended in a Couette flow, similar to a blade coating system. The effective viscosity of suspension flows is compared with the correlation results

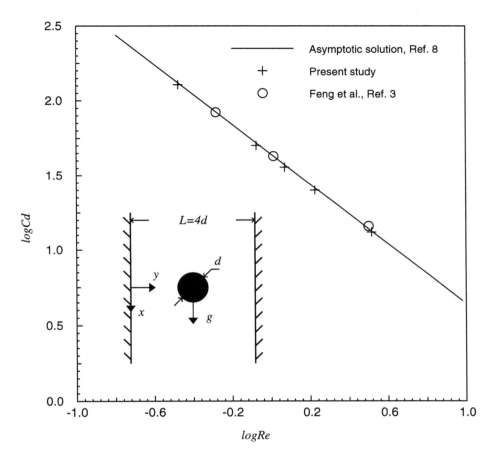

FIGURE 1: DRAG COEFFICIENTS OF A CIRCULAR PARTICLE SETTLING IN A 2-D CHANNEL AS COMPARED TO THE NUMERICAL RESULT OF FENG ET AL. (Ref. 3) AND THE EXACT RESULT OF THE ASYMPTOTIC SOLUTION (Ref. 8).

from experiments and other simulation results. In all cases, the agreements are very well.

## 2. RESULTS AND DISCUSSION

We first use the lattice Boltzmann method to simulate the sedimentation of a circular solid particle in a two-dimensional channel. A circular particle of diameter $d$ is released from different lateral positions, with zero initial velocity, in a channel of width $4d$. The coordinate system is shown in Figure 1. The density of the solid particle is two times larger than the fluid density. The inlet of the domain where zero velocity is applied uniformly is always $10d$ away from the moving particle, whereas the downstream boundary is $15d$ from the solid particle. The normal derivative of velocity are set to zero at the downstream boundary. Feng et al.[3] recently simulated the same problem by solving the Navier-Stokes equation for fluid phase and implementing Newtonian dynamics for the solid particle. Our simulation results of drag coefficients are compared in Figure 1 with the numerical results from Feng et al.[3], as well as the exact solution[8] of the Stokes equation. The settling trajectories of the circular particle released at two off centerline positions in the channel are presented in Figure 2 and Figure 3 for $Re = 1.03$ and $Re = 3.23$, respectively, along with the same results from Feng et al[3]. Due to the presence of inertia, the particle drifts horizontally toward the equilibrium centerline position with a counter-clockwise rotation. For the case of $Re = 3.23$, there is an overshot for the lateral migration. Our results are in good agreement with the simulation results of Feng et al. It shows that the lattice Boltzmann method can very effectively be used for simulation of solid particles suspended in fluid.

The extension of the lattice Boltzmann method and the boundary rules to three-dimensional cases is straightforward. With extension of our method to the three-dimensional cases, we have simulated the dynamics of a monolayer of spheres suspended in a Couette flow, similar to a blade coating system. The fluid is between two parallel surfaces, where one surface moves in the $x$ direction with a constant velocity, relative to the lower surface. The $y$-axis is vertically up. Periodic boundary condition is used in both $x$ and $z$ directions. In our simulations, the domain is either $128 \times 128 \times 16$ or $256 \times 256 \times 32$ lattices. Identical spherical particles are uniformly suspended in the domain. The number of particles ranges from 16 to 64. The results of the effective viscosity, relative to the pure fluid viscosity, versus volume fraction are shown in Figure 4. For comparison, Figure 4 also shows the results of Stokesian Dynamics[9,10], and the correlation results from experiments[11,12]. In the low to intermediate concentration regions, the agreements are very good. Besides the transport quantities, lattice Boltzmann simulations also give the detailed dynamics for both fluid phase and suspended particles, such as pressure distributions for fluid and trajectory for each particle.

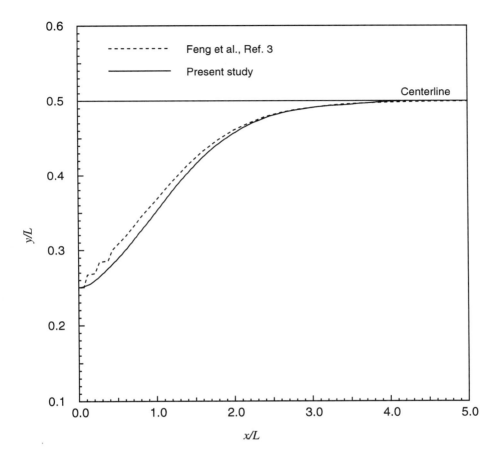

FIGURE 2: SETTLING TRAJECTORIES FOR PARTICLES WITH REYNOLDS NUMBER OF 1.03
RELEASED FROM DIFFERENT INITIAL POSITIONS. COMPARISON WITH THE SIMULATION
RESULT OF FENG ET AL. (Ref. 3) IS ALSO SHOWN HERE.

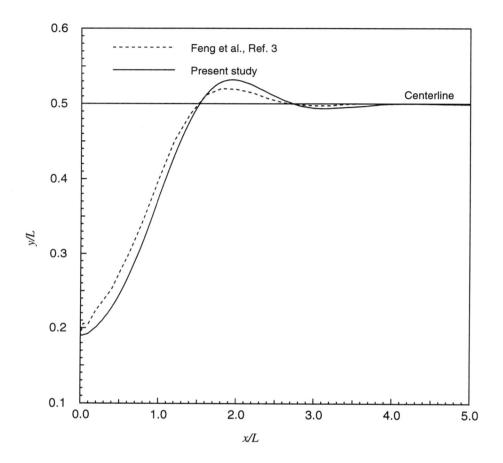

FIGURE 3: SETTLING TRAJECTORIES FOR PARTICLES WITH REYNOLDS NUMBER OF 3.23 RELEASED FROM DIFFERENT INITIAL POSITIONS. COMPARISON WITH THE SIMULATION RESULT OF FENG ET AL. (Ref. 3) IS ALSO SHOWN HERE.

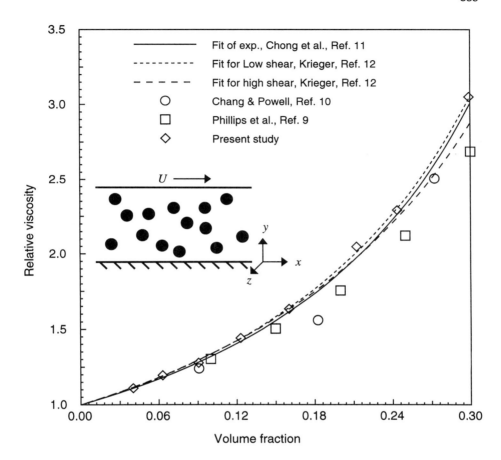

FIGURE 4: RELATIVE VISCOSITY OF MONODISPERSED SPHERICAL SUSPENSIONS IN A COUETTE FLOW AS A FUNCTION OF VOLUME FRACTION. THE SIMULATIONS ARE CARRIED OUT FOR THE SYSTEMS OF 128x128x16 OR 256x256x32 LATTICE NODES. THE NUMBER OF PARTICLES RANGES FROM 16 TO 64.

In conclusion, we have developed a new method for transportation of solid particles suspended in fluid. Lattice Boltzmann simulations have been carried out for the 2-D and 3-D parallel channel flows with moving particles. Results of the lattice Boltzmann simulations are in good agreement with experiments and other simulation results. The new method can very effectively be used for dynamical simulations of suspendsion flows in coating systems.

## ACKNOWLEDGMENTS

This study has been supported by the National Science Foundation's Young Investigator Award (CKA) through grant CTS-9258667, and by industrial matching contributions. The calculations were conducted, in part, using the facilities of National Center for Supercomputing Applications at Urbana-Champaign and Pittsburg Supercomputer Center at University of Pittsburg, both of which are being funded by the National Science Foundation.

## REFERENCES

1. Brady, J.F., and Bossis, G. Stokesian Dynamics, *Ann. Rev. Fluid Mech.*, **20**, 111 (1988).

2. Hu, H.H., Joseph, D.D., and Crochet, M.J. Direct simulation of fluid particle motions, *Theoret. Comput. Fluid Dyn.*, **3**, 285 (1992).

3. Feng, J., Hu, H.H., and Joseph, D.D. Direct simulation of initial value problems for the motion of solid bodies in a Newtonian fluid, *J. Fluid Mech.*, **261**, 95 (1994).

4. McNamara, G., and Zanetti, G. Use of the Boltzmann Equation to Simulate Lattice-Gas Automaton, *Phys. Rev. Lett.*, **61**, 2332 (1988).

5. Hou, S., Zou, Q., Chen, S., Doolen, G. and Cogley, A.C. Simulation of cavity flow by the lattice Boltzmann method, *J. Comp. Phys.*, **118**, 329 (1995).

6. Ladd, A.J.C. Numerical simulations of particulate suspensions via a discretized Boltzmann equation, *J. Fluid Mech.*, **271**, 285 (1994).

7. Aidun, C.K., and Lu, Y. Lattice Boltzmann simulation of solid particles suspended in fluid, *J. Stat. Phys.*, **81**, 49 (1995).

8. Faxen, H., *Proc. Roy. Swedish Inst. Eng. Res.*, No. 187 (1946).

9. Phillips, R.J., Brady, J.F., and Bossis, G. Hydrodynamis transport properties of hard-sphere dispersions, *Phys. Fluids*, **31**, 3462 (1988).

10. Chang, C., and Powell, R.L. Dynamic simulation of bimodal suspensions of hydrodynamically interacting spherical particles, *J. Fluid Mech.*, **253**, 1 (1993).

11. Chong, J.S., Christiansen, E.B., and Haer, A.D. Rheology of Concentrated Suspensions, *J. Appl. Polymer Sci.*, **15**, 2007 (1971).

12. Krieger, I.M. Rheology of Monodisperse Lattices, *Advan. Colloid Interface Sci.*, **3**, 111 (1972).

# TRIBOLOGICAL PROPERTIES OF CONDUCTING POLYMER FILMS FOR APPLICATION IN NANOTECHNOLOGY

C. BERIET, P.N. BARTLETT
Department of Chemistry
University of Southampton, Highfield, Southampton, SO17 1BJ (UK)

and

D.G. CHETWYND, J.W. GARDNER , X. LIU
Department of Engineering
University of Warwick, Coventry, CV4 7AL (UK)

In this paper we present results of a study of the friction and wear properties of conducting polymer films for application in micromechanics and nanotechnology. Polymeric bearings are commonly used as thin film bearings because of their low friction coefficient values. However, they are difficult to deposit uniformly and require simple bearing shapes. In contrast conducting polymer films are very easy to deposit electrochemically on small devices even on complex bearing shapes, the thickness of the film can be easily controlled and they also have good thermal and electrical properties. The friction coefficients and wear rates were measured using a pin-on-disc apparatus. We report the study of the deposition and morphology of smooth conducting polymer films with low friction coefficients which for optimum film thicknesses are as good as PTFE and with wear rates which are slightly better than PTFE.

## 1. INTRODUCTION

Thin film bearings have been used for many years. It is well known that under correct conditions, thin polymeric film bearings sliding on a glass counterface can provide smooth low friction bearings for applications requiring precise motion[1, 2]. Typical examples of low friction bearings include PTFE or composites like PTFE with graphite incorporated or PTFE/lead composite in a bronze matrix. Usually, bearing pads are produced by sintering operation, which greatly restricts the geometry to which they can applied. Otherwise these types of coatings require complex deposition techniques such as chemical vapour deposition for which the thickness of the film is difficult to control. There is therefore interest in finding new types of thin film coatings with good tribological properties. Conducting polymer films provide interesting properties as they are easy to deposit electrochemically even on complicated shape bearings and the thickness of the film can be easily controlled depending on the time and rate of deposition chosen[3]. It is also well known that the morphology of the electrodeposited film can be controlled by the choice of the solvent and electrolyte[4,5]. These conducting polymer films also have good thermal and electrical properties[6] which may be of use. We therefore believe that these materials have enormous potential in the fields of micromechanics and nanotechnology.

## 2. EXPERIMENTAL

### Reagents and materials

Pyrrole (Aldrich) was purified through an alumina column before preparing each solution. Methane phosphonic acid (MPA) and n-decane phosphonic acid (DPA) were purchased from Lancaster and used as received. Toluene sulfonic acid sodium salt (TSA) and dodecylbenzene sulfonic acid sodium salt (DBSA) were purchased from Aldrich and also used as received.

### Devices

Plano-convex lenses of 13 mm diameter were used as the substrates for testing the wear rate and friction coefficient of the various polymer films. 4 gold pads of 2000 Å with an underlayer of chromium (200 Å) were deposited on the lens using vacuum deposition. The lenses were then cleaned using iso-propanol in an ultrasonic bath. The gold pads were then individually connected so that a polymer film could be grown separately on each.

### Poly(pyrrole) growth

The poly(pyrrole) films incorporating various counter ions were prepared by electrochemical polymerization of pyrrole ($0.1$ mol dm$^{-3}$) from an aqueous solution containing $0.1$ mol dm$^{-3}$ of the corresponding counter ion. The films were deposited under potential control on the gold pads using a commercial potentiostat (EG&G) and a three electrode cell. The temperature was controlled at 20°C except for the solution containing DPA which had to be heated to 30 °C and the solution containing DBSA which had to be heated to 25°C.

### Pin-on-disc apparatus

The friction properties and wear rates of the various conducting polymer films were measured using a pin-on-disc apparatus shown in figure 1. It consists of a rotating optically flat glass disc, a sample assembly holding the substrate (lens) under test, a vertical probe for monitoring the height variation which therefore measures the wear of the polymer film and a spring flexure for the friction measurements. The sample assembly is attached tightly to the

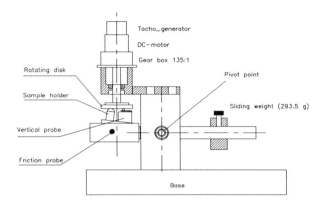

**Figure 1** : Schematic diagram of the slow speed test system.

moving platform of the spring flexure which is a notch-hinge linear spring mechanism and is centrally pivoted with a graduated rod on the other side. A counter mass of 293.5g can be moved along the rod to give a required loading force. The sample assembly has a holder which allows the lens to be turned so that one of the polymer pads comes in contact with the disk counterface. All experiments were performed in a temperature controlled laboratory, normally at 20 °C with average diurnal fluctuations of approximately 2°C. Typical measurements were obtained with a sliding speed of 5 mm s⁻¹ and a normal load of 2 N. Further details on the instrumentation may be found elsewhere[7].

## 3. RESULTS AND DISCUSSION

### 3.1 : Electrochemical deposition and study of the morphology of the films

After some preliminary studies[8] carried out on various conducting polymer films like poly(N-methylpyrrole), poly(aniline), poly(carboxindole) and poly(pyrrole) we concentrated this study on poly(pyrrole) films which gave the most promising results in terms of friction coefficients and wear rates. Four different poly(pyrrole) films were studied : poly(pyrrole) grown with MPA, DPA, TSA or DBSA. The films were grown under potential control either by stepping the potential to a value corresponding to a low polymerization current density (typically 0.6 to 0.65 V vs SCE) or double potential step, that is stepping the potential to a higher value for a short period of time where the rate of deposition is fast in order to create a large number of nucleation sites and then stepping the potential to a lower value to continue the growth.

After deposition, the film was conditioned by stepping to a lower potential at which the polymer was stable but still corresponding to the conducting state of the polymer film. If the rate of nucleation is slow, the double potential step technique is obviously a better choice in order to obtain more uniform films. Figures 2a and 2b show typical transients obtained with the two techniques. For all the counter ions used in this study, the rate of nucleation was fairly fast and so, for a given counter ion, no significant differences in film morphology were observed between films grown using the different techniques.

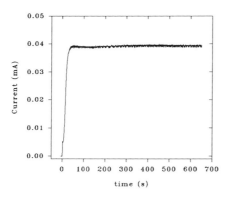

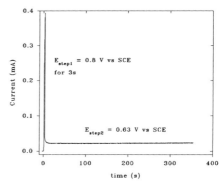

**Figure 2a :** Transient corresponding to the growth of poly(pyrrole) film with TSA. Solution 0.1 mol dm$^{-3}$ pyrrole + 0.1 mol dm$^{-3}$ TSA in water at 20 °C. $E_{step} = 0.6$ V vs SCE.

**Figure 2b :** Transient corresponding to the growth of poly(pyrrole) film with MPA by double potential step. Solution 0.1 mol dm$^{-3}$ pyrrole + 0.1 mol dm$^{-3}$ MPA in water at 20°C.

The poly(pyrrole) films grown with MPA, TSA and DBSA appeared smooth, shiny and uniform ranging in colour from green (MPA and DPA) or purple (DBSA) to black with increasing film thickness. Because the DPA growth solution was very viscous, the corresponding films were less uniform due to the poor conductivity of the solution and distortion due to iR drop. The poly(pyrrole) films grown with the phosphonic acid counter ions were very stable in contact with air and stayed smooth and shiny. The poly(pyrrole) films grown with TSA and DBSA were not as good. The adhesion of all of the films on the gold substrate was also fairly good but the films could easily be scratched with a finger nail. It is well known that the friction coefficient increases as the roughness of the film increases therefore the morphology of the various poly(pyrrole) films was studied using atomic force microscopy mainly because the resolution obtained with the scanning electron microscope was insufficient. Very smooth films were obtained with height variation less than ± 1 2 n m f o r poly(pyrrole) films grown with MPA (as shown in figure 3) and DPA, and height variation less than ± 7 5 n m f o r poly(pyrrole) films grown with TSA. The poly(pyrrole) films grown with DPA and MPA were both extremely smooth but had a completely different morpho- l o g y . T h e

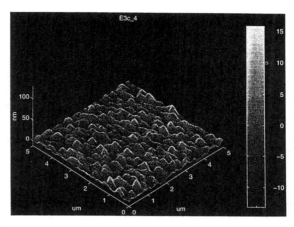

**Figure 3** : AFM image of poly(pyrrole) film grown with MPA. $I = 0.22$ mA.cm$^{-2}$ ; $Q = 183$ mQ.cm$^{-2}$

grain size for poly(pyrrole) films grown with DPA was smaller. Whenever the gold substrate was scratched, less uniform films were obtained. The scratches seemed to be favourable sites for nucleation leading to isolated mounds of polymer.

### 3.2 : Friction coefficients and wear rates of the conducting polymer films

The friction coefficients and wear rates of the various poly(pyrrole) films were measured using the pin-on disc apparatus. For comparison, the measurement was also carried out on a thin PTFE film bearing. Figure 4 shows typical transients obtained for PTFE and poly(pyrrole) film.

Three parameters are of interest, the friction coefficient at the start of the sliding is associated to the static friction coefficient and would be of importance if the bearing was used in a high precision displacement system for example. The dynamic friction coefficient is the average friction coefficient over a period of time that we would like as low as possible. And the fluctuation of the dynamic friction coefficient is important due to the stick-slip behaviour that we also want as low as possible. As seen in figure 4, the transients obtained for PTFE and poly(pyrrole) film are very similar. The dynamic friction coefficient was slightly lower for the conducting polymer film shown but the fluctuations were slightly larger.

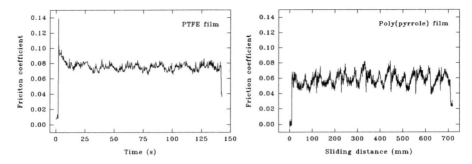

**Figure 4 :**    Transients showing the friction coefficient as a function of the sliding distance for a thin film of PTFE and for a conducting polymer film.

The reproducibility of the measurement was checked by depositing 12 presumably identical poly(pyrrole) films grown with TSA. For a thickness of ~0.15 $\mu$m a friction coefficient of 0.10 ± 0.01 was measured with a fluctuation of 0.012 ±0.005.

Figure 5 compares the dynamic friction coefficient obtained for four different poly(pyrrole) films and a thin film of PTFE. Poly(pyrrole) films grown with MPA, DPA have friction coefficients lower than 0.2, the values obtained for poly(pyrrole) films grown with TSA were slightly higher. However, the three polymer films have friction coefficient as good as PTFE (< 0.1). The friction coefficients measured for poly(pyrrole) films grown with DBSA were much higher (> 0.16). This film was not as smooth as the poly(pyrrole) films grown with MPA, DPA or TSA. It

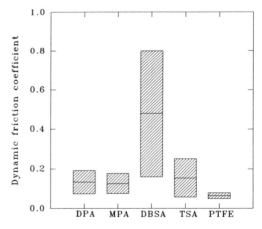

Poly(pyrrole) films with various counter ions

**Figure 5 :**    Range of dynamic friction coefficients measured for poly(pyrrole) films and PTFE.

was difficult to correlate the variation in the friction coefficient with the roughness of the films measured by AFM. For poly(pyrrole) films grown with MPA, DPA and TSA we found that the difference in the roughness of the films, from 10 nm to about 150 nm did not seem to affect significantly the measured friction coefficients. However, the important scatter in the friction coefficient measured for poly(pyrrole) films grown with DBSA could well be due to the roughness of the film and the irreproducibility in the growth of the film coming from scratches of the gold.

Figure 6 shows the influence of the thickness of the film on the dynamic friction coefficient. It is well known that the friction coefficient reaches a minimum value for an optimum film thickness but increases for both thinner and thicker films. For very thin films, the asperities of the substrate break through the film and the shear stress increases the friction coefficient. For thicker films, the contact area and ploughing of the film increases therefore the friction coefficient should also increase and tend to a constant value.

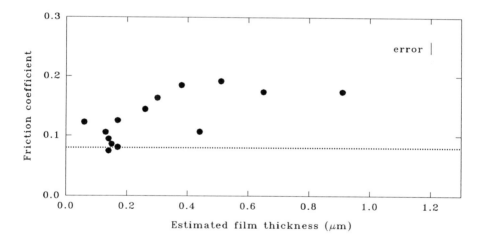

**Figure 6 :** Plot of the friction coefficient as a function of the thickness of the film for poly(pyrrole) films grown with DPA. ..... Typical friction coefficient value for a PTFE film.

The results plotted in figure 6 seems to follow quite well that theory with a low friction coefficient for an optimum thickness of ~0.15 $\mu$m. Similar trends were obtained for poly(pyrrole) films grown with MPA and TSA.

Another very important parameter in tribology is the wear of the film. The wear rates of the conducting polymer films were measured using the vertical probe fitted to the pin-on-disc apparatus. Figure 7 shows a plot of the friction coefficient as a function of the wear rate for three poly(pyrole) films grown with MPA, DPA or TSA. Some of the films were baked in an oven over night at a temperature varying between 120 °C to 180 °C. A significant improvement in the wear rate was observed for the films which were heat treated.

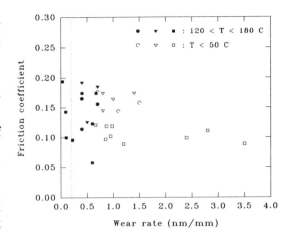

**Figure 7 :**    Plot of friction coefficient as a function of wear rate for poly(pyrrole) films grown with ● MPA, ▼ DPA, ■ TSA. ..... Typical value of wear rate for PTFE.

## 4. CONCLUSION

Very smooth and stable thin conducting polymer films could be deposited electrochemically and reproducibly. The poly(pyrrole films grown with the phosphonic acids counter ions have very interesting low friction coefficient. Friction coefficients lower than 0.2 could be produced reproducibly with values as good as PTFE for an optimum thickness. The wear a the polymer films seem to depend strongly on the temperature treatment carried out. For an optimum temperature, wear rates better than PTFE were obtained.

# REFERENCES

1. K. Lindsey, S.T. Smith and C.J. Robbie, Sub-nanometre surface texture and profile measurement with Nanosurf 2, Ann. CIRP, 37(1), (1988) 519-522.

2. S.T. Smith, S. Harb and D.G. Chetwynd, Tribological properties of polymeric bearings at the nanometre level, J. Phys. D., 25 (1A) (1992) 240-248.

3. G.K. Chandler and D. Pletcher, The electrochemistry of conducting polymers, Specialist Periodical Reports, Electrochemistry, 10 (1985) 117.

4. L.F. Warren, J.A. Walker, D.P. Anderson, C.G. Rhodes, A study of conducting polymer morphology, J. Electrochem. Soc., Vol 136 (1989) 2286-2295.

5. F.T.A. Vork, B.C.A.M. Schuermans, E. Barendrecht, Influence of inserted anions on the properties of poly(pyrrole), Electrochimica Acta, Vol 35 (1990) 567-575.

6. S. Roth, H. Bleier and W. Pukacki, Charge transport in conducting polymers, Faraday Discuss. Chem. Soc., 88 (1989) 223-233.

7. X. Liu, D.G. Chetwynd, J.W. Gardner, S.T. Smith, C. Beriet, P.N. Bartlett, Measurement of friction at light loads and low speeds in poly(pyrrole) film bearings, Proc. of ASPE 10th Annual Meeting, Texas, October 1995 and to be published in Precision Engineering.

8. S.T. Smith, S. Harb, V. Eastwick-field, Z.Q. Yao, P.N. Bartlett, D.G. Chetwynd and J.W. Gardner, Tribological properties of electroactive polymeric thin film bearings, Wear, 169 (1993) 43-57.

# SIMULATION OF VISCOELASTIC FLOWS

O.G. Harlen

*Department of Applied Mathematical Studies,*
*University of Leeds, Leeds, UK LS2 9JT.*

## SUMMARY

This paper describes a novel numerical method for simulating time-dependent flows of viscoelastic fluids. The method uses Lagrangian finite elements that move with the fluid, and is particularly well suited to problems involving liquid-air interfaces, such as those found in many coating applications. This method is used study the stretching of a viscoelastic filament.

## 1. INTRODUCTION

An important recent development in coating technology is the addition of polymeric additives to coating fluids to improve the quality and robustness of the finished coat. Coating fluids that contain polymer additives are viscoelastic, and so in order to be able to model the coating process numerically, it is necessary to develop numerical schemes for viscoelastic flows that can handle the kind of complex flow geometries found in coating applications. These typically involve both solid boundaries and liquid-air interfaces, and in addition the flow may be time-dependent.

Viscoelastic fluids differ from Newtonian fluids or generalised Newtonian fluids (such as power law or Carreau fluids) in that viscoelastic fluids have 'memory'. The fluid stress in a generalised Newtonian fluid depends only upon the instantaneous local value of the rate-of-strain and pressure, whereas the stress in a viscoelastic fluid depends upon the rate-of-strain at previous times, as well as its current value. This strain-rate history is associated with a particular fluid element and moves with it along particle paths. Consequently the calculation of the stress in a viscoelastic fluid is naturally Lagrangian, (*i.e.* it is more naturally expressed in a frame of reference moving with the fluid, rather than one fixed in space).

The vast majority of numerical codes for simulating either viscous or viscoelastic fluids adopt a frame of reference that is fixed in space. This is particularly advantageous for problems with solid boundaries where the edge of the fluid domain remains constant. However, the position of a fluid-air interface moves with the fluid and so, like the calculation of the viscoelastic stress, is a naturally Lagrangian problem. Thus for viscoelastic flows with fluid-air interfaces the advantages of using a Lagrangian scheme may outweigh those of using a fixed grid.

In this paper we outline a novel numerical method for transient viscoelastic flows that uses Lagrangian elements, and discuss its application to the problem of the stretching of fluid filament. Although this problem is not directly related to a coating process, it shares many of the features found in coating flow geometries in having both solid boundaries and liquid-air interfaces.

## 2. GOVERNING EQUATIONS

The governing equations for a viscoelastic fluid are comprised from the standard equations of conservation of mass and momentum

$$\nabla \cdot \mathbf{u} = 0 \tag{1}$$

$$\rho \frac{D\mathbf{u}}{Dt} = -\nabla \cdot \boldsymbol{\sigma} \tag{2}$$

(where $\mathbf{u}$ and $\boldsymbol{\sigma}$ are respectively the fluid velocity and stress) together with a constitutive equation for the stress $\boldsymbol{\sigma}$.

Unlike Newtonian fluids, where there is a single undisputed constitutive law relating the stress to the rate-of-strain, the complex microstructure of polymeric fluids makes it impossible to formulate an exact constitutive relation. Instead a plethora of approximate models have been proposed (see for example Larson [4]). A simple constitutive model, but one that encapsulates the essential underlying physics in a polymer solution, is the Oldroyd-B fluid. Although this model has a number of short-comings, it is able to provide reasonable quantitative agreement between several different experiments for a restricted class of polymeric fluids[3].

Polymer molecules are highly flexible structures and can lie in a very large number of different configurations. In solution an equilibrium distribution of configurations is set up in which the polymers are on average coiled up in a random-walk (the precise details depend upon the quality of the solvent). If a flow is now applied to the solution the velocity gradients will try to stretch out the polymers (see figure 1), disturbing this equilibrium. This is opposed by molecular relaxation due to thermal fluctuations, which tries to restore the equilibrium distribution. It is this competition that gives rise to the elasticity of the fluid. A simple analogy is think of polymers as behaving like microscopic elastic bands.

A simplified model of the gross distortion of the polymer may be obtained by replacing the polymers with elastic dumbbells. The stretching of the polymer by the flow is now represented by the drag on the beads, while the spring represents the relaxation of the polymer. The configuration distribution at each point in space may now be expressed as a symmetric second-rank tensor $\mathbf{A}$, which denotes the ensemble average of $\mathbf{RR}$ (where $\mathbf{R}$ is the dumbbell length).

After an appropriate non-dimensionalisation there are just two non-Newtonian parameters; $c$, a measure of the concentration of polymer; and the Weissenberg number, We, the ratio of the velocity gradient to the relaxation-rate. If the

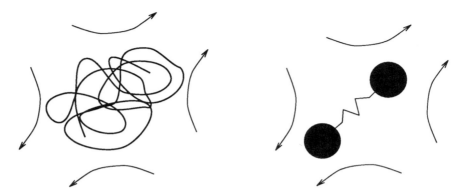

Figure 1: The gross deformation of the polymer is modelled as the stretching of an elastic dumbbell

Weissenberg number is small, the flow is too weak to deform the polymer structure and the fluid is only weakly elastic. On the other hand, when We is large the flow can produce large deformations of the polymer generating large elastic stresses.

For this model, the dimensionless fluid stress is given by

$$\boldsymbol{\sigma} = -p\mathbf{I} + \boldsymbol{\nabla}\mathbf{u} + \boldsymbol{\nabla}\mathbf{u}^\dagger + \frac{c}{\text{We}}\left(\mathbf{A} - \mathbf{I}\right) \tag{3}$$

where $p$ is the fluid pressure and $\dagger$ denotes transpose. The structure tensor $\mathbf{A}$ evolves according to the equation

$$\frac{\partial \mathbf{A}}{\partial t} + \mathbf{u} \cdot \boldsymbol{\nabla}\mathbf{A} - \mathbf{A} \cdot \boldsymbol{\nabla}\mathbf{u} - \boldsymbol{\nabla}\mathbf{u}^\dagger \cdot \mathbf{A} = -\frac{1}{\text{We}}\left(\mathbf{A} - \mathbf{I}\right) \tag{4}$$

This last equation describes the stretching and relaxation of the polymers as they are advected with fluid, and is a hyperbolic equation where the characteristics are the particle paths.

Equation (4) can be greatly simplified by noting that the terms on left-hand side correspond to the time-derivative in a frame of reference that moves and deforms with the fluid. This eliminates all the spatial gradients from the equation, and reduces the computation of $\mathbf{A}$ to the solution of an ordinary differential equation at each point in space.

## 3. NUMERICAL METHOD

The central idea behind our numerical method is to solve the momentum and continuity equations (1) and (2) in an Eulerian frame, but switch to frame deforming with the fluid for the calculation of the structure. The numerical method consists of a two stage algorithm, in which we first find the flow corresponding to the current values of the structure **A** and then step forward in time to find new values for **A**.

The solution of the conservation equations (1), (2) together with (3) for the current values of the structure tensor **A**, is obtained using a standard finite element technique. We use triangular elements in which the velocity and pressure nodes are the vertices of the triangles. The resulting system of algebraic equations is solved iteratively using Preconditioned Conjugate Residuals.

In order to calculate the evolution of the structure in a codeforming frame, we treat the nodes of the triangles as material points. This means that each triangle represents a volume of fluid, and so the triangle sides behave as lines and can be used as base vectors for a coordinate system based in the fluid. To solve equation (4) we first transform **A** into a coordinate system based on two of the triangle sides at the beginning of the timestep. We then solve the equation

$$\frac{d\tilde{\mathbf{A}}}{dt} = -\frac{1}{\text{We}} \left( \tilde{\mathbf{A}} - \tilde{\mathbf{I}} \right) \tag{5}$$

where $\tilde{\mathbf{A}}$ are the components of **A** in triangle-side coordinates. Finally we transform back to a coordinate system based on fixed axes using the triangle sides at the end of the timestep.

A major problem with using Lagrangian elements is that their shape can become highly distorted by velocity gradients within the fluid. This will ultimately degrade the accuracy of the finite element solution, and so it is necessary to limit the distortion of the mesh in some way. The solution we have adopted is to retain the nodes as material points, but reconnect them in the way that produces the "best" possible triangulation. What we wish to avoid are very acutely angled triangles and so the optimal choice is the Delaunay triangulation [1], which maximises the minimum angle in any triangle. After each timestep the mesh is reconnected, where necessary, to obtain the Delaunay triangulation.

Further details of the numerical method can be obtained from the paper by Harlen et. al.[2]

## 4. STRETCHING FILAMENT

The stretching filament experiment was proposed by Sridhar[5] as a way to measure the extensional properties of polymeric materials. A small sample of fluid is placed between two plates. The plates are then moved apart so that the distance between them increases exponentially in time (see figure 2), so that the average extension-rate over the length of the filament is constant in time. A force transducer measures the force required to stretch the filament. If the

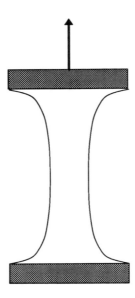

Figure 2: Sketch of the stretching filament device.

fluid was able to slip at the plates, the sample would experience a perfectly uniform uniaxial extension. However, since the fluid must obey no slip at the plate surface, the column of fluid thins initial thins more in the middle, into a shape similar to a wine glass stem.

Although the effects of gravity and surface tension can be incorporated into the numerical method, for simplicity the calculations shown here neglect both effects, so that the filament-air boundary is a free-surface. Surface tension provides an additional mechanism for squeezing the filament in the middle, while gravity introduces a slight up-down asymmetry by thickening the base of the filament and thinning the top. Fluid inertia is also neglected as the Reynolds' number is less than one tenth.

We can non-dimensionalise this problem by choosing the initial separation of the plates as the unit of length and the reciprocal of the average extension-rate as the unit of time. For a Newtonian filament, there is a single dimensionless parameter, the initial aspect ratio of the sample, which controls the subsequent dynamics. For a viscoelastic fluid there are two additional parameters; $c$ a measure of the polymer concentration; and the Weissenberg number, We, which is the ratio of the average extension-rate of the filament to the relaxation rate of the polymer. For a given fluid, $c$ is fixed, but We can be varied by changing the extension-rate in the experiment.

If the Weissenberg number is less than one half, (*i.e.* if the filament is

Figure 3: Comparison between the shape of the filament of a Newtonian (left) and an Oldroyd B fluid (right) when the filament is 2.56 times its original length.

stretched slowly), the extension-rate is too weak to overcome the relaxation of the polymer and the elastic stress remains small. If, on the other hand the Weissenberg number is greater than one half, then extension wins over relaxation and the polymers can stretch indefinitely given sufficient time, and so the elastic stress continues to grow as the filament stretches.

In the simulations we fixed the initial aspect ratio (plate separation: sample diameter) as 1:3.5, the value used in one experimental set-up. The parameter $c$ determines the relative magnitude of the elastic and viscous stresses, so we choose $c = 8$ to give a strongly elastic fluid.

The shape of the filament for a Newtonian fluid and the viscoelastic fluid at a Weissenberg number of unity is compared in figures 3 and 4. These show the filament profile when the filament length is respectively 2.56 and 24.5. In both figures the right-hand picture shows the viscoelastic fluid, and the shading shows Tr**A** the extension of the polymer. At early times (figure 3), both the Newtonian and viscoelastic profiles are similar as the elastic stress has not had sufficient time to build up.

As the plates are pulled further apart, the Newtonian filament continues to thin most in the middle (figure 4). Since the total force is constant across any cross-section, the stress and therefore the rate-of-strain must be proportional to inverse square of radius, so that

$$\frac{dr}{dt} \propto -\frac{1}{r^2} \tag{6}$$

However, in the viscoelastic fluid the elastic stress grows as the square of the polymer extension and so it is largest where the filament width is least. This resists further extension in the middle part of the filament. The result is that the filament cross-section is almost constant along its entire length, except for a small foot-region near the plates. This foot region is much smaller than in the Newtonian filament as the higher tension in the filament tends to draw the fluid out of the foot. Qualitatively at least these profiles match those seen in experiment, though further work is required to make a quantitative comparison.

The force required to stretch the Newtonian filament decreases as its stretches due to the thinning of the filament. In the viscoelastic filament the axial com-

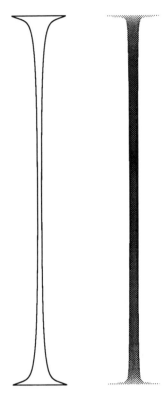

Figure 4: Comparison between the shape of the filament of a Newtonian (left) and an Oldroyd B fluid (right) when the filament is 24.5 times its original length.

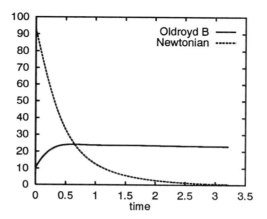

Figure 5: Plot of the dimensionless force in the filament as a function of time for an Oldroyd B fluid at We = 1 and a Newtonian fluid.

ponent of the structure tensor, $A_{zz}$, satisfies (from equations (4) and (1))

$$\frac{d}{dt}A_{zz} + \frac{4}{r}A_{zz}\frac{dr}{dt} = -\frac{1}{\text{We}}(A_{zz} - 1) \quad (7)$$

so that when $A_{zz} \gg 1$

$$A_{zz}r^4 \propto \exp\left(-\frac{t}{\text{We}}\right)$$

and hence the tension force, $F$, in the filament should ultimately scale

$$F \propto \exp\left[\left(1 - \frac{1}{\text{We}}\right)t\right].$$

Thus, if the Weissenberg number exceeds unity, the force should increase with time and if We is equal to unity the force should tend to a constant. Figure 5 shows a plot of the dimensionless force (scaled on the viscosity in steady shear flow) in the filament as function of time for a Newtonian filament and a viscoelastic filament at We = 1.

## 5. CONCLUSION

The numerical method outlined above is motivated from the idea that momentum and constitutive equations should be solved in their natural frames of reference. This is achieved by using Lagrangian finite elements that move and deform with the fluid. Our method is particularly well suited to flows with moving boundaries such as air-liquid interfaces, where the shape of the fluid domain changes with time. Consequently, it should be capable of handling the complex flow geometries found in coating processes.

## 6. REFERENCES

1. Delaunay, B.N. 1934 Sur la sphere vide, *Bull. Acad. Sci. USSR VII: Class. Sci. Math.*, 793–800

2. Harlen, O.G., Rallison, J.M., and Szabo, P. 1995 A split Eulerian-Lagrangian method for simulating transient viscoelastic flows, *J. non-Newtonian Fluid Mech.*, **60**, 81–104.

3. Keiller, R.A. 1992 Modeling of the extensional flow of the M1-fluid with the Oldroyd equation. *J. non-Newtonian Fluid Mech.*, **42**, 49–64.

4. Larson, R.G. 1988 *Constitutive equations for polymer melts and solutions* Butterworth, Boston

5. Tirtaatmadja, V., Sridhar, T. 1993. A filament stretching device for measurement of extensional viscosity. *J. Rheology*, **37**, 1081–1102

# SECTION 7

# Novel Coating and

# Thin Film Problems

# The chemistry of thin film deposits formed from hexamethyldisiloxane and hexamethyldisilazane plasmas.

S.M. Bushnell-Watson*, M. R. Alexander, A. P. Ameen, W. M. Rainforth,
R. D. Short and F. R. Jones
Department of Engineering Materials, University of Sheffield, Sir Robert Hadfield
Building, Mappin Street, Sheffield, S1 3JD.
W. Michaeli, M. Stollenwerk, G. Mathar and J. Zabold.
Institut Für Kunststoffverarbeitung, RWTH Aachen, D-52056 Aachen.

## Summary

Organosilicon films have been deposited from microwave sustained plasmas.
This plasma polymerisation deposition technique allows films of several microns
thickness to be readily formed on a variety of substrates. The deposition
parameters can be used to alter the chemistry of the film from an organic to an
inorganic nature. The films described here were obtained from
hexamethyldisiloxane - oxygen or nitrogen and from hexamethyldisilazane -
nitrogen or ammonia plasmas. Two different microwave plasma systems were
used, one of which employs an electron cyclotron discharge. The chemistry of the
resulting films is discussed with reference to data obtained by X-ray photoelectron
spectroscopy and infra-red spectroscopy.

## Introduction

The use of plasma polymerisation to produce thin film coatings is a technique
with potential applications in many areas of technology. The description, plasma-
enhanced chemical vapour deposition (PECVD), is sometimes used to describe the
process and emphasises its similarity with chemical vapour deposition (CVD)
processes.[1] Both deposition processes use gaseous starting materials which are
subsequently decomposed, by thermal energy for the CVD process and by electron
impact for the PECVD process. This technique is applicable to a wide range of
organic gases and vapours, many of which are not suitable for conventional
polymerisation processes. The resulting films are different to those formed
through conventional polymerisation in that the regular repeats units are not
present.[2]

Plasma deposition has been used for many years in the microelectronics
industry but thin films deposited by these techniques have potential use in many
areas of modern technology. For example, Wróbel[3] suggests that plasma
polymerised organosilicon thin films may find use as protective coatings, as

dielectrics in microelectronics, as anti-reflection coatings in conventional optics and as highly biocompatible materials in medicine owing to their outstanding thermal and chemical resistance and their remarkable electrical, optical and biomedical properties. Probably the largest single industrial application of plasma polymer films is the deposition of protective coatings of plasma polymerised hexamethyldisiloxane on the evaporated aluminium mirror film of car headlights to prevent corrosion of the aluminium layer.[4]

Morosoff [2] details certain characteristics of plasma polymerised films which makes plasma polymerization an attractive technique. These include the ability to form films which are often highly coherent and adherent to a variety of substrates including conventional polymer, glass and metal surfaces. These coatings are pinhole-free and highly crosslinked and multilayer deposits or films with grading of chemical or physical characteristics are easily made.

In recent years the use of microwave plasma polymerisation techniques have been developed. Microwave plasmas yield more polymerizable species than those generated by r.f frequencies and therefore lead to a higher deposition rate.[5] Deposition rates in the order of several hundred angstroms per second can be achieved which makes the process industrially attractive. The ability to tailor the film structure and properties by control of the deposition parameters such as substrate temperature, flow rates and the power density of the plasma means that a variety of films of organic or inorganic nature can be deposited.

Recently, electron cyclotron resonance (ECR) plasma sources [6] have been developed and are now widely used for thin film deposition. In these systems the microwave energy is coupled to the natural frequency of the electron gas in the presence of a static magnetic field. Such systems are usually operated at relatively low pressures to minimise the loss of energy through collisions and so the transfer of energy will be maximised. The presence of the axial magnetic field will reduce the plasma loss to the chamber walls. ECR plasma sources are therefore capable of producing very high plasma density at relatively low pressures and therefore allow deposition at low pressures, low temperatures and with potentially low damage to substrates and the deposited coating.

The work discussed in this paper describes the chemistry and morphology of films deposited from two different organosilicon monomers subjected to a microwave plasma. In the first instance, coatings obtained from hexamethyldisiloxane / oxygen plasmas, produced in two different apparatus, are compared. The second coating system utilises an ECR plasma source rather than an ordinary microwave sustained plasma. The effect of changing the ECR gas to nitrogen rather than oxygen is noted.

Secondly, the nature of films deposited from hexamethyldisilazane subjected to either nitrogen or ammonia ECR plasma is investigated.

In all these systems the effect of varying process parameters such as microwave power, flow rate of the reactant gases and the substrate temperature will be discussed.

## Experimental

Plasma polymers were prepared from two different monomers, hexamethyldisiloxane [$(CH_3)_3Si-O-Si(CH_3)_3$] and hexamethyldisilazane [$(CH_3)_3Si-NH-Si(CH_3)_3$]. Two different microwave sustained plasma reactors were used to deposit these films, one of which uses an ECR system. Schematic diagrams of both apparatus are given below.

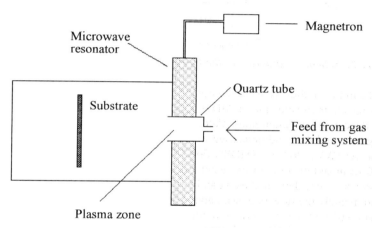

**Figure 1.** *Schematic diagram of the microwave plasma deposition unit.*

In the reactor geometry described in Figure 1, the monomer vapour and the reactant gas are introduced together, i.e. are subjected to the microwave excitation together. The resultant excited species then pass into the reactor chamber where deposition of the film occurs on a substrate.

A schematic diagram of the ECR plasma deposition unit is shown in Figure 2. In this system in the region of the ECR coils the electrons are under the influence of both a magnetic field and an electromagnetic field oscillating at the electron cyclotron frequency. The electrons absorb energy from the electromagnetic field through the ECR. The reactant (ECR) gas is fed into this discharge zone and microwave excitation is applied. The excited species then pass out of the discharge zone into the plasma processing region. The monomer vapour is injected into the process chamber where a secondary plasma is induced and film deposition occurs on the substrate.

Both apparatus operate under vacuum conditions, details of which are given in the experimental section.

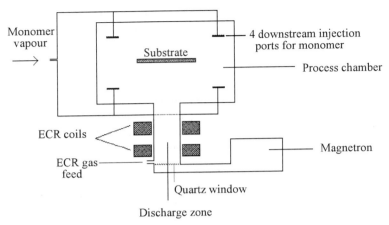

**Figure 2.** *Schematic diagram of the ECR plasma deposition unit.*

The initial studies were carried out on a series of plasma polymers deposited from hexamethyldisiloxane, (HMDSO), and oxygen plasmas. Films, 5 μm thick, were deposited onto aluminium plate. A schematic of the apparatus used is shown in Figure 1 and the conditions used were as follows: total pressure (P) fixed at 50 Pa, power (W) at 200 W, HMDSO flow at 20 sccm*, oxygen flow varied between 0-200 sccm and the microwave generator at 2.45 GHz.

For the films deposited using an ECR discharge several parameters were varied, namely the flow rate of monomer, the flow rate of the ECR gas and the power density of the microwave source. A standard Astex 2.45 GHz microwave source was used with ECR coils producing an axial magnetic field of 875 Gauss which is required for electron cyclotron resonance to occur. The base pressure in the system was in the region of 10$^{-5}$ Pa and the process pressure varied between 0.1 and 0.4 Pa.

Films were produced from HMDSO and oxygen using a monomer flow of 23 sccm, ECR gas flow of 4.5 sccm and power varied from 350-1000 W. The power and ECR flow were then held constant and the HMDSO flow reduced to 4 sccm. Finally the HMDSO flow was increased to 8 sccm and the oxygen flow to 19.4. Two films were deposited from HMDSO with nitrogen as the ECR source. In both cases a flow rate of 8 sccm HMDSO and 25.3 sccm nitrogen was used but powers of 600 and 1000 W were applied.

For the coatings deposited from hexamethyldisilazane, (HMDSZ), the parameters described above were varied and the effect of substrate heating was also introduced. A range of powers were investigated but only results obtained at 700 W will be discussed. The actual conditions used are summarised in Table 1.

---

* standard cubic centimetres per minute

| HMDSZ flow | ECR flow | Substrate heating conditions |
|---|---|---|
| 2 * | $N_2$ / 20 | Heater not used |
| 4.2 | $N_2$ / 50 | Not used, 350 °C |
| 9.8 | $N_2$ / 50 | Not used, 350 °C |
| 17.6 | $N_2$ / 50 | 350 °C |
| 24.6 | $N_2$ / 50 | 350 °C |
| 4.2 | $NH_3$ / 50 | Not used, 350 °C, 450 °C |
| 9.8 | $NH_3$ / 25 | Not used, 450 °C |

All at 700 W except where marked *, 650 W used; flow in sccm

**Table 1.** *Deposition conditions for the HMDSZ plasma polymer films.*

The HMDSO plasma polymers were deposited onto both stainless steel and aluminium foil. The HMDSZ plasma polymers were deposited onto either stainless steel or pressed potassium bromide discs. Samples produced from the latter were characterised by infra-red spectroscopy. The films obtained were all analysed by X-ray photoelectron spectroscopy (XPS) to determine the elemental composition of the surface. A VG CLAM 2 electron analyser and Mg kα radiation source with a sample take off angle of 30° was used, giving a calculated analysis depth of ~ 2.1 nm.

**Results**

*Chemical analysis by X-ray photoelectron spectroscopy.*

XPS analysis of films produced from the plasma polymerisation of HMDSO - $O_2$ plasma were similar in all cases and a typical example of a widescan spectrum is shown in Figure 3.

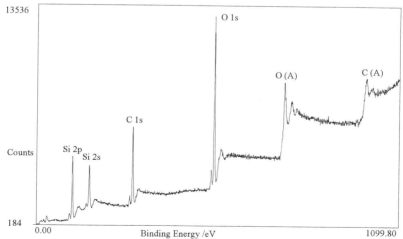

**Figure 3.** *XPS widescan of an HMDSO and oxygen plasma polymer.*

Within the detection limits of this technique, the only elements observed are carbon, oxygen and silicon. Elemental compositions of the films produced from the first series of coatings were calculated and are shown in Table 2. For all of these a power of 200W and HMDSO flow of 20 sccm was used.

| Oxygen flow/sccm | $O_2$:HMDSO | Silicon | Carbon | Oxygen |
|---|---|---|---|---|
| 0 | 0 | 26.80 | 50.31 | 22.89 |
| 10 | 0.5 | 27.74 | 39.93 | 32.33 |
| 20 | 1.0 | 28.88 | 35.57 | 35.56 |
| 40 | 2.0 | 28.63 | 29.94 | 41.43 |
| 100 | 5.0 | 29.77 | 22.66 | 47.57 |
| 200 | 10.0 | 30.50 | 18.74 | 50.76 |

**Table 2.** *Elemental composition of the HMDSO and oxygen plasma polymers.*

The calculated elemental composition of the HMDSO - $O_2$ plasma polymer films deposited from the ECR plasma source are given in Table 3.

XPS analyses of the two deposits produced with nitrogen as the ECR gas gave spectra which were similar to those produced with oxygen. However, a peak due to nitrogen was observed at the higher power and smaller peaks attributable to oxygen were seen. The elemental compositions were calculated as before and are included in Table 3.

| HMDSO flow | ECR flow | ECR gas: HMDSO | Power | Si | C | O | N |
|---|---|---|---|---|---|---|---|
| 23 | $O_2$, 4.5 | 0.2 | 350 | 25.4 | 53.1 | 21.6 | - |
| " | " | " | 700 | 26.5 | 50.6 | 23.0 | - |
| " | " | " | 1000 | 24.4 | 50.1 | 25.5 | - |
| 4 | $O_2$, 4.5 | 1.13 | " | 25.0 | 52.8 | 22.3 | - |
| 8 | $O_2$, 19.4 | 2.4 | " | 27.7 | 45.8 | 26.6 | - |
| 2 | $O_2$, 4.5* | 8.7 | 500 | 27.9 | 42.5 | 29.7 | - |
| 8 | $N_2$, 25.3 | 12.7 | 600 | 29.1 | 57.9 | 13.1 | 0.0 |
| " | " | " | 1000 | 25.8 | 57.5 | 15.2 | 1.6 |

* also used argon carrier gas

**Table 3.** *Calculated elemental composition for the ECR deposited HMDSO films.*

For all the HMDSO - $O_2$ films deposited from the ECR plasma, the composition determined from deposits on both the steel and the aluminium substrate were very similar. This indicated that a homogenous deposition was obtained over the whole of the substrate area. For ease of comparison an average value of composition is presented rather than data for both substrates.

XPS spectra were obtained from the HMDSZ with nitrogen or ammonia films. All show peaks attributable to silicon, carbon nitrogen and oxygen. The calculated elemental atomic composition of the HMDSZ plasma polymers is given below in Table 4. All were deposited at a power of 700 W, with the exception of the first coating listed in the table when 650 W was used.

| HMDSZ flow | ECR flow sccm | Substrate heater not used | | | | Substrate heater used | | | |
|---|---|---|---|---|---|---|---|---|---|
| | | Si | C | O | N | Si | C | O | N |
| 2 sccm | $N_2$, 20 | 25.3 | 57.8 | 7.1 | 9.8 | | | | |
| 4.2 sccm | $N_2$, 50 | 26.8 | 58.7 | 5.3 | 9.2 | 35.3 | 41.0 | 10.6 | 13.1 |
| 9.8 sccm | $N_2$, 50 | 24.0 | 58.0 | 7.3 | 10.7 | 32.4 | 32.2 | 15.7 | 19.7 |
| 17.6 sccm | $N_2$, 50 | | | | | 36.0 | 42.3 | 5.8 | 15.9 |
| 24.6 sccm | $N_2$, 50 | | | | | 35.1 | 45.0 | 5.1 | 14.8 |
| 4.2 sccm | $NH_3$, 50 | 26.7 | 46.6 | 15.8 | 10.9 | 28.5 | 17.4 | 37.4 | 16.8 |
| | | | | | | 24.1 | 17.7 | 45.2 | 13.1 |
| 9.8 sccm | $NH_3$, 25 | 27.5 | 55.6 | 3.3 | 13.6 | 35.0 | 23.9 | 15.7 | 25.4 |

Substrate heating of 350 °C used except for the HMDSZ - $NH_3$ at 4.2 /50 where 450 °C was also, used and the 9.8 / 25 where 450 °C only was used.

**Table 4.** *Calculated elemental composition for the ECR deposited HMDSZ - $N_2$ and HMDSZ - $NH_3$ films.*

*Infra red spectroscopy.*

Certain of the HMDSZ films were deposited on to potassium bromide substrates which were then used to obtain infra-red spectroscopy data. Spectra typical of plasma polymerised HMDSZ were obtained.[3,7] Peaks identified showed the presence of Si-$CH_3$, N-H and Si-N-Si bonds. These data show that the original structural groups of the monomer are incorporated in the coating. Peaks indicating Si-O-Si, Si-O-C bonds are also observed and these obscure any contribution from the $CH_2$. Little variation in data is seen for films deposited under different flow or power regimes but the spectra obtained using a heated substrate are markedly different. In this case one broad peak is observed, with few other features visible. These spectra are characteristic of a strongly crosslinked material with a high content of inorganic structure,[4] showing that the use substrate heating is altering the chemistry of the film.

**Discussion**

*Films deposited from hexamethyldisiloxane.*

The series of films produced in the conventional microwave plasma system show a clear change in the elemental composition as the oxygen flow rate is increased. As the flow is increased from 0 to 200 sccm, the oxygen concentration is more than doubled (Table 2). The O:Si ratio of these films varies from 0.85 to 1.66 and so begins to approach that found in silica ($SiO_2$) of 2.0. The shortfall is consistent with the retention of 19% carbon, which is probably bonded to the silicon.

Films produced from HMDSO and oxygen using an ECR plasma do not show a pronounced a change in the plasma polymer composition as the ratio between the HMDSO and oxygen is varied. Some small decrease in the amount of carbon is seen but the composition changes are much smaller than those described above. Direct comparison of the composition of the coatings can not be made as different deposition systems were used and different processing parameters employed. However, as similar oxygen to HMDSO flow ratios were used some comparisons can be made. For example, comparing the film deposited with no oxygen and that from the ECR system where an $O_2$ : HMDSO ratio of 0.2 was used, shows that very similar compositions were obtained. Moving to $O_2$ : HMDSO ratios of 2.0 or 2.4 (ECR plasma) shows a large difference in the calculated percentages of carbon and oxygen. These data are compared in Table 5.

| $O_2$ : HMDSO | Power | Silicon | Carbon | Oxygen |
|---|---|---|---|---|
| 0 | 200 | 26.8 | 50.3 | 22.9 |
| 0.2 † | 700 | 26.5 | 50.6 | 23.0 |
| 2.0 | 200 | 28.9 | 35.6 | 35.6 |
| 2.4 † | 700 | 27.7 | 45.8 | 26.6 |

† deposited from ECR microwave plasma

**Table 5.** *Comparison of the compositions of HMDSO - $O_2$ films produced from each microwave plasma system at similar $O_2$ : HMDSO ratios*

For the second series of coatings, when the ECR gas is changed to nitrogen there is only evidence of a small amount being incorporated into the film surface. Using a higher flow rate of nitrogen ECR gas than was used for the oxygen, only 1.7% nitrogen was detected in the surface of the film at a microwave power of 1000 W. The carbon levels in this film are a little higher and those of oxygen are reduced. The latter element is not present in the monomer but is incorporated either during the deposition cycle, or as is more likely, by reaction with the atmosphere when the sample is removed from the process chamber.

The effect of varying the flow rates of the monomer and the reactant gas are therefore found to be different in each of the microwave plasma used. When the monomer and reactant gas are injected and subjected to the microwave excitation together, as in the first series of coatings, the gas is incorporated into the film surface. Increasing the proportion of oxygen injected into the system resulted in a very significant increase in the oxygen content of the film. When the ECR plasma source is used, i.e. when the plasma is formed in the reactant gas which in turn excites the monomer vapour, the incorporation of the reactant gas into the film surface is only small. For the ECR deposited coatings examined in this study, there was little change observed in the composition of the resulting film when the gas flow ratios were varied.

*Films deposited from hexamethyldisilazane.*

The compositions of the films deposited from HMDSZ with nitrogen as the ECR gas again show little change with the ratios of the flow of monomer vapour and reactant gas or with the power density used. A significant change in composition is seen when the substrate is heated during the deposition. The amount of carbon decreases whilst the percentage of silicon, nitrogen and oxygen all increase. As stated above, the oxygen is most likely incorporated into the surface by reaction of remaining active species with the atmosphere after removal from the process chamber. This change is also reflected in the IR spectra of the films where data typical of a strongly crosslinked material containing a high inorganic matter are seen when substrate heating is applied. The intensities of the peaks do not change significantly with power as reported elsewhere [4] but as discussed above, little compositional change was observed in these ECR deposited systems, when this parameter is varied.

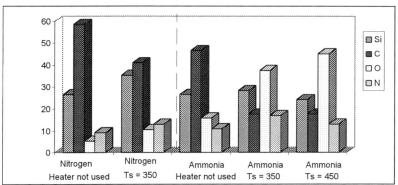

**Figure 4.** *Effect of varying ECR gas and substrate temperature, Ts, on the composition of HMDSZ films deposited at 4.2 sccm monomer and 50 sccm ECR gas flow.*

When ammonia is used as the reactive gas, a significant change in the chemistry of the film is again seen. In this instance, the amount of silicon is relatively unchanged but the decrease in carbon content and the increase in oxygen content are more pronounced. The effect on the composition of the resulting film when the ECR gas is changed from nitrogen to ammonia is to decrease the carbon content and increase the percentage of oxygen. These variations are illustrated in Figure 4.

**Conclusions**

The use of plasma polymerisation to produce thin film deposits is a technique which has several advantages. The ability to tailor the film properties by

386

manipulation of the process parameters is particularly valuable. Variables include flow rates of reactive species, substrate temperature and power density.

This study has shown that for a microwave sustained plasma deposition of HMDSO and oxygen, altering the ratio of the reactant gases has a significant effect on the composition of the film. When an ECR plasma source is used, the same effect is not observed. In the latter case there is shown to be little incorporation of the reactant gas into the film surface. The use of a heated substrate is shown to have the most significant effect on the composition of coatings deposited from the ECR plasma source.

## Acknowledgements

The data presented here were obtained from work carried out on two projects. The ECR plasma deposition study is funded as a LINK / DTI / EPSRC project within the Enhanced Engineering Materials Programme, {GR/J19078}. The other microwave deposited films work is a Brite EuRam funded project {BRE2-0453} and also involves groups within Department of Metallurgy and Materials Engineering, KU Lueven, Belgium and the Department of Materials and Production Engineering, Luleå University of Technology, Sweden.

## References

1. Jansen, F. 'Plasma Deposition Processes', *Plasma Deposited Thin Films*, Ed. Mort, J. and Jansen, F., CRC Press, Florida, (1986) 1-20
2. Morosoff, N. 'An Introduction to Plasma Polymerisation', *Plasma Deposition, Treatment, and Etching of Polymers*, Ed. R. d'Agostino, Academic Press Limited, London, (1990) 1-94
3. Wróbel, A. M., Klemberg, J. E., Wertheimer, M.R. and Schreiber, H.P. Polymerisation of organosilicones in microwave discharges. II Heated substrates. *J. Macromol. Sci. -Chem. A,* **15,** 197-203 (1981)
4. Wróbel, A.M. and Wertheimer, M.R. 'Plasma Polymerised Organosilicones and Organometallics', *Plasma Deposition, Treatment, and Etching of Polymers,* Ed. R. d'Agostino, Academic Press Limited, London, (1990) 163-268
5. Biederman, H. and Osada, Y. *Plasma Technology, 3: Plasma Polymerisation Processes* Elsevier Science Publishers B.V., Amsterdam, (1992)
6. Asmussen, J. Electron cyclotron resonance microwave discharges for etching and thin-film deposition. *J. Vac. Sci. Technol. A,* 7, 883-893, (1989)
7 Inagaki, N., Kondo, S., Hirata, M. and Urushibata, H. Plasma Polymerisation of Organosilicon Compounds. *J. Appl. Polym. Sci.,* **30,** 3385-3395 (1985)

# THE ONSET OF RIBBING RECONSIDERED :
# AN EXPERIMENT

Decré M., J.M. Buchlin,

*von Karman Institute for Fluid Dynamics*
*72, Chée de Waterloo, B-1640 Rhode-St-Genhse, Belgium,*

J. Schmidt ,

*DGH, Dept. Fluid Mechanics*
*Building 404, PJ 2800 Lyngby, Denmark,*

and

M. Rabaud

*F.A.S.T., bât 502*
*Campus Universitaire,*
*91405 Orsay Cedex, France.*

## SUMMARY

Experiments are reported to measure the onset of ribbing by means of direct, quantitative measurement of the destabilizing meniscus, yielding better than 5% accuracy in the determination of the onset. The onset of ribbing is measured under asymmetric operating conditions, namely at varying speed ratio for constant gap ratio. Two set ups are compared : a roll coating rig and a journal bearing rig. The measurements show that these set ups are genuinely differing in the features of the onset curves. The findings are also in good agreement with existing theoretical and experimental results.

## 1. INTRODUCTION

Ribbing is a well known instability of the interface in roll coating processes. It is characterized by a sinusoidal destabilisation of the previously straight meniscus, along the cylinder generatrix (for illustration of the phenomenon, see [1]). This instability is impairing the final product in roll coating processes, and therefore many studies have been conducted to understand its origin and, if possible, to avoid its occurence.

Authors have early identified the dimensionless parameters characterizing roll coating processes: the gap ratio $G$, related to the gradient of the distance separating the cylinders, and the capillary number $Ca = \mu U/\sigma$ quantifying the ratio between viscous and capillary forces, where $\mu$ is the fluid dynamic viscosity, $U$ some tangential velocity typical of the flow and $\sigma$ is the surface tension of the liquid/air interface [2,3,4]. The occurence of ribbing can, like any other instability, be described by means of a marginal stability curve in the parameter space. This has been done experimentally for the plate/cylinder case [2,5,6,7,8,1,9]. In this case, $G = R/h_0$, with $h_0$ for the gap distance and $R$ for the roll radius, and $Ca$ is determined by the tangential velocity of the roll, $U$. In the symmetric forward roll coating experiment, the marginal stability curve has been measured as a function of $(G, Ca)$, where $G = R/h_0$ is obtained when $R$ is the radius of both rolls, and $Ca$ is computed by means of $U$, their tangential velocity [3,10,11,12,13,14]. All these works have been put together [9], and share the general features that, at constant $G$, ribbing occurs for capillary numbers higher than a critical value $Ca^{crit}$, and that this critical value is a decreasing function of increasing gap ratio $G$.

In the asymmetric case, $G = R/h_0$ with $1/R = 1/2R_1 \pm 1/2R_2$, where the sign should be negative in the journal bearing case, and it is necessary to use one capillary number describing the independent rotation of each cylinder $Ca_i = \mu U_i / \sigma$.

So far, it seems that results, together with models, of the asymmetric operation of forward roll coating rigs have only been reported in two works [15,16]. The experimental results in these papers are quite scattered, probably due to the detection techniques. The asymmetric mode of operation being quite widely used, it is desirable to obtain more accurate measurements of the onset of ribbing under these conditions. Such an approach has been designed by direct visualisation of the destabilizing meniscus through transparent cylinders in a journal bearing set up[17], an accurate method that guarantees to observe the onset of the instability where it takes place.

In the present communication, we report asymmetric experiments for the measurement of the critical capillary number pairs $(Ca_1, Ca_2)$. These experiments are performed concurrently on the journal bearing rig[17], and on an original forward roll coating set up with one transparent cylinder, in order to obtain accurate measurements of the onset of ribbing in asymmetric forward roll coating. We then compare these set ups with respect to their onset curves, and to the mathematical model[16].

## 2. EXPERIMENTAL SET UPS

The concept of each set up used in the present work is depicted in fig..

The transparent journal bearing set up used here has been extensively described elsewhere[17,18,19]. We will therefore not describe it in detail here. Suffice to say that the inner cylinder has length $L_1 = 380$ mm and radius $R_1 = 33,0$ mm, the outer cylinder has length $L_2 = 420$ mm and radius $R_2 = 50,0$ mm, the gap can be set from 0.1 to $2 \pm 0.01$ mm and the tangential speeds of the cylinders can vary independently from 0 to $280.0 \pm 1.0$ mm/s. The liquid used in this rig is silicone oil RHODORSYL 47V100 (Rhône-Poulenc) with dynamic viscosity $\mu = 0.0965$ Pa.s, specific mass $\rho = 970$ kg/m$^3$ and surface tension $\sigma = 0.021$ N/m.

The roll coating rig has been originally developped for the present experi-

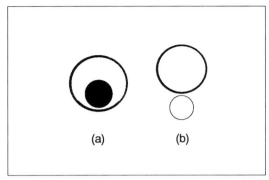

Figure 1: set up concepts: (a) journal bearing, (b) roll coating

ments, on the basis of the previously described journal bearing rig. It consists of a hollow, plexiglass upper cylinder of outer radius $R_1 = 65.0$ mm and length $L_1 = 310$ mm, put on top of a metallic lower cylinder of radius $R_2 = 25.0$ mm and length $L_2 = 260$ mm. The gap separating the cylinders can be set from 0.5 mm to 2 mm, with measured fluctuations of the plexiglass radius leading to $h_0$ being determined within 0.07 mm. The hollow, plexiglass cylinder is fitted with an inner, movable mirror inclined at 45 degrees with the horizontal, for direct observation of the meniscus in a window 8 cm long and 6 cm across. The independent tangential speeds of the cylinders can vary from 7 to 500 mm/s. The liquids used in this rig are silicone oils RHODORSYL 47V100 (see above), and 47V500 (Rhône-Poulenc) with dynamic viscosity $\mu = 0.57$ Pa.s, specific mass $\rho = 970$ kg/m$^3$ and surface tension $\sigma = 0.021$ N/m.

The technique used for the determination of the onset is identical for both rigs: the meniscus is filmed with a CCD camera for several operating conditions spanning the visually detectable onset, yielding a precision of better than 5% on the onset of ribbing. The magnification of the CCD camera allows to detect ribbing amplitudes above 0.13 mm. Each image is processed to determine the existence of periodic waves, their amplitude and wavelength. An illustration of the measurement of onset is presented in fig.2, for the roll coating rig, $G = 36.1$, $Ca_1 = 0.76$ and varying $Ca_2$.

A remark must be done here about the comparison between measurements performed on both rigs. For things to be tested against $G$-similitude, identical $G$ should be obtained. This is however very difficult to achieve, for the following reason. The radii to be used for the determination of $G$ in each rig are $R_J = 194.1$ mm for the journal bearing rig and $R_R = 36.1$ mm for the roll coating rig, respectively. One must therefore work with a gap 5.4 times smaller for the roll coating case than for the journal bearing case. Since the smallest achievable gap for the roll coating rig is 0.5 mm, the corresponding gap for the journal

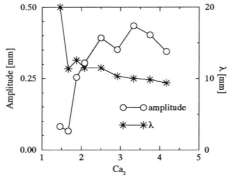

Figure 2: amplitude (○) and wavelength evolution (*) as function of $Ca_2$. $G = 36,1$ et $Ca_1 = 0,76$

bearing set up should be no less than 2.7 mm, a value which is impossible to reach, because of feeding, filling and meniscus shape differences at one and the same time. This is made clear by estimating the capillary length $\sqrt{\sigma/\rho g} \approx 1.5$ mm $< h_0 = 2.7$ mm. Since this is the typical length along which surface tension forces act, it is evident here that a gap larger than 1 mm will cause menisci much different from usual coating situations. The closest values of $G$ which are attained in the present work are consequently $G_J = 194.1$ and $G_R = 72.2$. It should be also noted that, in the journal bearing case, the system is closed and the feeding is always maintained, while in the roll coating case, the lower cylinder is fed only. This leads to limitations in the allowed capillary numbers for the roll coating rig, $Ca_1/Ca_2 >\approx 1$.

## 3. RESULTS

Figure 3 compares the measurements obtained on the open set up with existing results[15,16]. In their study, Benkreira et al.[15] do not observe any influence of the gap number when the results are presented as in fig. 3, i.e. as a function of modified capillary numbers $Ca_i\sqrt{G}$. Carter and Savage[16] do not comment specifically on that matter, but the model they propose does not scale as a function of $Ca_i\sqrt{G}$ (see their equation (42)). From fig. 3, it is clear that the present, precise measurements of the marginal stability curve distinguish between different gap configurations, even when the capillary numbers are scaled with $\sqrt{G}$. Otherwise, the present results compare well with previous ones. Moreover, as it will be seen in the next section, it is possible to compare reasonably well these experimental data with the theory developed by Carter and Savage[16].

The comparison made in fig. 4 between the onset curves obtained for the roll coating and the journal bearing set ups is more striking: the curves are

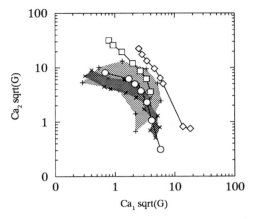

Figure 3: onset of ribbing in roll coating : × ref.15, + ref.16, this work: $G =$ circles: 72.2, : squares: 48.1, diamonds: 36.1

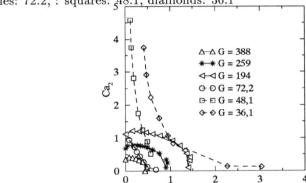

Figure 4: comparison of all onset curves : $G > 100 =$ journal bearing

totally different, and it is impossible to find some continuous transformation, even qualitative, from one case to the other, as $G$ continuously changes. A major feature of the curves obtained in the journal bearing case is their robustness. The curves presented here are by all means analogous to the ones reported previously for journal bearing[17,19,18]. Moreover, their shape appears to be very insensitive to modifications of the conditions, like modifying the total amount of fluid in the experiment, or inclining the set up with respect to gravity, as it is shown in fig. 5. In fig. 6, the curves for the journal bearing set up have been scaled with respect to their extreme values, demonstrating the similitude of the curves through the whole range of $G$ values. It has however not been possible to derive any scaling law for $G$.

The striking difference between the marginal stability curves of the roll coating and journal bearing cases clearly calls for different theoretical descriptions.

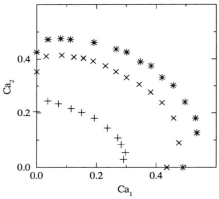

Figure 5: journal bearing: gravity effects. $*$:20.9 deg., $\times$:0 deg., $+$: -20.2deg.

Since the material contained in the following section supports the existing theoretical description of roll coating processes[16], one has to look for major deviations of the journal bearing case with respect to the assumptions of roll coating descriptions. A recent theoretical work[20] sheds light on that matter, by providing the first predictions displaying qualitative agreement with existing journal bearing onset curves for ribbing. This work[20] contains two important features. Firstly, it takes into account the fact that the amount of liquid contained in the set up is constant. Secondly, it describes the film formation region with great mathematical detail. Both features seem to achieve the goal of describing satisfactorily the onset curves, but the latter is so exacting on the $Ca$ domain of validity of the mathematical expansions, that no description so far extends to $G$ and $Ca$ values encountered in practice or in existing literature, including the present one. The results obtained by Reinelt[20], together with the present measurements of the onset displaying such robustness in the journal bearing case, support the assumption that the constant amount of liquid is a property that is very constraining on the global flow, and should be contained in any further description of it.

## 4. COMPARISON WITH THEORY

In this last section, the experimental results of the present work are compared to the theoretical predictions of Carter and Savage[16]. This paper being restricted to the presentation of experimental results, the authors will not extensively comment this model. The interested reader will refer to existing literature about the lubrication theory of the base flow[7,8,21] underlying the perturbation theory of ribbing developed in the original article[16].

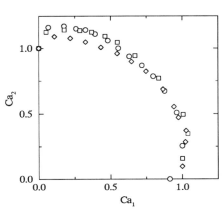

Figure 6: journal bearing, normalized curves for $G = 388, 259, 194$

For the sake of clarity, the major equations will be repeated here:

$$\frac{-\cos^2\theta}{2\sqrt{G}Ca_m} = \left(\theta + \frac{\pi}{2} + \frac{\sin 2\theta}{2}\right)$$
$$- \left[\frac{\left(1 + S^{3/2}\right)}{(1 + S)(1 + \sqrt{S})\cos^2\theta}\right]\left(\frac{\theta}{2} + \frac{\pi}{4} + \frac{\sin 2\theta}{3} + \frac{\sin 4\theta}{24}\right) \tag{1}$$

$$Ca_2^{crit} = b\left(\frac{d}{b} + \left(\frac{k}{C}\right)^2\right)\left[\frac{9\mathcal{F}}{3a\mathcal{F} - 4e}\right] \tag{2}$$

where $a, b, d$ and $e$ are functions of the gap ratio $G$, the speed ratio $S$ and the dimensionless streamwise position of the meniscus $X_M = x_M\sqrt{Rh_0} = \tan\theta$, and $k$ is the wavenumber of the periodic perturbation[16].

Two editorial errors should be noted here about the original article: equation (28) in should read as equation 1 above, i.e. as equations (3.6) in Savage(1977)[7] and (2.26) in Savage(1982)[21], and equation (39) for the constant $d$ should read $d = X_M\sqrt{G}$ These modifications are algebraically correct and are necessary to reproduce the original results. They are therefore solely due to typing errors.

Equation 1 provides the solution for the position of the meniscus $X_M$ which is needed to obtain the constants of equation 2, that yield the critical capillary number $Ca_2^{crit}$ for given operating conditions $(G, Ca_1/Ca_2)$, when looking for the minimum of equation 2 using $dCa_2/dk$. This is done here and displayed in fig., for the three values of $G$ investigated experimentally with the roll coating set up. Taking into account that fig. is not logarithmic, the agreement between experiment and theory is very good. The model proposed by Carter and Savage[16] can thus be relied upon for the prediction of the onset of ribbing in asymmetric roll coating. Remark that the distance between theory and experiment increases with decreasing $G$, the gap ratio. This can be attributed to a

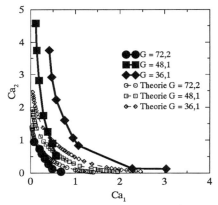

Figure 7: roll coating: comparison between theory (empty symbols) and experiment

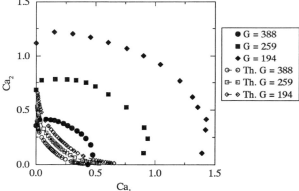

Figure 8: journal bearing: : comparison between theory (empty symbols) and experiment

light 'starving' effect of the feeding conditions at smaller gap ratios, observed in previous works[22,23]. When the gap is less well fed, the meniscus is located closer to the gap, a situation that delays the onset of ribbing.

Figure confirms, on the other hand, the inappropriateness of roll coating descriptions for journal bearing ribbing.

## 5. CONCLUSIONS

Experiments have been performed to determine the onset of ribbing in asymmetric operating conditions of both a roll coating and a journal bearing set up, principally by varying the speed ratio of the cylinders. ¿From the results, it is clear that the onset of ribbing has different features in each set up, calling for different assumptions in the describing models. The precision (5%) obtained by the direct observation of the destabilizing meniscus allows to resolve effects of the gap ratio upon the onset of ribbing, and demonstrate that the onset does not scale as $\sqrt{G}$, a phenomenon that was as yet not reported. The results for roll coating are otherwise in very good agreement with both existing experiments[15,16] and theory[16].

## 6. REFERENCES

1. V.HAKIM, M.RABAUD, H.THOMI & Y.COUDER 1990, Directional growth in viscous fingering. in 'New Trends in Nonlinear Dynamics and Pattern Froming Phenomena ', Eds. P.Coullet and P.Huerre, Plenum Press, New York, 327–337.

2. J.R.A.PEARSON 1960, The instability of uniform viscous flow under rollers and spreaders. J. Fluid Mech. **7**, 481–500.

3. E. PITTS & J. GREILLER 1961, The flow of thin liquid films between rollers. J. Fluid Mech. **11**, 33–50.

4. G.I.TAYLOR 1963, Cavitation of a viscous fluid in narrow passages. J. Fluid Mech. **16**, 595–619.

5. L.FLOBERG 1965, On hydrodynamic lubrication with special reference to sub-cavity pressures and number of streamers in cavitation regions. Acta Polytech. scand., **ME 19**.

6. T.BAUMAN, T.SULLIVAN & S.MIDDLEMAN 1982, Ribbing instability in coating flows: effect of polymer additives. Chem. Eng. Sc. **14**, 35–46.

7. M.D.SAVAGE 1977a, Cavitation in Lubrication. Part 1. On boundary conditions and cavity-fluid interfaces. J. Fluid Mech. **80**, 743–756.

8. M.D.SAVAGE 1977b, Cavitation in Lubrication. Part 2. Analysis of wavy interfaces. J. Fluid Mech. **80**, 757–767.

9. K.ADACHI, T.TAMURA & R.NAKAMURA 1988, Coating flows in a nip region and various critical phenomena. AIChE J. **34**(3), 456–464.

10. J.C.HINTERMAIER & R.E.WHITE 1965, The splitting of a water film between rotating rolls. Tappi J. **48**(11), 617–625.

11. C.C.MILL & G.R.SOUTH 1967, Formation of Ribs on Rotating Rollers. J. Fluid Mech. **28**, 523–529.

12. J.GREENER, T.SULLIVAN, B.TURNER & S.MIDDLEMAN 1980, Ribbing Instability of a two-roll coater : newtonian fluids. Chem. Eng. Commun. **5**, 73–83.

13. M.D.SAVAGE 1984, Mathematical model for the onset of ribbing. AIChE J. **30**(6), 999–1002.

14. D.J.COYLE 1984, The fluid mechanics of roll coating : steady flows, stability, and rheology. PhD Thesis, University of Minnesota, Minneapolis.

15. H.BENKREIRA, M.F.EDWARDS & W.L.WILKINSON 1982b, Ribbing instability in the roll coating of Newtonian fluids. *Plast. Rubb. Process. and Appl* **2**(2), 137–144.

16. G.C.CARTER & M.D.SAVAGE 1987, Ribbing in a variable speed two-roll coater. *Math.Engng Ind.* **1**(1), 83–95.

17. M.RABAUD, S.MICHALLAND & Y.COUDER 1990, Dynamical regimes of directional viscous fingering: spatiotemporal chaos and wave propagation. *Phys. Rev. Lett.* **64**(2), 184–187.

18. Y.COUDER, S.MICHALLAND, M.RABAUD & H.THOMI 1990, The Printer's Instability : the dynamical regimes of directional viscous fingering, in *Nonlinear evolution of spatio-temporal Structures in Dissipative Continuous Systems.* F.H.Busse & L.Kramer Edt., Plenum Press, New York, 487–497; M.RABAUD,Y.COUDER & S.MICHALLAND 1991, Wavelength Selection and Transients in the One-dimensional Array of Cells of the Printer's Instability. *Europ. J. of Mech. B/Fluids* **10** (2)-Suppl.,253–260. S.MICHALLAND & M.RABAUD 1992, Localised phenomena during spatio–temporal intermittency in the printer's instability, *Physica D* **61**, 197–204; S.MICHALLAND, M.RABAUD & Y.COUDER 1993, Transition to chaos by spatio-temporal intermittency in directional viscous fingering. *Europhys. Lett.* **22**(1), 17–22;

19. S.MICHALLAND 1992, Etude des différents régimes dynamiques de l'instabilité de l'imprimeur. Thèse de doctorat, Université de Paris VI.

20. D.A.REINELT 1995, The primary and inverse instabilities of directional viscous fingering. *J. Fluid Mech.* **285**, 303–327.

21. M.D.SAVAGE 1982, Mathematical models for coating process *J. Fluid Mech.* **117**, 443–455.

22. M.DECRÉ (1994), "Etude expérimentale des comportements de l'interface dans l'enduisage par rouleaux", thèse de doctorat, U. Paris VI.

23. M.DECRÉ, E.GAILLY AND J.-M.BUCHLIN (1995), "Meniscus shape experiments in forward roll coating", *Phys. Fluids* **7** (3), 458–467.

# EFFECTS OF EVAPORATION DURING SPIN-COATING

J.H. Lammers & M.N.M. Beerens

*Philips Research Laboratories,*
*Prof. Holstlaan 4, 5656 AA Eindhoven, The Netherlands.*

and

S.B.G.M. O'Brien

*Department of Mathematics,*
*University of Limerick, Limerick, Ireland.*

## 1. INTRODUCTION

Spin coating can be used as a wet-chemical process for applying a thin layer on a flat substrate. A coating solution is spun on a rotating substrate to obtain a uniform liquid film, which is subsequently dried. The deposition of layers on displays, silicon wafers and optical disks are typical examples.

The coating can either be dried while still spinning, or in a separate process step, e.g. in an oven. Industrial applications usually require that drying is performed as quickly as possible. We will therefore discuss the case where spinning is continued until the coating is dry.

Non-uniform evaporation, e.g. due to turbulence or edge effects then leads to a coating defect: thickness variations over the substrate. An example is given by D.E. Bornside et al.[1].

This contribution focuses on the prediction of the coating profile for a known evaporation rate distribution. We will discuss the following topics.
1. The Meyerhofer model[3], which can be solved exactly.
2. The decay of a non-uniform initial condition.
3. The coating non-uniformity caused by evaporation differences over the substrate.

## 2. SOLUTION OF THE MEYERHOFER MODEL.

The Meyerhofer model predicts the coating thickness for uniform evaporation, small Peclet number and uniform initial conditions. The model can be solved analytically for arbitrary dependence of the liquid parameters on the solute concentration. The result is

$$S_f = \tilde{h} \left( 3 \int_{c_0}^1 \frac{\mathrm{d}c'}{c'^4 \nu(c') e_v(c')} + \frac{\tilde{h}^3}{(c_0 h_0)^3} \right)^{-\frac{1}{3}}. \tag{1}$$

with

$$\tilde{h} = \left( \frac{3\mu_l(c_0)e_v(c_0)}{2\rho_l(c_0)\omega^2} \right)^{\frac{1}{3}}. \tag{2}$$

For the typical height $\tilde{h}$, the decrease in liquid height due to spinning and evaporation are approximately equal. It can be calculated from the initial values of the liquid viscosity $\mu_l$, the density $\rho_l$ and the evaporation rate $e_v$, which depend on the initial solute concentration $c_0$. The initial liquid height is $h_0$, and the rotation rate $\omega$. The functions $\nu(c)$ (kinematic viscosity) and $e_v(c)$ have been non-dimensionalised with their respective initial values.

Since the integrand in (1) decays quickly due to the inverse fourth power of the concentration it suffices to integrate to the value $c = 3\,c_0$ instead of $c = 1$. After the concentration $c = 3\,c_0$ has been reached, the decrease in coating thickness due to spinning is no more than about 1%. This is a practically very important new result, since it means that one needs to measure the fluid properties only in the interval between $c_0$ and $3 \cdot c_0$, with emphasis on the lower end of this interval. This reflects the fact that spinning effectively ceases, i.e. $h$ becomes small compared to $\tilde{h}$ before the concentration has risen considerably.

For dilute coating liquids based on a single solvent the fluid properties hardly change in this interval of concentration, so the fluid properties can then even be regarded as constant. If initially enough coating liquid is dispensed, we obtain the result

$$S_f = c_0\tilde{h}, \tag{3}$$

which we will refer to as the 'Meyerhofer formula'. It can be derived by dividing the spin-coating process in two conceptually distinct phases. First, spinning dominates and the concentration hardly changes. Later, evaporation dominates, flow has effectively ceased and the coating thickness hardly decreases.

## 3. THE DECAY OF A NON-UNIFORM INITIAL CONDITION.

The method of characteristics can be used to study the effects of the way in which the coating solution is introduced on the substrate. This was done by Emslie et al.[2] for spin-coating without evaporation. The method can be extended to the process with significant evaporation. One result is shown in figure 1 for the flow of a coating liquid with constant fluid properties.

## 4. EVAPORATION NON-UNIFORMITY.

A perturbation method is used to calculate the thickness variations due to small non-uniformities in the evaporation rate distribution. Using the method of characteristics we have been able to calculate the Green's function due to an

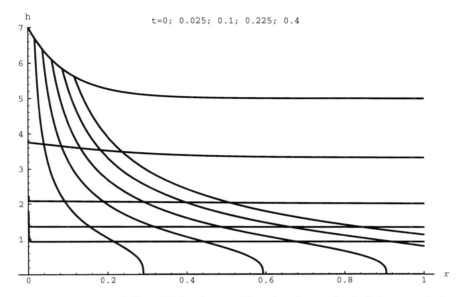

Figure 1: Solution of the initial value problem by the method of characteristics. The characteristics run top to bottom, the liquid level profiles (radius $r$) at different times are the nearly horizontal lines. The liquid height $h$ has been non-dimensionalized with $\tilde{h}$ and time $t$ has been non-dimensionalised with $\tilde{h}/e_v(c_0)$.

isolated evaporation non-uniformity

$$e_v(r) = e_v(0)\left(1 + \epsilon\delta(r - r')\right) \tag{4}$$

at $r'$. It is given by

$$S_f(r) = c_0\tilde{h}(1 + \epsilon S_1(r/r')), \tag{5}$$

with

$$S_1(x) = \hat{S}_1\delta(x - 1) + \frac{4}{3}\frac{x^5}{1 - x^6} - \frac{16}{9}\int_x^1 \frac{q^{\frac{5}{3}}x^{\frac{7}{3}}}{(1 - q^6)^{\frac{1}{3}}(1 - x^2q^4)^{\frac{5}{3}}}dq, \tag{6}$$

which is shown in figure 2. The Green's function may be used to calculate the result of a small but arbitrary evaporation non-uniformity by convolution with the evaporation rate profile. In particular if the evaporation rate profile along a radial line can be obtained in the form of a Taylor series

$$e_v(r) = e_v(0)\left(1 + \epsilon\sum_{n=1}^{\infty} e_{v,n}r^n\right) \tag{7}$$

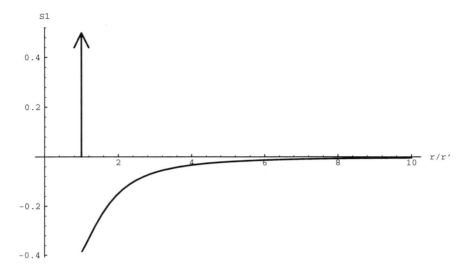

Figure 2: The Green's function $S_1$, eq(6), for the coating thickness at radius $r$ due to a localized evaporation non-uniformity of strength unity at $r'$. The arrow indicates the strength of the resulting coating non-uniformity at $r = r'$.

the coating profile is given by

$$S(r) = c_0 \tilde{h} \left( 1 + \epsilon \sum_{n=1}^{\infty} s_n e_{v,n} r^n \right) \tag{8}$$

and

$$s_n = \int_0^1 x^n S_1(1/x) \mathrm{d}x. \tag{9}$$

The first few numbers $s_n$ given in table 1. Since these coefficients do not deviate too much from $\frac{1}{3}$, the exponent in (2), one may as a rough first estimate calculate the coating profile by inserting the local value of the evaporation rate into the Meyerhofer formula (3).

## 5. CONCLUSIONS

We have analyzed a few problems related to the spin-coating process, in the version with significant evaporation.

Firstly, the analysis by Meyerhofer[3] of the effect of changes in the solute concentration is completed. His approximate reasoning, viz. that the process can be separated into two conceptually distinct phases, with either spinning or evaporation being important, yields a remarkably accurate result for the deposited coating thickness, eq.(3). From the exact solution, eq.(1) it is possible

Table 1: Dependence of shape factor $s_n$ on exponent $n$.

| $n$ | $s_n$ |
|-----|--------|
| 0 | 0.3333 |
| 1 | 0.3784 |
| 2 | 0.4049 |
| 3 | 0.4222 |
| 4 | 0.4343 |
| 5 | 0.4432 |
| 6 | 0.4500 |
| $\infty$ | 0.4989 |

to assess the influence of changes in the fluid properties due to the increase of solute concentration during the process. The process is such that spinning effectively ceases to be important before the concentration increases dramatically. Hence, one needs to measure the fluid properties only in the range between $c_0$ and $3c_0$, where $c_0$ denotes the initial solute concentration. This knowledge leads to considerable savings on the experimental effort to obtain the properties of coating liquids during the spinning process.

Finally, we have been able to calculate the Green's function due to an isolated evaporation rate non-uniformity. By convolution one can calculate the effect of small but arbitrary shaped evaporation non-uniformities. As a rough estimate, one can use the local evaporation rate in the Meyerhofer formula to calculate the local coating thickness.

## 6. REFERENCES

1. Bornside D.E. & Brown R.A. 1993 The effects of gas phase convection on mass transfer in spin coating, *J. Appl. Phys.*, **73**, 585-600.

2. Emslie A.G., Bonner, F.T. & Peck L.G. 1958 Flow of a viscous liquid on a rotating disk, *J.Appl.Phys.*, **29**, 858.

3. Meyerhofer D. 1978 Characteristics of resist films produced by spinning, *J.Appl.Phys.*, **49**, 3993-3997.

4. Wimmers O.J., Beerens M.N.M. & O'Brien S.B.G.M. 1994 Spin coating of oxidic layers on 27" CRT screens to form an antireflection coating. Contribution to *7th International Coating Process Science & Technology Symposium*, Atlanta

# List of Delegates

**Dr. S.J. Abbott**
Autotype International Limited, Grove Road, Wantage,
Oxon OX12 7BZ, United Kingdom
Tel: 01235 771111 Fax: 01235 771191 Email: steven@abbott.demon.co.uk

**Prof. K. Adachi**
Department of Applied Chemistry, Kyushu Institute of Technology
Tobata, Kitakyushu 804, Japan
Tel: 81 93 884 3311 Fax: 81 93 884 3300 Email: k.adachi@post.isct.kyutech.ac.jp

**Mr. J.A. Addison**
Jesus College, Oxford, OX1 3DW, United Kingdom
Email: jaddison@jesus.ox.ac.uk

**Prof. C. K. Aidun**
Institute of Paper Science and Technology, Georgia Institute of Technology
500 10th Street, N.W., Atlanta, GA 30318-5794, USA
Tel: 404 853 9777 Fax: 404 853 9510 Email: cyrus.aidun@ipst.gatech.edu

**Mr. M. Barnard**
3M UK Plc, Gorseinon Road, Penllergaer
Swansea SA4 1GD, United Kingdom

**Dr. S. Barnard**
Coates Coating International, Station Lane
Whitney, Oxon OX8 6XZ, United Kingdom
Tel: 01993 707400 Fax: 01993 775579

**Dr. D. Beck**
Bayer AG, D-51368, Bayer Leverkusen
MD-IM-FA, Germany

**Dr. H. Benkreira**
Dept. Chem. Engineering, University of Bradford
Bradford BD7 1DP, United Kingdom
Tel: 01274 383721 Fax: 01274 385700 Email: H. Benkreira@bradford.ac.uk

**Dr. C. Beriet**
Dept. of Chemistry, University of Southampton, Highfield
Southampton SO17 1BJ, United Kingdom
Tel: 01703 594169 Fax: 01703 676960 Email: cb@soton.ac.uk

**Dr. T.D. Blake**
Surface and Colloid Science Group, Kodak Limited, Headstone Drive
Harrow, Middx. HA1 4TY, United Kingdom
Tel: 0181 427 4380

**Dr. M.F.J. Bohan**
Department of Mechanical Engineering, University of Wales, Swansea
Singleton Park, Swansea SA2 8PP, United Kingdom
Tel: 01792 205678 ext. 4105

**Prof. P. Bourgin**
Université Louis Pasteur, Institut de Mecanique des Fluides
2 rue Boussingault, 67000 Strasbourg, France
Tel: 33 88 41 65 78 Fax: 33 88 61 43 00 Email: bourgin@imf.u-strasbg.fr

**Mr. F. Brotz**
BASF-AG, ZET/EA - L544, 67056 Ludwigshafen, Germany

**Mr. B. Broughton**
Rexam Custom, Warrington Road, Manor Park, Runcorn
Cheshire WA7 1SN, United Kingdom
Tel: 01928 579111 Fax: 01928 579222

**Prof. J-M. Buchlin**
von Karman Inst. for Fluid Dynamics, Chaussee de Waterloo 72
1640 Rhode-St-Genese, Belgium
Tel: 02 358 1901 Fax: 02 358 2885 Email: buchlin@vki.ac.be

**Dr. S. Bushnell-Watson**
Dept. of Engineering Materials, The University of Sheffield
Sir Robert Hadfield Building, Mappin Street, Sheffield S1 3JD,
United Kingdom
Tel: 01742 825477 Fax: 01742 754325

**Mr. H. Bussmann**
Agfa-Gevaert AG, D-51368, Leverkusen, IT-1, Geb. E1, Germany

**Mr. P. Campbell**
ICI Films, PO Box 90, Wilton, Middlesbrough, Cleveland
TS90 8JE, United Kingdom
Tel: 01642 454144 Fax: 01642 432762

**Mr. M.S. Carvalho**
Dept. of Chem. Eng. and Mat. Science, University of Minnesota
151 Amundson Hall, 421 Washington Avenue SE,
Minneapolis MN 55455-0132, USA
Tel: 612 625 1313 Fax: 612 626 7246

**Mr. E. Catot**
Glaverbel SA, Centre de Recherche et Developpement
Rue de l'Aurore, B. 6040 Jumet, Belgium

**Dr. D. C-H Cheng**
191 Ickneild Way, Letchworth, Herts SG6 4TT, United Kingdom
Tel: 01462 672146 Fax: 01462 672146

**Dr. A. Clarke**
Surface and Colloid Surfaces, Kodak Ltd., Headstone Drive
Harrow, Middx HA1 4TY, United Kingdom
Tel: 0181 424 5681 Fax: 0181 424 3750 Email: clarke_a@kodak.com

**Dr. O. Cohu**
Laboratoire de Rheologie, BP 53X, 38041 Grenoble, Cedex France
Tel: 33 76 82 5155 Fax: 33 76 82 5164

**Mr. H.A. Coulson**
Sensitisers Research Ltd., 1 Blackdown Road, Camberley
Surrey GU16 6SH, United Kingdom

**Mr. F. Cunha**
von Karman Inst. for Fluid Dynamics, Chaussee de Waterloo 72
1640 Rhode-St-Genese, Belgium
Tel: 02 358 1901 Fax: 02 358 2885

**Mr. N. Daniels**
Dept. of Applied Mathematical Studies, University of Leeds
Leeds LS2 9JT, United Kingdom
Email: N.Daniels@leeds.ac.uk

**Mr. J.F.H. de Dreu**
Heideweg 10, 5971 DR, Grubbenvorst, The Netherlands

**Mr. M.J. de Ruijter**
Dalialaan 31, B-3191 Boortmeerbeek, Belgium
Tel: 065 37 34 57 Fax: 065 37 30 54 Email: michel@gibbs.umh.ac.be

**Dr. M. Decré**
von Karman Inst. for Fluid Dynamics, Chaussee de Waterloo 72
1640 Rhode-St-Genese, Belgium
Tel: 02 358 1901 Fax: 02 358 2885

**Mr. M. Douglas**
3M UK Plc, Gorseinon Road, Gorseinon
Swansea SA4 1GD, United Kingdom

**Miss L. Dunford**
3M Atherstone, Ratcliffe Road, Atherstone
Warwickshire CV9 1PJ, United Kingdom

**Mr. H.G. Eggels**
Heistraat 78, 6071 SK Swalmen, The Netherlands
Tel: 077-592416 Fax: 077-595499

**Mr. R. Ellison**
3M Company, 3M House, PO Box 1, Bracknell, Berkshire RG12 1JU
United Kingdom
Tel: 01344-858732

**Mr. J. P. Flett**
ICI Films, Melinex PD and E, Semi-Tech 3, Wilton Centre
PO Box 90, Wilton, Middlesbrough TS90 8JE, United Kingdom

**Mr. J. Frodin**
Applied Computing and Engineering Ltd., Genesis Centre
Science Park South, Birchwood, Warrington WA3 7BH, United Kingdom
Tel: 01925 830085, Fax: 01925 826460

**Mr. E. Gailly**
von Karman Inst. for Fluid Dynamics, Chaussee de Waterloo 72
1640 Rhode-St-Genese, Belgium
Tel: 02 358 1901 Fax: 02 358 2885

**Mr. A. L. Gardner**
Dupont Printing and Publishing, Coal Road
Leeds LS14 2AL, United Kingdom
Tel: 0113 273 7475 Fax: 0113 251 4030

**Dr. P.H. Gaskell**
Department of Mechanical Engineering, University of Leeds
Leeds LS2 9JT, United Kingdom
Tel: 0113 233 2201 Fax: 0113 242 4611 Email: P.H.Gaskell@leeds.ac.uk

**Mr. B. Goetmaeckers**
Dept. of Coating and Drying Research, Agfa-Gevaert N.V.
Septestraat 27, B 2640 Mortsel, Belgium
Tel: 3 444 8610 Fax: 3 444 7492

**Mr. E. W. Grald**
Coating Analysis Group, Fluent Inc., 10 Cavendish Court
Lebanon NH 03741, USA
Tel: 603 643 2600 Fax: 603 643 3967 Email: ewg@fluent.com

**Mr. F. Gurcan**
Dept. of Applied Mathematical Studies, University of Leeds
Leeds LS2 9JT, Uniied Kingdom
Email: F.Gurcan@leeds.ac.uk

**Dr. R. Haas**
Hoechst Ag, BUDRUCK, KDTT, Rheingaustrasse 190
D-65174 Wiesbaden, Germany

**Mr. H. Hardegger**
Bachofen + Meier AG, Maschinenfabrik, Feldstrasse 60
8180 Bulach, Switzerland
Tel: 41 1 860 45 45 Fax: 41 1 861 0171

**Dr. O.G. Harlen**
Dept. of Applied Mathematical Studies, University of Leeds
Leeds LS2 9JT, United Kingdom
Tel: 0113 233 5189 Fax: 0113 242 9925 Email: oliver@amsta.leeds.ac.uk

**Dr. O.J. Harris**
Theory and Modelling, Corporate Technology, Courtaulds
PO Box 111, Lockhurst Lane, Coventry CV6 5RS United Kingdom
Tel: 01203 582153 Fax: 01203 582136 Email: owen.harris@corp.courtaulds.co.uk

**Mr. H. Hofmann**
Agfa-Gevaert AG, D-51368, Leverkusen IT-1, Geb. E1, Germany

**Dr. G.M. Homsy**
Dept. of Chemical Engineering, Stanford University, Stanford
CA 94305-5025, USA
Tel: 1 415 723 2419 Fax: 1 415 725 7294 Email: bud@chemeng.stanford.edu

**Prof. A. T-L. Horng**
Department of Applied Mathematics, Feng Chia University, Taichung
Taiwan, ROC
Tel: 886 4 4517250 Ext. 5206, Fax: 886 4 4510801 Email: tlhorng@math.fcu.edu.tw

**Dr. G. Hultzsch**
Hoechst Ag, BUDRUCK, KDTT, Rheingaustrasse 190, D-65174 Wiesbaden,
Germany

**Mr. B. Ikin**
Ilford Ltd., Town Lane, Mobberley, Knutsford,
Cheshire WA16 7JL, United Kingdom
Tel: 01565 650000

**Dr. G. E. Innes**
Dept. of Mechanical Engineering, University of Leeds
Leeds LS2 9JT, United Kingdom
Tel: 0113 233 2202 Fax: 0113 242 4611 Email: G.E.Innes@leeds.ac.uk

**Mr. G. Jackson**
3M United Kingdom Plc, Ratcliffe Road, Atherstone
Warwickshire CV9 1PJ, United Kingdom
Tel: 01827 710247 Fax: 01827 710271

**Dr. M.N. Kadiramangalam**
Coating Analysis Group, Fluent Inc., 10 Cavendish Court,
Lebanon NH 03766, USA
Tel: 603 643 2600 Fax: 603 643 3967 Email: mnk@fluent.com

**Dr. S. Kalliadasis**
Department of Chemical Engineering, University of Leeds,
Leeds LS2 9JT, United Kingdom

**Mr. N. Kapur**
Dept. of Mechanical Engineering, University of Leeds
Leeds LS2 9JT, United Kingdom
Email: N.Kapur@leeds.ac.uk

**Mr. I. Kliakhandler**
School of Mathematical Sciences, Tel-Aviv University
Tel Aviv 69978, Israel
Tel: 972 3 6408043/4 Fax: 972 3 6409357

**Mr. J. Knew**
3M United Kingdom Plc, Ratcliffe Road, Atherstone
Warwickshire CV9 1PJ, United Kingdom
Tel: 01827 710247 Fax: 01827 710271

**Mr. S. Krauss**
Dept. of Fluid Mechanics, University of Erlangen-Nurnberg
Cauerstrasse 4, D-91058 Erlangen, Germany
Tel: 49 9131 859487 Fax: 49 9131 859503

**Dr. J. H. Lammers**
Philips Research Laboratories, Building WA - Box WA01
Prof. Holstlaan 4, 5656 AA Eindhoven, The Netherlands
Tel: 31 40 742072 Fax: 31 40 743352 Email: lammersj@prl.philips.nl

**Mr. U. Lange**
Dept. of Fluid Mechanics, University of Erlangen-Nurnberg
Cauerstrasse 4, D-91058 Erlangen, Germany
Tel: 49 9131 859485 Fax: 49 9131 859503 Email: ulange@lstm.uni-erlangen.de

**Mr. S. Leefe**
BHR Group, Cranfield, Bedford, MK43 0AJ, United Kingdom
Tel: 01234 750422 Fax: 01234 750074

**Dr. K. Looney**
ICI Films, PO Box 90, Room G415, Wilton Centre
Middlesbrough, Cleveland TS90 8JE, United Kingdom
Tel: 01642 454144 Fax: 01642 432762

**Dr. M. Lott**
Horsell Graphic Industries Ltd., Howley Park Estate
Morley, Leeds, LS27 0QT, United Kingdom

**Mr. E. Middelman**
Akzo Nobel Central Research, Postbus 9300
6800 SB Arnhem, The Netherlands
Tel: 31 85 66 4722

**Mr. M. Minale**
Dept. of Chemical Engineering, University Federico II of Naples
P. Le Tecchio, 80, 80125 Napoli, Italy
Tel: 39 81 2391800 Email: dic@irc.na.cnr.it

**Mr. P. Moussalli**
ICI Filmsm PO Box 90, Room G413, Wilton Centre
Middlesbrough, Cleveland TS90 8JE, United Kingdom
Tel: 01642 454144 Fax: 01642 432762

**Mr. A. Munch**
Institute for Applied Mathematics, TU Munchen, Postfach 20 10 32
80010 Munich, Germany
Tel: 089 5522 48 25 Fax: 089 5522 48 19 Email: amue@appl-math.tu-muenchen.de

**Dr. T. Myers**
Mathematical Institute, University of Oxford, 24-29 St. Giles
Oxford OX1 3LB, United Kingdom
Tel: 01865 273525 Fax: 01865 273583 Email: myers@maths.ox.ac.uk

**Dr. S. O'Brien**
Department of Mathematics, University of Limerick, Ireland
Tel: 353 61 202644, Fax: 353 61 202572 Email: obrien@ul.ie

**Ir. W. Overdiep**
Akzo Nobel Central Research, Dept. RGH, Postbus 9300
6800 SB Arnhem, The Netherlands
Tel: 31 85 66 2956 Fax: 31 85 66 5464

**Dr. B. Peppard**
Horsell Graphic Industries Ltd., Howley Park Estate
Morley, Leeds LS27 0QT, United Kingdom

**Mr. R. Peterson**
Dept. of Applied Mathematical Studies, University of Leeds
Leeds LS2 9JT, United Kingdom

**Mr. C. Powell**
Dept. of Applied Mathematical Studies, University of Leeds
Leeds LS2 9JT, United Kingdom
Email: amtcap@sun.leeds.ac.uk

**Prof. M. Rabaud**
Laboratoire FAST, bat. 502, Campus Universitaire
91405 Orsay Cedex, France
Tel: 33 1 60 19 01 16 Fax: 33 1 60 19 34 90 Email: rabaud@fast.u-psud.fr

**Mr. S. Rees**
Pilkington Technology Centre, Hall Lane, Lathom
Nr. Ormskirk, Lancs, L40 5UF, United Kingdom
Email: ReesSJ@challenger.pilkington.co.uk

**Mr. C. Richardson**
Dept. of Applied Mathematical Studies, University of Leeds
Leeds LS2 9JT, United Kingdom
Email: amt5cr@sun.leeds.ac.uk

**Mr. N. Roberts**
Autotype International Limited, Grove Road, Wantage
Oxon OX12 7BZ, United Kingdom
Tel: 01235 771111, Fax: 01235 771191

**Dr. K.J. Ruschak**
Emulsion Coating Technology, Eastman Kodak Company
Rochester, NY 14652-3701, USA
Tel: 716 477 6830 Fax: 716 722 4012

**Dr. M.D. Savage**
Dept. of Applied Mathematical Studies, University of Leeds
Leeds LS2 9JT, United Kingdom
Tel: 0113 233 5118 Fax: 0113 242 9925 Email: M.D.Savage@leeds.ac.uk

**Dr. P.M. Schweizer**
Ilford AG, Industriestrasse 15, CH-1701 Fribourg, Switzerland
Tel: 41 37 214 111 Fax: 41 37 215 212

**Mr. R. Shaw**
Dept. of Applied Mathematical Studies, University of Leeds
Leeds LS2 9JT, United Kingdom
Email: amtrs@sun.leeds.ac.uk

**Dr. Y.D.Shikhmurzaev**
Institute of Mechanics, Moscow University, 119899 Moscow, Russia
Tel: 7 95 939 2339 Fax: 7 95 939 0165 Email: yulii@inmech.msu.su

**Dr. Y. Shnidman**
Eastman Kodak Company, RL 83-2, 1999 Lake Avenue, Rochester
NY 14650-2216, USA
Email: shnidman@kodak.com

**Mr. G. Sterzi**
3M Italia s.p.A., via della Liberta 57, Ferrania - 17016 Cairo
Montenotte (SV), Italy
Tel: 019 5223051

**Mr. S. Storey**
Dept. of Applied Mathematical Studies, University of Leeds
Leeds LS2 9JT, United Kingdom
Email: amt5ss@sun.leeds.ac.uk

**Prof. M. Sugawara**
Dept. of Electrical Engineering, Hachinohe Institute of Technology
88-1, Myo, Ohbiraki, Hachinohe, Aomori-Ken, Japan
Tel: 81 178 25 3111 Fax: 81 178 25 1430 Email: sugawara@hi-tech.ac.jp

**Dr. J.L. Summers**
Department of Mechanical Engineering, University of Leeds
Leeds LS2 9JT, United Kingdom
Tel: 0113 233 2151 Fax: 0113 242 4611 Email: j.l.summers@leeds.ac.uk

**Mr. G.H. Sunderhauf**
Dept. of Fluid Mechanics, University of Erlangen-Nurnberg
Cauerstrasse 4, D-91058 Erlangen, Germany
Tel: 49 9131 859487 Fax: 49 9131 859503

**Dr. H.M. Thompson**
Department of Mechanical Engineering, University of Leeds
Leeds LS2 9JT, United Kingdom
Tel: 0113 233 2110 Fax: 0113 242 4611 Email: h.m.thompson@leeds.ac.uk

**Mr. G. Van Sever**
Human Resources Department, UCB Chemicals
Anderlechtstr. 33, B-1620 Drogenbos, Belgium
Tel: 32 2 371 47 86 Fax: 32 2 371 45 11

**Mr. G. Vancoppenolle**
Dept. of Coating and Drying Research, Agfa-Gevaert N.V.
Septestraat 27, B 2640 Mortsel, Belgium
Tel: 3 444 8610 Fax: 3 444 7492

**Ir J. Vrancken**
Hoechst Holland NV, Wetering 20, 6002 SM Weert, The Netherlands
Tel: 31 4950 76508 Fax: 31 4950 76753

**Mr. D. Walker**
Dept. of Applied Mathematical Studies, University of Leeds
Leeds LS2 9JT, United Kingdom
Email: men5djw@sun.leeds.ac.uk

**Mr. J. Walker**
Horsell Graphic Industries Ltd., Howley Park Estate
Morley, Leeds LS27 0QT, United Kingdom

**Dr. C.M. Warwick**
CIBA Composites, Duxford, Cambridge CB2 4QD, United Kingdom
Tel: 01223 838219 Fax: 01223 838469

**Dr. S.K. Wilson**
Department of Mathematics, University of Strathclyde
Livingstone Tower, 26 Richmond Street, Glasgow G1 1XH, United Kingdom
Tel: 0141 552 4400 Ext. 3820 Fax: 0141 552 8657 S.K.Wilson@strath.ac.uk

**Mr. J.A.H. Wilson**
CIBA Composites, Duxford, Cambridge CB2 4QD, United Kingdom
Tel: 01223 838219 Fax: 01223 838469

**Mr. M. Wilson**
Dept. of Applied Mathematical Studies, University of Leeds
Leeds LS2 9JT, United Kingdom
Email: amtmw@sun.leeds.ac.uk

**Dr. A. Young**
Department of Mechanical Engineering, University of Leeds
Leeds LS2 9JT, United Kingdom
Email: menaey@sun.leeds.ac.uk